Environmental Earth Sciences

Series editor

James W. LaMoreaux, Tuscaloosa, AL, USA

Environmental Earth Sciences encompass mulitdisciplinary studies of the Earth's atmosphere, biosphere, hydrosphere, lithosphere and pedosphere and humanity's interaction with them. This book series aims to provide a forum for this diverse range of studies, reporting on the very latest results and documenting our emerging understanding of the Earth's system and our place in it. The type of material published traditionally includes:

- proceedings that are peer-reviewed and published in association with a conference;
- post-proceedings consisting of thoroughly revised final papers; and
- research monographs that may be based on individual research projects.

The Environmental Earth Sciences series also includes various other publications, including:

- tutorials or collections of lectures for advanced courses;
- contemporary surveys that offer an objective summary of a current topic of interest; and
- emerging areas of research directed at a broad community of practitioners.

More information about this series at http://www.springer.com/series/8394

Mark Anglin Harris

Geobiotechnological Solutions to Anthropogenic Disturbances

A Caribbean Perspective

Springer

Mark Anglin Harris
Department of Biology, Chemistry
and Environmental Science
Northern Caribbean University
Mandeville
Jamaica

ISSN 2199-9155 ISSN 2199-9163 (electronic)
Environmental Earth Sciences
ISBN 978-3-319-30464-9 ISBN 978-3-319-30465-6 (eBook)
DOI 10.1007/978-3-319-30465-6

Library of Congress Control Number: 2016940812

Printed on acid-free paper

This Springer imprint is published by Springer Nature
The registered company is Springer International Publishing AG Switzerland

The author dedicates this book to a small group of inspirational students from his classes at Northern Caribbean within 2002–2016. He spontaneously portrays them as follows:

Raffique (2002), Lloyd Wiggan, LaSonya Fletcher (2009): competent with a staunch work ethic (biogeography); Bryson Brown, Charles Koomson, Yvette Berry (2007–2009): highly driven graduate researchers (environmental science); Jacqueline Brown (2007–2009): graduate student—magnificent written work of the highest standard; Joan Black (2003): top 1 % of undergraduates—absolutely flawless writing; Valden Simpson and Mr. Sappleton (2009): well-prepared, capable with total commitment, topped class in graduate environmental chemistry; Diana Dayes (2004), Cathy-Ann Woodburn (2009): steadfast workers, consistently scoring highest marks in respective classes; Kevor Nicholls-Thomas (2010–12) graduate student (environmental Science): attentiveness to detail produced excellent preparation for tests (2010–12); Antoinette Hawthorn-Ingleton (2014–): thorough, high standard of work, never flinches from required tasks.

Foreword

Degradation of soil materials poses a great threat to agricultural production, environmental quality, human health, and economic activities all over the world. Soil salinization in southern Mesopotamia and in several parts of the Tigris-Euphrates valley destroyed the ancient societies that had successfully thrived for several centuries. Salts of different types and origin deposited into soils have affected soil quality hindering plant growth and water use efficiency. In addition to the natural processes, anthropogenic activities have accelerated these problems in recent times. More recently, global activities of using waste waters for irrigation and dumping of untested waste materials into soils have contaminated the soils with toxic metals and organic compounds. Mining activities in every continent have produced harmful residues in large quantities making it difficult to rehabilitate the environment. The factors involved in the degradation processes are complex and are specific to each region and even, each site. Scientific investigations are necessary to identify these factors in order to develop economically viable solutions to make productive use of the affected soils and mining residues.

Dr. Mark Harris has dedicated his efforts in search of geotechnical solutions to the anthropogenic disturbances, focusing on the Caribbean region. His research activities in Australia and Jamaica, collaborating with a number of colleagues, have resulted in a number of publications. They include the reclamation of sodic soils with gypsum, green manure, and decomposable phytogenic materials; processes of water movement in a hydrocarbon-contaminated soil in Trinidad; reclamation of bauxite waste and red mud using phyto-organics; geotechnical stability of carbonated lithified bauxite red mud wastes; quenching of phosphorus fixation using root exudates; bioremediation of acid drainage using plant materials; and the use of different organic amendments in the restoration of mining wastes. These papers published in different journals are very useful in developing innovative solutions for environmental degradation, depending on the factors involved.

I am sure this book which is based on the compilation of a number of his papers rewritten with more input, though focused on Caribbean, will be useful to scientists, students, and the managers of environmental remediation, all over the world.

Dr. Pichu Rengasamy
Senior Research Fellow
The University of Adelaide
Australia

Preface

Mining, which supplies materials thought essential for modern living, has serious environmental impacts to land, sea, and air. Open-cast mining impacts all ecosystem components. This book has two objectives: (a) To provide an environment-friendly approach to reducing toxicity caused by mining, and (b) to discourage the expansion of mining to hitherto unmined and pristine areas in the Caribbean. Hence, in addition to catering to the academic audience, the aim is to increase the readability for users such as stakeholders in government departments, private industry, and other concerned individuals. Thus, the use of technical jargon is reduced as far as is feasible; explanations for abbreviations are included, and it is hoped that several color photographs of the impacts of mining increase the awareness of the problems.

Without efficient methods to neutralize toxic mine wastes, the impacts on the environment can be long-lasting. Groundwater has the longest recovery times from pollution due to low flow, dispersion and dilution, colder temperatures, and lower bacteria (degrading) levels. It may take thousands of years to cleanse itself of degradable wastes, and non-degradable waste is permanent. Through peer-reviewed updated research articles, the author studies sustainable geobiotechnological modes of amelioration, remediation, and recycling of toxic mine wastes. The ensuing chapters in this volume are comprised of thirteen, updated and/or new, journal articles, published during the last 15 years. Below the updated articles in the table of contents are the titles of the original articles on which they are based.

The following Caribbean countries are engaged in open-cast, or other forms of mining: Jamaica, Guyana, Virgin Islands, and Haiti (bauxite); Guyana, Dominican Republic, Jamaica and Cuba (one or more of gold/silver/lead/zinc); and Trinidad (Petroleum, mined in deep wells but also forming surface seepages, thereby impacting soil and water) and Barbados (petroleum). Mining, by increasing slope cuts and degrading soils, increases the likelihood of landslips, water erosion, and dust production. An integrated management strategy of topography, surface cover, and vegetation is required (Part I: Open-Cast Mining: Some Problems and Solutions).

The waste from mines creates disposal problems. Some of these wastes contain materials which can help to increase efficiencies of certain industrial products. Chapters titled: "Geotechnical Stability of Two Gypsum-Treated Bauxite Red Muds and 'Marl' as a Road Base Under Submerged Conditions" and "Uniaxial Compressive Strength of Some Un-calcined Red Mud Mortars: Geotechnical Implications" (Part II) examine the use of red muds as road surfaces and as an ingredient in concrete. Hydrocarbon contamination can create water repellence in soils. This often results in high infiltration rates which can cause rapid spreading of aqueous toxicants. This section includes a study of the use of bauxite waste to reduce hydraulic conductivity via clay dislocation, using field soils taken from a hydrocarbon-contaminated area in Trinidad.

Aqueous toxic contaminants such as acid mine drainage inflict arguably the greatest damage on aquatic systems. The Camaguey region of Cuba and Pueblo Viejo in Dominica Republic produce vast quantities of acid mine drainage, polluting several freshwater sources. Part III presents non-toxic bioremediation methods to neutralize acid mine drainage.

The caustic soda used in the Bayer process to extract aluminum oxide remains in the ground, disaggregating (hard-setting) soils and increasing alkalinity and the sodium concentration in ground waters. Part IV (the last five chapters) examines the following: (1) the ability of gypsum with or without the addition of N-rich decomposable phyto-organics to increase soil aggregation in poorly structured sodic subsoils, and, by extension, sodic bauxite waste; (2) the role of phyto-organics and biosolids in reclaiming acid sulphate soils; (3) the preparation of a mine capping for disused mine sites; and (4) phosphorus quenching.

Mark Anglin Harris

Acknowledgements

According to Twidale and Campbell (2003),[1] Isaac Newton said that if he had seen further, it was by standing on the shoulders of giants. The author similarly expresses his gratitude to several colleagues. Dr. Pichu Rengasamy, a world-class soil scientist and Senior Research Fellow at Adelaide University, gave invaluable supervision during the author's training in graduate soil science. Dr. Rowl Twidale, former chair for the Department of Geology and Geophysics, and arguably Australia's foremost geomorphologist, presented to the author a copy of his masterpiece [book] on Australian geomorphology which was gratefully received. Under his able directorship, the author served as a tutor at the University of Adelaide in 1992–1996. Twidale honed the concept of the working hypothesis. His input, at the author's request in 2007, into the research study on landslips in northeastern Jamaica increased the value of that publication of 2009. The experience of Dr. Victor Gostin, also a former chair for the Department of Geology and Geophysics, University of Adelaide, and expert on planetary geology (who presented the author with a copy of his very readable book—on environmental geology), and under whose directorship the author also served as a tutor, is appreciated. Dr. Martin Williams, foundation professor of environmental science at the University of Adelaide, and Dr. Santo Ragusa of the Centre for Groundwater Studies, CSIRO (South Australia), gave useful and game-changing comments during the author's Ph.D. research in environmental aqueous geochemistry.

Last but not least, the foresight of the author's parents who encouraged correct speech and study in the early years is gratefully appreciated.

Based to some extent on the advice of the above-mentioned giants, the author was able to acquire the following:

Recipient of the *Distinguished Faculty Award* at Northern Caribbean University (NCU) on four occasions (unprecedented) for scholarship (2003) and for research (2007, 2011, and 2016), twice chosen to deliver the *C.J. Thomas*[2] *Memorial Lecture*

[1]C.R. Twidale and E.M. Campbell (2005) Australian Landforms: understanding a low, flat, arid and old landscape. Rosenberg Press.

[2]Dr. C.J. Thomas, Professor Emeritus Department of Biological Sciences (NCU) 1967–2000.

(the lead presentation) at the NCU Annual Science Symposium (2013 and 2016), delivering the *W.D. Carter*[3] *Memorial Lecture* (2006), and receiving three annual certificates of commendation: "*In Recognition of Outstanding Service to the (NCU) University*" in 2011, 2012, and 2014.

[3]W.D. Carter, lecturer at West Indies College prior to the changeover to NCU in 1996.

Contents

Part II Mine Waste: Utilization and Recycling

Part III Acidic Mine Water: Bioremediation

Abbreviations

AD	Acid drainage
AHD	Australian height datum
AMD	Acid mine drainage
ANC	Acid neutralizing capacity
ANZECC	Australia and New Zealand Environment Conservation Council
ASS	Acid sulphate soil
ASSMP	Acid Sulphate Soil Management Plan
BD	Bulk density—the ratio of unit mass to unit volume of soil
BE	Base exchange (or cation exchange)
BS	Base saturation
BSR	Bacterial sulphate reduction
BTEX	Benzene, toluene, ethyl benzene, and xylenes
CE	Cation exchange
CEC	Cation exchange capacity
Clay	Detrital mineral particle of any composition having a diameter less than 1/256 mm (4 μm)
DNAPL	Dense non-aqueous phase liquid
DO	Dissolved oxygen
ESP	Exchangeable sodium percentage
HSD	Honestly significant difference
ICPAES	Inductively coupled plasma atomic emission spectroscopy
MCLs	Maximum contaminant levels
$mSv\ yr^{-1}$	Millisievert per year
NAPL	Non-aqueous phase liquid
ORP	Oxidation–reduction potential
PPM	Parts per million
PAH	Polyaromatic hydrocarbon
pH	Acidity (or basicity) level
PSA	Particle-size analysis
PTE	Potentially toxic elements
RM	Red mud

SAR	Sodium absorption ratio
SG	Specific gravity
SOM	Soil organic matter
SRB	Sulphate reducing bacteria
SST	Sea surface temperature
TDS	Total dissolved solids
TOC	Toxic organic compounds
USEPA	United States Environment Protection Agency
VOC	Volatile organic compounds

Part I
Open-Cast Mining: Some Problems and Solutions

Post-mining Nutrient Depletion of Bauxite Overburdens in Jamaica: Storage Methods or Subsoil Dilution?

Mark Anglin Harris and Samson N. Omoregie

Abstract Rapid degradation of disturbed soil from a karst bauxite mine in Jamaica was recorded. Substantial macro-nutrient losses occurred during a short (1 month) or long (12 months) storage of the replaced top-soils during frequent wet/dry changes. The results suggested very high rates (>70 % in the first year) of soil degradation from storage, alongside moderate rates (30 %) within the same storage dump. However, higher levels of soil organic matter (SOM) were indicated just below the surface, compared with the surface horizons. It was unlikely that under a high leaching humid tropical rainfall regime, natural degradation processes could have re-emplaced such material firmly intact in the 15–30 cm zone. It was therefore concluded that these SOM anomalies were due to mechanical dilution of surface soil with sub-soil material during overburden removal and emplacement rather than from long storage. Increasing the soil organic content during storage could be one corrective approach. However, it is far less costly to exercise greater care to apply more precise overburden removal and emplacement techniques initially, than it is to correct the results of top-soil contamination with sub-soil. Although this study was limited to one mine, in the context of imminent large-scale mining expansion and current practices, further investigations are needed to accurately ascertain the proportion of similar sub-soil contamination in other bauxite-mined sites.

Keywords Bauxite mining · Karst bauxite · Mine-capping · Overburden · Soil storage · Jamaica · West Indies

Original article: **Post-mining deterioration of bauxite overburdens in Jamaica: storage methods or sub-soil dilution?** *Environmental Geology March 2008, Volume 54, Issue 1, pp 111–115*

M.A. Harris, *Geobiotechnological Solutions to Anthropogenic Disturbances*, Environmental Earth Sciences, DOI 10.1007/978-3-319-30465-6_1

Highlights

- After overburden removal, very high rates (>70 % in the first year) of soil degradation from storage, alongside moderate rates (30 %) were found within the same top-soil storage dump.
- Soil organic matter anomalies were due to mechanical dilution of surface soil with sub-soil material during overburden removal.

1 Introduction

Bauxite mining in Jamaica disturbs approximately 63 ha of land annually, being mined by the open pit method using a variety of earth-moving equipment (Neufville 1993). Prior to mining, the land is cleared of all vegetation and roughly 30 cm of top-soil removed (Neufville 1993). However, the organic fraction in such soils is often much closer to the surface than 30 cm. This matter is discussed later. The sub-soil (bauxite) is then mined until the underlying limestone is exposed, using a combination of earth moving machines. The excavations result in irregularly shaped pits of varying depths, with side slopes ranging from gentle to vertical (Lyew-Ayee and Stewart 1982). Before mining, bauxite soils in Jamaica are deep, friable, and contain few rock fragments (Greenberg et al. 1996). Habitat destruction is a major negative physical impact which affects not only the topography of the land, but flora and fauna, through de-vegetation and the removal of top-soil (Neufville 1993).

In the humid tropics, deep tropical weathering and decomposition of rocks have caused excessive solution and leaching of bases, leaving largely the insoluble Fe and Al sesquioxides (Fig. 1). With an area the size of Rhode Island, and rapid increases in bauxite mining globally in recent years (Hai et al. 2014), the rate of Jamaica's loss of agricultural land to bauxite mining is set to be continue. Top-soil materials, when they can be salvaged, are typically much lower in rock content than mine spoils and therefore have better water retention characteristics (Nicolau 2002; Morenode las Heras et al. 2008). However, if current methods of the removal and replacement of top-soil (overburden) during mining operations are applied, their inherently low base saturation and low fertility could be exacerbated (Fig. 1).

In an effort to determine the causes of poor yields and restricted growth on mined-out bauxite lands, comparisons can be drawn between the nutrient status and history of undisturbed and mining-disturbed soils on one hand, and on the other, characteristic effects of mixing top-soils with sub-soils. These are investigated below.

Fig. 1 In this active karst bauxite mine pit, the apparent differentiation between top-soil horizons and sub-soil zone is not easily observed or detected. This can cause inadvertent mixing of top-soil with sub-soil. Source: usgs.gov

1.1 Nutrient Status and History

Two problems have emerged in recent years as being critical to the sustainability of top-soil removed to facilitate open-cut mining: Firstly, soil deterioration varies directly with the period of storage after removal (Hannan 1978). Secondly, these impacts include loss of vegetation, destruction of soil structure, an initial increase in soil organic matter mineralization, compaction, dilution of nutrients, and loss of soil biota (Dangi et al. 2011) including the majority of the seeds and other plant propagules such as rhizomes, lignotubers, roots etc. (Ghose 2001) (Fig. 2).

1.2 Physical Effects

Thus the effects of organic matter on the physical condition of soils are far in excess of its proportion in the soil, and very small losses of the order of <0.5 % in organic content can cause breakdown of some physical properties of the soil (Brady 1974). Laffran (1994) refers to slaking as the breakdown of a lump of soil into smaller fragments on wetting, caused by swelling of clay and the bursting out of entrapped air. A subsequent comminution of soil aggregates into smaller entities increases the likelihood of erosional losses by water and wind but organic matter reduces slaking by binding mineral particles and by slowing the rate of wetting (Stefanson 1971; Oades 1984; Beare et al. 1994; Baldock et al. 1994; Laffran 1994; Rowell 1995; Denef et al. 2001).

Post-mining landscaping should leave a final landform visually compatible with the surrounding natural landscapes, ensure that the land is stable and will not

Fig. 2 Utility of plant propagules seen on the top-soil is contrasted with low activity of relatively barren sub-soil layers. Source: usgs.gov

erode, and provide an adequate substrate for revegetation (SVDP 2004). Macro-aggregate stability is largely responsible for macro porosity, which determines soil drainage rate and aeration (Sheoran et al. 2010). Aggregate structure breaks down as successive layers of soil are removed and stockpiled elsewhere on the site when mining begins. The resulting compaction reduces water holding capacity and aeration. During open-cut mining, the area is completely stripped of vegetation and overburden to uncover the bauxite ore. This process is similar to conventional tillage practices, which disrupt soil aggregates, exposing more organic matter to microbial attack (Beare et al. 1994; Brady and Weil 2002). Changes that occur in soil include increased bulk density, decreased water holding capacity, chemical changes, reduced nutrient cycling, reduced microbial activity, and loss or reduction of viable plant remnants and seeds (Mummey et al. 2002).

Losses of organic matter

Organic matter at various stages of decomposition, including cells and tissues of soil organisms and the substances synthesized by them is the chief stabilizing factor in most soils (Oades 1984). Organic matter maintains soil structural integrity and acts as a bonding agent linking the soil particles in an open network, reducing the wetting rates through its inherent hydrophobicity (Oades 1984), and by providing nutrients for the breakdown activities of soil microflora, micro- and macro-fauna.

Furthermore, the shaking and agitation by heavy equipment such as draglines and mechanical shovels involved in the overburden removal process breaks up organic structures and bonds, creating macro-pores which let extra air into the soil, leading to rapid volatilization of soil organic matter. This cycle of removing, dumping, removal, then replacing a soil layer using mechanical shovels in the open-cast process applies physical stresses to the soil not unlike those accompanying the use of tillage instruments (Ghose 2001). Top-soil thus removed becomes biologically sterile if is not preserved properly (Kundu and Ghose 1997; Mummey et al. 2002). In a study of soil microbial activity in British open-cast coal mine restorations that utilized stockpiling techniques, Harris and Birch (1989) found less carbon stored in the soil, and carbon decreased further after re-spreading of the soil. One of the most immediate consequences of stockpiling is the loss of organic carbon levels (Sheoran et al. 2010). Moreover, mining operations often involve the storage of top-soil in stockpiles for long periods of time, often several continuous years. The natural process of soil development can take hundreds of years. Stockpiled top-soil becomes highly degraded the moment this long-term structure is disturbed (Sheoran et al. 2010). Research studies show that there is one time-frame where the most damage occurs. This time frame is when top-soil is initially stripped from the ground (Sheoran et al. 2010).

Dilution with sub-soil

Clay enrichment of bauxite overburdens which often happens when deep mechanical mixing of top-soil with sub-soil or "inverted horizons," occurs during overburden removal (Harris and Omoregie 2008) or during the placement of the stored top-soil. Differential horizontal zones of organic matter (hence varying fertility) therefore occur in the same overburden within short distances (often less than 2 m). Further, such horizontal inconsistencies of nutrients exacerbate the problem of treating phosphorus fixation by requiring large applications of additives to quench the hidden high-fixing spots, compared with rates for less heterogeneous overburdens.

Secondly, top-soil mixed with sub-soil leads to the degradation of the top-soil by increased dispersion due to decreased SOM and lowered porosity (Baldock et al. 1994). Jamaica bauxites have remained unburied by sediments since the times of accumulation (Bardossy 1984), thus alterations from secondary processes have not been significant (Bardossy 1984); soil profiles on the highly weathered red bauxites are consequently weakly developed with no colour gradations. This presents difficulties in visually demarcating the true depth of top-soil and boundaries with sub-soil during overburden removal prior to mining. The likelihood of mechanical mixing of top-soil with sub-soil in the early stages of mining is therefore high. Because mining operations involve removal and storage of both layers (i.e., top-soil and overburden), mixing of sub-soil and top-soil layers can create plant establishment problems. The sub-soil layers lack the organic and microbial organisms necessary to sustain plants.

Nevertheless a contrary view is advanced by Abdul-Kareem and McRae (1984) who state that while it is clear that adverse effects due to storage and earthmoving equipment exist, the extent of deterioration of soil in stockpiles has been greatly overestimated. They say that their studies in England show that there is no reason why soils should not continue to be stockpiled although with greater care given to minimize compaction and mixing of top-soil with sub-soil. Studies by Williamson and Johnson (1990) as well as Widdowson et al. (1982), conducted in Wales and New Zealand, have shown that the soil pH, and the mineral content of stockpiled soils are not affected, as long as the soil is not stored for long periods of time in deep stockpiles. These studies (Mummey et al. 2002) have also shown that the soil biology of stockpiled top-soil recovers relatively quickly once the soil is re-spread.

1.3 Traditional Method in Jamaica Bauxite Mines

Sheoran et al. (2010) state that three to four feet of loose, non-compacted soil material are required to hold enough water to sustain plants through prolonged droughts (Sheoran et al. 2010). He reasons that severely compacted (bulk density >1.7 g/cc) mine soils, particularly those with less than two feet of effective rooting depth, shallow intact bedrock and the presence of large boulders in the soil simply cannot hold enough plant-available water to sustain vigorous plant communities through protracted drought.

In Jamaica, bauxite is extracted down to the underlying limestone bedrock. The bedrock is often dramatically undulating, and areas narrower than 3 meters are not mined. When mining is complete, piles of limestone rubble mixed with bauxite are used to fill in the lowest areas in the pit. Then the top-soil is spread back over the whole disturbed area and grass is usually planted. Greenberg et al. (1996) note that reclaimed soils are shallower, have weaker structure, and exhibit higher bulk density, pH (6.5–8.0), and limestone fragment content than pre-mined soils. Reclaimed bauxite soils could therefore be classified as Typic or Lithic Udarents depending on the depth to bedrock (Greenberg et al. 1996).

This investigation seeks to determine the effects of removal, storage, and replacement, on soil organic matter (SOM) and macro-nutrients (N, P, K) of two red soils which have been disturbed by bauxite mining.

2 Materials and Methods

2.1 Study Site

Jamaica is constructed on a Cretaceous age basement of mainly volcanic rocks, plus some granodiorite intrusions, sedimentary conglomerates and minor limestones, and metamorphic rocks including schists and minor serpentinite, all of which are sporadically exposed at the surface of the island (Wacaster 2008). About two-thirds of the island is covered by Tertiary age platform and reef complex limestone, and a thick sequence of red bed conglomerates lies between the basement and the limestone cover. Limestone is often overlain by bauxite, which is the country's primary mineral export (Wacaster 2008).

The site used in this study is located on the Manchester Plateau in south-western Jamaica (Fig. 3), and comprised of red kaolinite clay soils rich in bauxite, autochthonously derived in resistant.

Eocene–Miocene limestone several miles thick. To the north, this limestone extends into an almost impenetrable crystalline dissected plateau. Normally, weathering of rock starts with brecciation (Kesse 1984). In the present case, brecciation resulted from down-faulting of parts of central plateau (Hill and Ostojic 1984), and a feature which distinguishes carbonate from siliceous rocks is that

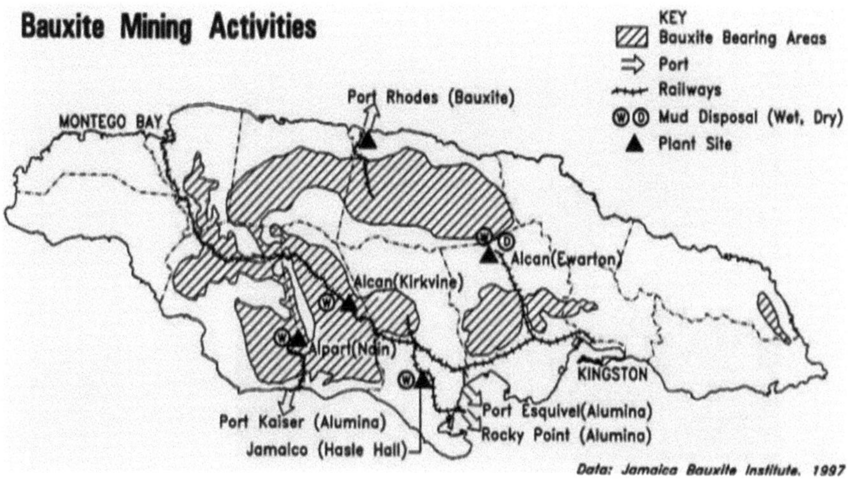

Fig. 3 Location of study area can be seen at Kirkvine, amongst other karst bauxite deposits

they have a greater tendency to fracture and fault, rather than fold and buckle, under stress (Reading et al. 1995). In Jamaican karsts, typically well jointed (Twidale and Campbell 2005) with the development of a series of vertical cylindrical pipes, and being within the humid tropics, high rates of solution and collapse has led to irregular karstic hills (hummocks) and deep, open basins (dolines) called cockpits, hence the area being called the Cockpit Country (Twidale and Campbell 2005). The hills are remarkably uniform in height (100–160 m) (Fig. 4) with base diameters around 300 m (Faniran and Jeje 1983). It is believed that the dissolution of the limestone left residues in the down-faulted enclosures (Hill and Ostojic 1984), resulting in diagenesis of red bauxite. Deposits range in size from a few hundred to millions of metric tons with a thin soil overburden (Wacaster 2008).

The increased surface area of such rough terrain weakens the North East Trade Winds which blow diagonally across the east–west axis of the Cockpit Country. Moreover, the lack of a sufficiently high land barrier here often excludes rainfall from these moisture-laden winds. Orographic precipitation in south-western Jamaica is therefore minimal. However, the combination of relatively long distance from the sea, low elevation and low latitude causes frequent, regular summer afternoon convectional thundershowers. Between late August and November, this normally occurs for at least 2 days of every week. As all these events coincide with the northern hemisphere hurricane season, the annual average rainfall total of about 1250 mm occurs mostly between July and November. This contrasts with >3000 mm in the northeastern foothills of the >2000 m Blue Mountain Range less than 150 km east. The annual dry seasons are therefore more distinct than those in north-eastern Jamaica, and often prolonged, causing frequent droughts in south-western locations.

(a)

(b)

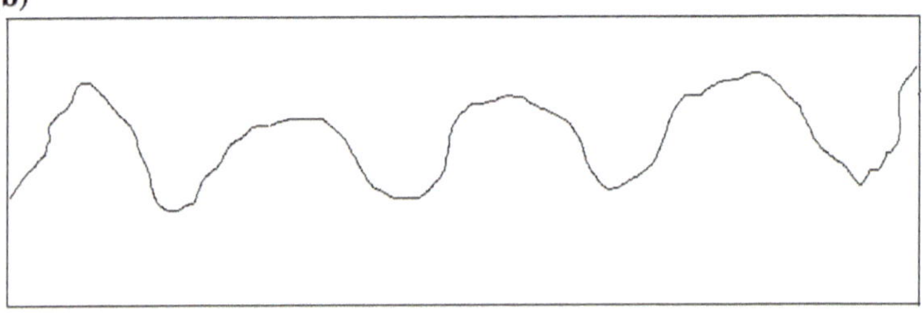

Fig. 4 **a** Aerial view of the karst landscape in the Cockpit Country. In this relatively young karst, the magotes (hills) dominate the dolines (depressions), which contain alluvial deposits of bauxite. Source: Jamaica Bauxite Institute. **b** Transverse sketch of the Cockpit Country. Due to the "young" geological age of this white limestone plateau, in size, the magotes (hills) always dominate the cockpits (dolines). This arrangement is the obverse of the older, more eroded bauxite-yielding karsts such as those of Yugoslavia, for example, where the dolines are much wider. This is caused by erosion of magotes, a process linking several dolines at ground level into larger geologic depressions called Poljes

2.2 Top-Soils: Description and Analysis

Jamaican karst bauxite is soft, earthy and friable (Hill and Ostojic 1984), of high porosity and highly permeable (Bardossy 1984). Humid tropical weathering has depleted the soils of alkalis and alkaline earth minerals, and more prolonged leaching of silica has decreased plasticity (stickiness) in the top-soils, rendering them very porous (just as the extraction of gluten, the sticky network trapping gases, decreases the rising of wheat flour by failing to trap gases). Rainfall thus sinks readily into the soils. Soil samples were taken from Mount Nelson Mines, a deeply weathered red karst soil. Samples from undisturbed sites at the edge of the mines were also taken in the same manner, to be used as reference soils. SOM was measured by the Walkley and Black method (Piper 1942). Soil N analysis was

Table 1 Some properties of an un-mined soil from Mount Nelson, Jamaica

Organic matter	4.2 %
pH	5.3
P (ppm exchangeable ions)	29
K (ppm exchangeable ions)	112
Cu (ppm exchangeable ions)	4.1
Mn (ppm exchangeable ions)	305
Fe (ppm exhangerable ions)	75
Kaolinite (%)	15
Gibbsite (tri-hydrate bauxite) %	<3
Haematite %	>10

done by titrimetric acid digestion (Kjeldahl; Bremner 1960). Potassium was measured by a 1-M ammonium acetate extraction, using EDTA titration. Phosphorus was assessed by the Truog method, using the Milton-Roy Spectrometer. The micro-nutrients Fe, Mn, Cu, Zn were determined by 1 M ammonium acetate–0.1 M EDTA extraction and the atomic absorption spectrophotometer. Sampling was carried out randomly, using a soil auger, to a depth of 30 cm. The characteristics of the undisturbed soil from Mount Nelson are shown in Table 1.

3 Results

Table 2 depicts the results from the processing of samples from the Mount Nelson Mine after storage for 1 year (see also Figs. 5–8). The most striking aspect is the presence of almost three times the level of organic matter in the lower level (15–30 cm) compared with that of the surface zone (0–15 cm). This is highly unusual because firstly, organic matter, however minimal, is usually greater in the top layer of any undisturbed soil profile, and, as pointed out by Brady (1974), represented by a thin surface layer in strongly leached tropical red soils; not the other way around. Secondly, it is of interest, although not surprising, that the highest concentrations of macro-nutrient ions are found also in the second level (15–30 cm), i.e. where there is greatest SOM. For example, in the 0–15 cm level and the 15–30 cm level, respectively, K levels are 0 and 26 ppm, and N levels

Table 2 Changes in mined Mount Nelson soil after I year of storage

	SOM	N	P	K	Zn	Cu	Mn	Fe
Unmined soil (cm depth)								
0–15	42000	2660	29	111	7	4.2	325	76
15–30	38000	2220	29	112	8	3.9	288	75
Mined soil (cm depth)								
0–15	12000	820	27	0	2	1.4	110	50
15–30	34000	1960	23	32	111	2.5	313	46

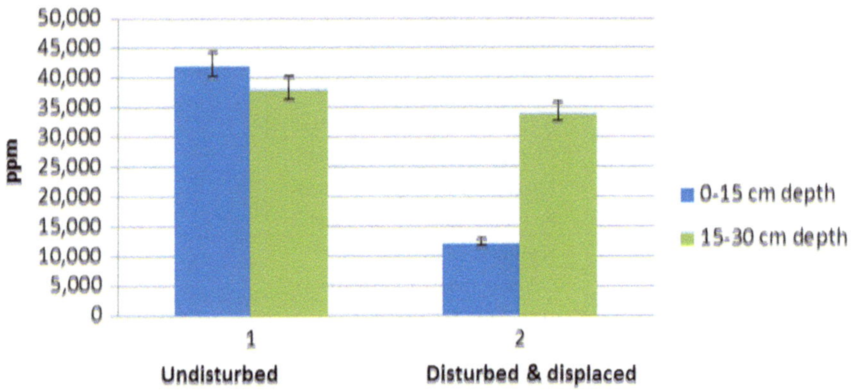

Fig. 5 Organic C in undisturbed-, as compared to that of disturbed & displaced, bauxite overburden after one year

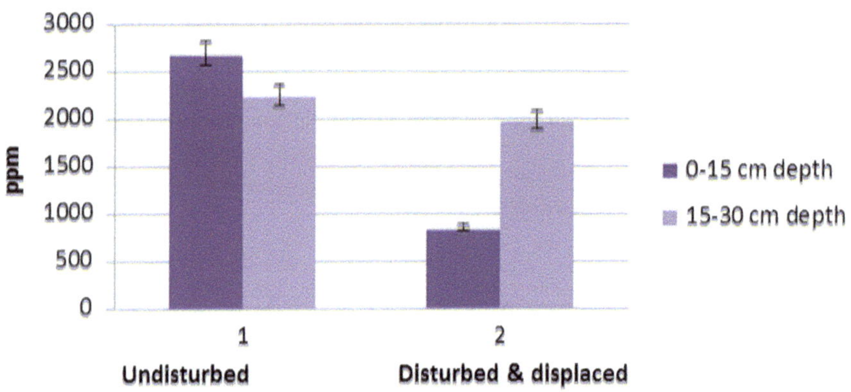

Fig. 6 Nitrogen in undisturbed-, compared to that of disturbed & displaced, overburden after one year

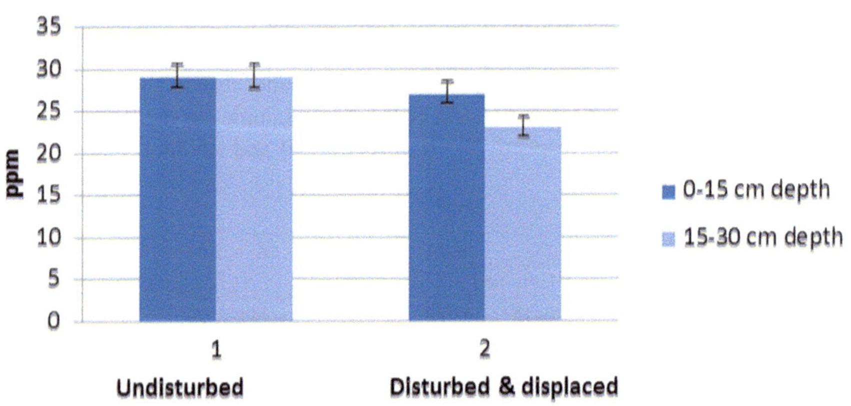

Fig. 7 Phosphorus in undisturbed-, compared to that of disturbed & displaced, overburden after one year

have been doubled (although P has changed in the opposite direction; this is discussed later). Being strongly leached, deep, well-drained base-depleted soils with a homogenous structure and no clear horizons prior to disturbance, it would be expected that after further loosening by removal and storage for 1 year, some depletion of organic matter and macro-nutrient ions would have occurred from the upper surface layer (0–15 cm). Yet for the reference soil in comparison, no such changes have occurred (Table 2). The losses of SOM and macro-nutrients, however, amounting to 60 % in stored top-soils, as were observed in this study after 12 months, are unusual. However, there are several ways in which such heavy losses could have been accomplished. Allison (1967) pointed out that in top-soil, SOM is largely the source of nutrient ions. This is the case especially in the humid tropics. Therefore, ion concentrations vary directly with SOM, being the sole source of macro-nutrients in deforested tropical red karst soils. Because the removal of vegetative coverage exposes the top layer of the soil to the weather, SOM losses would normally be expected to occur most prevalently in the upper horizon at the interface with the weather. Further illuviation would be expected as pulverization of the soil by earth-moving machinery would have opened up the soil fabric, increasing the surface area of exposure. Clay skins, found in many well-aggregated soils, are repositories of organic matter and nutrient ions (Allison 1967). Allison (1967) further found that any disturbance of the soil as by wetting and drying, freezing and thawing, and even microbial action caused fragmentation of (nutrient-rich) clay skins in the surface soil. He concluded that less disturbance at lower soil depths permitted more stability of clay skins there. As the red soils of this study were found to be open textured and well drained, material from the upper horizons thus exposed would be expected to pass unhindered towards lower levels. Considering these factors, it seems possible that leaching was responsible for macro-nutrient and SOM depletion in the 0–15 cm level of the soil.

In northern India (within similar latitudes to those of Jamaica), Ghose (2001) also observed heavy SOM losses for a stored top-soil after 12 months, although smaller (47 %) than those in Jamaica. He attributed the deficiency of nutrients in stockpiled soil to reduction in soil microbes, as did Rana et al. (2007), and excessive leaching, i.e., temporarily lodging subjacently but subsequently elluviated.

However, the climatic changes of India are more sharply compartmentalized than those under Tropical Marine Caribbean type conditions with more evenly distributed annual rainfall totals. Even though droughts are frequent in western Jamaica, they are of shorter duration than those in northern India. Thus, it seems reasonable to conclude that there are more frequent wet/dry cycles occurring in the study area of Jamaica within the same time period (causing a greater number of leaching periods) compared to anywhere on the Deccan. But even if such a climatic regime could adequately explain the more rapid disappearances of SOM from the stored top-soils in Jamaica compared with those in India, it does not account for the high SOM anomaly at just 15–30 cm below the surface in Jamaica. In fact, on the assumption that the high frequency wet/dry cycle explanation is accepted, the accumulation of SOM in India should as a result be even greater than that in Jamaica, as fewer leaching cycles on the subcontinent should facilitate greater accumulation of organic colloids out of the soil to cause greater accumulation of SOM. Yet Ghose (2001) in India found no accumulation of SOM in horizontal layers anywhere below the soil surface. It is therefore unlikely that organic colloids were effectively held back in a 15–30 cm subsurface zone in Jamaican red bauxite soils by an open soil network under a frequently heavy rainfall regime. As stated earlier, Bardossy (1984) found all red bauxite soils to be of high porosity and excellent permeability. Being microporous (<0.05 mm) and porous (0.05–0.5 mm), he observed, at exposed cuts, water gushing off the limestone basement rocks underlying the red Jamaican bauxites >200 cm below the surface during rains. Therefore, despite all the above arguments in favour of exposure and leaching as an explanation, the nature of red soils in the study area would not facilitate such a relatively large accumulation—a veritable "logjam" of SOM, N, and soluble base ions (relative to the 0–15 cm level: i.e. 30-fold K, 5-fold Zn, 2-fold Cu, 3-fold Mn) so close under the depleted (0–15 cm), heavily leached 0–15 cm zone. Harris and Rengasamy (2004) found that green manure incorporated into a soil of similar pH and allowed to decompose caused significant increases in N, P, K, and Cu. The increased levels of such ions could similarly have solubilized from large amounts of SOM emplaced at the 15–30 cm level of the present study. A plausible explanation therefore is that at some stage between overburden removal and its emplacement, the original top layer (0–15 cm) was physically switched with the underlying (15–30 cm) layer in several instances during the earth-moving process. The 0–15 cm surface layer may have been therefore accidentally buried under a deeper layer by the filling machines, or simply turned upside down. The former sub-soil layer (now "top layer" in its "wrong" position) would therefore now be lower in organic matter and base ions, thereby explaining the highly depleted characteristics now on the surface.

4 Conclusions

Although the rates of losses of SOM and macro-nutrients in stored red soils were substantially larger than had been expected, the unusual vertical sequence of SOM after storage and the lack of a sound scientific explanation suggest that human-induced mechanical misplacement caused a shift in natural position of SOM. This would have exacerbated the effects of soil storage per se. Even though this research was done on one mine, it may be indicative of conditions in other mines. Considering the planned expansion of bauxite mining in Jamaica, and the consequent potential for soil deterioration, more incisive evaluations of the impact of current methods of overburden removal and storage on soils seem necessary. In another study, mixing top-soil and spoils produced a lower saturated hydraulic conductivity (Mentis 2006).

5 Geobiotechnological Applications

Considering that only one small mine was investigated in this study, and assuming (a) similar mining practices currently prevail and (b) the imminence of large-scale expansions of open-cast mining in the small island of Jamaica (11,500 km^{-2}), more deliberate planning of overburden removal and storage seems imperative for the sustainability of top-soil resources. The following specific recommendations are suggested to pre-empt rapid top-soil deterioration in other local mines:

- SVDP (2004) recommends the following chronological steps prior to mining: (1) a full botanical survey should be carried out before mining, and seeds collected, (2) top-soil is stripped and placed in properly sequenced layers where it is stockpiled (3) rehabilitation of earthworks recreates the sub-soil surface, and top-soil is then replaced and mulched and a cover crop is sown with native seeds incorporated. The cover crop protects germinating seedlings. Thus, suitable plant-growth media are required for the revegetation of mined and/or disturbed areas. This objective can be achieved by the selective handling of top-soil during the mining process. Retention of soil is the key to revegetation. Top-soil, together with leaf and plant litter, should be removed and stockpiled separately from underlying material. Depending on the soil profile, up to 300 mm of top-soil may need to be stripped, and the Rehabilitation Plan should indicate the location and depth of soil stripping (SDVP 2004). During overburden removal, ensure that the upper 0–15 cm of soil is not mixed with sub-soil at any stage. This cannot be accomplished conventionally (Fig. 5).
- This step could be achieved by using calibrated scrapers similar to those used by commercial "landscapers" where possible, while ensuring that soil is replaced "right side up".
- If possible, storing soil only when unavoidable, and for the shortest time necessary (as suggested by Ghose 2001).
- As grass adds more SOM in the top 0–15 cm through root death, decay and re-growth than any other vegetation (Laffran 1994), and land under pasture grass

Fig. 8 Scooping of overburden causing mixing of top-soil with sub-soil unintentionally dilutes the quality of the top-soil. Specially calibrated equipment is required (source: usgs.gov)

retained its SOM, integrity and durability for longer periods than land under other vegetation (Harris and Rengasamy 2004), SOM levels during storage would increase if soils were left under actively grazed pasture grass for a prolonged period prior to mining disturbance. Grass has further effectiveness because of (a) its high root:shoot ratio, and (b) its root fineness (exceeding that of other plants), thereby permeating more soil particles with SOM. Weeds on the other hand, with tap root systems, are not as effective. If mining is to occur, grass seeds may be sown several months in advance. However, to prevent colonization by weeds, weed control operations should precede this operation. Thus to preempt infestations, as has been has been recommended by Ghose (2001), cleaning of the ground area before seeding to remove weeds and weed seeds should be carried out.

- Soil should be re-spread as soon as possible after stripping in order to maximize its fertility, seed viability and microbial activity (SVDP 2004). The Rehabilitation Plan should indicate the approximate depth of re-spread soil and its location. Consideration may also need to be given to the re-spreading techniques used, with a view to minimizing soil compaction that can inhibit later revegetation.
- With all of its deleterious effects, top-soil stockpiling is still better than attempting to establish plants in the soils that exist below the top-soil layer (SVDP 2004).
- Recent research shows that anaerobic conditions and poor nutrient cycling result from deep stockpiling, especially stockpiles deeper than one meter (Sheoran et al. 2010).
- Studies also show that long-term storage is detrimental to soil health.
- New techniques are being used with varying rates of success. For example, road construction projects use staging so that only limited parts of the road must be closed at any given time. Staging the project means that stockpiles are often re-spread in a very short timeframe, a few months or less. If mining operations could be staged in a similar manner, it is possible some of the effects of long term stockpiling could be reduced. Another challenge in both mining and civil engineering operations that use stockpiling is in the use of heavy equipment to remove and replace top-soil. Although research shows that use of heavy equipment results in compaction, no research was available on a better way for moving the large amounts of soil to be stockpiled.
- Stockpiling the top-soil followed by its re-laying would seem to potentially produce planting conditions that are closer to the pre-disturbance conditions than would planting on the exposed sub-soil layers (USDA Forest Service RM-447). When mining operations are complete, the overburden material is reapplied and levelled and then the top-soil is reapplied and spread over the overburden material to provide a planting medium (USDA Forest Service RM-447).
- Stockpiled soil reapplied quickly, with care to reduce the compaction inherent in the use of mechanical means for stockpiling, conserves production potential (Strohmayer 1999). Thus, ideally, the top-soil is removed first and stockpiled in one pile and the soil layer below is also removed and stockpiled separately. This sub-soil layer is sometimes referred to as the overburden (Strohmayer 1999). However, where there is no effort to separate top-soil from sub-soil,

"overburden" connotes all soil covering the ore layer. Here, significant loss of OM generally results from top-soil stripping and/or storage prior to mining as a result of increased microbial decomposition rates and dilution of OM through soil horizon mixing.

- Compacted zones may also perch water tables during wet weather conditions, causing saturation and anaerobic conditions within the rooting zone. Repeated traffic of wheeled mining machineries (loaders and haulers), and bulldozers to a lesser extent, form compacted zones in the mining dumps (Sheoran et al. 2010).
- Mine soil compaction can be minimized by grading when mine soil materials are dry and using small dozers with low ground pressures (Sheoran et al. 2010).
- When "top-soil" from the pre-mining forest is available for use in preparing the final surface, leaving any stumps, logs, or other woody debris that is in that top-soil on the final land surface will aid reforestation (Strohmayer 1999).
- The depth of the stockpile and the length of time it is stored affect the quality of the soil at replacement (Strohmayer 1999). Thus Harris and Birch (1989) noted that soil stockpiled in piles more than a meter deep, chemical accumulated ammonium, and anaerobic conditions occurred in the top-soil at the base of the pile. They found other detrimental biological effects including absence of propagules and decrease in viability of buried seeds.
- For optimum survival of mycorrhizal propagules, the depth of stockpiles should be restricted to the rooting depth of covering vegetation (Strohmayer 1999). If plant cover can be maintained with roots extending throughout the depth of the stockpile, nutrient cycling processes and microbial activity can continue while the stockpile is stored.
- Special attempts should be made to avoid stripping deeper soil horizons with the top-soil since they may have poor structure or high clay contents (Primus 2014). In areas where the Al horizon is not obvious, the top 100–300 mm of soil would be recovered.
- Double stripping the top-soil, where the top 50–100 mm of soil is removed, and returned separately and on top of the remaining top-soil, may be warranted, particularly since the aim is to restore the native flora. Most of the seeds are stored in this top layer of soil, and its removal and return as a thin layer on the surface will maximise the contribution of these seeds to the post-mining flora.
- Mine-site rehabilitation efforts should not be evaluated and awarded too quickly; in many cases the success of mine-site rehabilitation can only be judged well after mining has ceased (Lottermoser 2003).
- To protect the plants as well as the seed bank during construction activity, a modified stockpiling technique can be used. The top-soil is carefully stripped and stockpiled in low berms and replaced in less than a month so that the seed bank, the plants, and the mycorrhizal fungi necessary for their continued success would not be lost (SDVP 2004).
- Unmined bauxite soils have strong granular structure but are friable despite a high clay content. Recently reclaimed soils are compacted and have large blocky structural units (imported from the deep bauxite deposits) commonly admixed with the stockpiled top-soil. After decades under grass, the blocks begin to disaggregate into smaller units and granular structure slowly forms.

References

Abdul-Kareem AW, McRae SG (1984) The effects on top-soil of long-term storage in stockpiles. Plant Soil J 76:357–363

Allison FE (1967) Soil aggregation: some facts and fallacies as seen by a microbiologist. Soil Sci 106 (7):136–140

Baldock JA, Aoyama M, Oades JM, Susanto, Grant CD (1994) Structural amelioration of a South Australian red brown earth using calcium and organic amendments. Aust J Soil Res 32:571–594

Bardossy G (1984) Karst bauxites: developments in economic geology. Elsevier Scientific, Amsterdam, p 441

Brady NC, Weil RR (2002) The nature and properties of soils, 13th edn. Prentice Hall, New Jersey (Published by ASMR, 3134 Montavesta Rd., Lexington, KY 40502)

Beare MH, Cabrera ML, Hendrix PF, Coleman DC (1994) Aggregate protected and unprotected organic matter pools in conventional and no-tillage soils. Soil Sci Soc Am J 58:787–795

Brady NC (1974) The nature and properties of soils. McMillan, New York

Bremner JM (1960) Determination of nitrogen in soil by the Kjeldahl method. J Agric Sci Camb 55:1–33

Dangi Sadikshaya R, Stahl Peter D, Wick Abbey F, Ingram Lachlan J, Buyer Jeffrey S (2011) Soil microbiual community recovery in reclaimed soil on a surface coal mine site. Soil Sci Soc Am J 76(915–924):2011. doi:10.2136/sssaj.0288

Denef K, Six J, Bossuyt H, Frey SD, Elliott ET, Merckx R, Paustian K (2001) Influence of dry-wet cycles on the interrelationship between aggregate, particulate organic matter, and microbial community dynamics. Soil Biol Biochem 33:1599–1611

Faniran A, Jeje LE (1983) Humid tropical geomorphology. Longman's, London

Ghose MK (2001) Management of top-soil for geo-environmental reclamation of coal mining areas. Environ Geol 40:1405–1410

Greenberg WA, Wilding LP, McDaniel SC (1996) Infiltration and erosion potential on reclaimed bauxite mine lands in Jamaica (Scientific registration no: 1277, symposium no: 38, Soil and Crop Science Dept., Texas A&M Univ., College Station, TX 77843-2474, USA)

Hai LD, Khai NM, Quy TV, Huan NX (2014) Material composition and properties of red mud coming from alumina processing plant tanrai, Lamdong, Vietnam. Int J Res Earth Environ Sci. ISSN 2311-2484 http://www.ijsk.org/ijrees.html

Hannan JC (1978) Coal: rehabilitation of coal mines in the Hunter Valley of New South Wales. In: Rummery RA, Howes KMW (eds) Management of lands affected by mining. Division of land resources management, CSIRO, Australia, p 18–32

Harris JA, Birch P (1989) Soil microbial activity in opencast coal mine restorations. Soil Use Manage 5(4):155–160

Harris MA, Omoregie SN (2008) Post-mining deterioration of bauxite overburdens in Jamaica: storage methods or sub-soil dilution? Environmental Geology 54(1):100–115

Harris MA, Rengasamy P (2004) Treatment of a potential soil capping for open-cut mines (part 1): influence of prior management on responses to a phyto-organic amendment. Environ Geol 46(1):15–21

Hill VG, Ostojic S (1984) The characterization and classification of bauxites. In: Leonard J Jr (ed) Bauxite: proceedings of the 1984 bauxite symposium, Los Angeles, CA, 27 Feb–1 Mar 1984, p 31–48

Kesse GO (1984) The geology of the bauxites of Ghana. In: Leonard J Jr (ed) Bauxite: proceedings of the 1984 Bauxite Symposium, Los Angeles, CA, 27 Feb–1 Mar 1984, p 572–602

Kundu MK, Ghose MK (1997) Micobiological studies of soils in coal mining areas. Indian J Soil Water Conserv 25(2):110–113

Laffran J (1994) Soils and their management. NSW agriculture land management series, vol 3. Tocal, Paterson

Lottermoser, Bernd G (2003) Mine wastes: characterization, treatments, impacts. ISBN: 3-540-00-526-9, Springer Verlag Berlin Heidelberg, New York 2003 p 3–11
Lyew-Ayee PA, Stewart R (1982) Stratigraphic and compositional correlation between bauxites and their limestone hosts in Jamaica. J Geol Soc Jamaica p 19–35 (Proceedings of Bauxite Symposium No. V)
Mentis MT (2006) Restoring native grassland on land disturbed by coal mining on the Eastern Highveld of South Africa. S Afr J Sci (May/June) 102(5–6):193–197. ISSN 0038-2353
Morenode las Heras M, Nicolau JM, Espigares T (2008) Vegetation succession in reclaimed coal mining slopes in a Mediterranean-dry environment. Ecol Eng 34:168–178
Mummey DL, Stahl PD, Buyer JS (2002) Soil microbiological properties 20 years after surface mine reclamation: spatial analysis of reclaimed and undisturbed sites. Soil Biol Biochem 34:1717–1725
Neufville L (1993) Bauxite is mined by the open pit method using a variety of earth-moving equipment. Jamaican Geogr (9) p 6
Nicolau JM (2002) Runoff generation and routing on artificial l slopes in a Mediterranean–continental environment: the Teruel coal field. Spain Hydrol Process 16:631–647
Oades JM (1984) Soil organic matter and structural stability: mechanisms and implications for management. Plant Soil 76:319–337
Piper CS (1942) Soil and plant analysis. University of Adelaide, Adelaide, p 223–226
Primus V (2014) Tamakay mining project Prashad Nagar, Georgetown Guyana www.epaguyana.org/index.php/2014-02-10-20-21-58?task…id=26. Accessed 20 Dec 2014
Rana S, Stahl PD, Ingram LJ, Wick AF (2007) Soil microbial community structure and diversity under different vegetation in Wyoming minelands. Presented at the ASMR National Meeting, Gillette WY, 2–7 June 2007. In: Barnhisel RI (ed) Thirty Years of SMCRA and Beyond (Published by ASMR, 3134 Montavesta Rd., Lexington, KY 40502)
Reading AJ, Thompson RD, Millington AC (1995) Humid tropical environments. Blackwell, Oxford (The State of Victoria, Department of Primary Industries, 2004. Guidelines for Environmental Management in Exploration and Mining. State Government Victoria, Department of Primary Industries. ISBN 0 7306 9495 X)
Rowell DL (1995) Soil science: methods and applications. Wiley, New York
Sheoran V, Sheoran AS, Poonia P (2010) Soil reclamation of abandoned mine land by revegetation: a review. Int J Soil Sediment Water 3(2). ISSN: 1940-3259
Stefanson RC (1971) Effect of periodate and pyrophosphate on seasonal changes in aggregate stabilization. Aust J Soil Res 9:33–41
Strohmayer P (1999) Soil stockpiling for reclamation and restoration activities after mining and construction. Restor Reclam Rev 4(7)
SVDP (2004) The State of Victoria, Department of Primary Industries, ISBN 0 7306 9495 X2004
Twidale CR, Campbell EM (2005) Australian landforms: understanding a low, flat, arid and old landscape. Environ Geol 54:111–115 (Rosenberg Publishing Pty, NSW)
Wacaster S (2008) The mineral industries of the islands of the Caribbean Aruba, The Bahamas, Barbados, Dominican Republic, Jamaica, Trinidad and Tobago, and other islands. U.S. Department of the Interior, U.S. Geological Survey, Aug 2010
Widdowson JP, Gibson EJ, Healy WB (1982) Effects of stockpiling top-soils associated with opencast mining. NZ J Sci 25:287–294
Williamson JC, Johnson DB (1990) Determination of the activity of soil microbial populations in stored and restored soils at opencast coal sites. Soil Biol Biochem J 22:671–675

Dust Reduction in Bauxite Red Mud Waste using Carbonation, Gypsum & Flocculation

Mark Anglin Harris

Abstract Producing alumina by the Bayer Process creates fine air borne red dust in de-watered dumps which devalues property and causes irritation to the human respiratory system. Samples of red mud waste (1) treated 10 years before at the 0–15 cm depth zone with 40 t ha^{-1} of gypsum (2) from the subjacent 15–30 cm zone, were collected, crushed and passed through a 0.5 mm diameter sieve. Leaves from *Acacia senensis* (a legume) were finely chopped to <1 mm and thoroughly mixed with the sieved bauxite waste at 25- and 50 %, and the samples incubated for 6 weeks at ambient room temperatures, at 60 % soil water-holding capacity. To determine the fraction of potential dust, the treated samples were submerged in de-ionized water for several days until there was no change in discoloration (due to clay dislocation) of the water. The samples were removed from the water and the water evaporated and the residues dried and weighed. This submergence process was repeated. The first submergence produced 30- and 20 % w/w dust respectively from the 25- and 50 % phytogenic treatments, and 0 % dust from the non-phytogenic treatments. Submergence #2 released no clay (wet dust) particles from any sample. Finally, the same samples were air-dried, re-crushed and dry-sieved through a 100 μm sieve, because dust-size particles are <100 μm in diameter. Dry-sieving produced dust as follows: non-phytogenic-treated subjacent red-mud zone = 50 % w/w, non-phytogenic—0–15 cm depth = 30 % w/w, phytogenic—at 25 % = 15 % w/w, phytogenic at 50 % = 7 % w/w. In total, the dust-reducing capabilities of the treatments were: 50 % phytogenic >25 % phytogenic >0–15 cm depth non-phytogenic >15–30 cm-depth non-phytogenic. The 50 % phytogenic-treatment reduced potential particles of fugitive dust by 70 % over the untreated controls and 95 % over the crushed-only (subjacent red mud; no organics added) samples. The binding of clay particles into larger entities by organic gums produced agglomerates which (1) reduced the number of <100 μm particles and (2) hence, particles which would have resisted air movements. The initial increases in fine clay dislocation amongst phytogenic treatments was attributed to the increase of small negatively charged organic colloids from the decomposition process, which repelled similar-sized negatively charged clay particles. All in all, phyto-organics increased average particle size >100 μm by flocculation, thereby

M.A. Harris, *Geobiotechnological Solutions to Anthropogenic Disturbances*, Environmental Earth Sciences, DOI 10.1007/978-3-319-30465-6_2

creating stable agglomerates which resisted disintegration and breakdown under simulated rain. This treatment promises the use of gypsum + phyto-organics for reducing the emanation of surface dust from red mud waste sites onto surrounding areas.

Keywords Guyana bauxites · Gypsum-organic amendments · Jamaica WI · Organic remediation · Virgin islands

Highlights

- A phyto-organic treatment decreased dust generation from dried bauxite waste red muds.
- Crushing and incubating lithified impermeable red mud into sub 2-mm particles failed to increase permeability in the crushed beds, but increased dust generation.

1 Introduction

1.1 Red Dust: Origin and Damages

Red mud waste (RMW) surfaces (crusts) of bauxite mine spoils are potential sources of dust emission. During storage, the surface gets dried, even though the consistency may be still liquid at a few millimeters below (Graham and Fawkes 1992). A strong wind can then blow a red dust cloud over the surroundings at a considerable financial and environmental cost. Buildings have been defaced and vegetation damaged by dust accumulation on leaves.

Of four categories, it was the high porosity un-compacted bauxites that stained the strongest (Bardossy 1982), and this strongest-staining group includes Jamaican bauxite, where the RMW particles are <2 microns in diameter (Ly 2001) due to the Bayer process of extraction. In a table indicating the textural analysis for several bauxite wastes, Wehr et al. (2006) reported bauxite residues (sand (%) 2–0.02 mm, silt (%) 0.02–0.002 mm clay (%) <0.002 mm) respectively, from various locations as follows: Australia—Gove 13, 40, 47; Australia—Kwinana 30, 30, 40; Canada 0, 47, 36; Jamaica 6–10–75; Spain-12, 50, 38; USA-Texas 8, 66, 26. It can be seen that Jamaica bauxite waste contains, on average, at 75 %, i.e., at least twice as much clay as any of the other bauxites wastes. This suggests the highest concentration of potential dust.

One successful approach in some countries has been the covering of aged red mud with 200-mm of top-soil followed by revegetation (Graham and Fawkes 1992). Another approach uses fresh or salt water irrigation sprays (Graham and Fawkes 1992), but generally, these are economically impractical. Potential topsoil sources in Jamaica are either already under plantation crops such as sugar cane and bananas or protecting other subsoils. Indeed, in any country, removing and transferring topsoil to cover red-muds is destructive to the area from which the topsoil has been removed, and hence is counterproductive.

Already inadequate water resources in the karst bauxite areas are also unavailable for such purposes; large quantities of water diverted to the amelioration of bauxite waste may be viewed by local communities as an "unfair" usage of a scarce resource. Therefore, due to a shortage of topsoil and water, alternative strategies for the reduction of wind-blown dust have to be envisioned for, not only Jamaica, but wherever the Bayer process extracts alumina from bauxite.

1.2 Drawbacks of Gypsum Treatment

As a treatment for decreasing the alkalinity of red muds, gypsum has long been applied, but only after red muds had congealed and dried to a depth of at least 10–20 cm (Harris 2009) in the form of a thick crust, a process often requiring years of atmospheric exposure. Harris (2009) postulated that gypsum-treatment reduces dust by indurating the dried red muds, because gypsum-treatment after several years produces a thick crust throughout the layer so treated, due to subaerial carbonation from the air (Harris and Megahraj 2001). This facilitates the movement of heavy machinery on the surface to subsequently spread more of the gypsum. The solidifying process also permits mechanical comminution by tractors, which is an advantage if the gypsum is to be mixed in with the red mud waste. But such mixing was difficult to accomplish, and was not thorough when done by O'Callaghan et al. (1998). Difficulties of ploughing were attributed to the semi-viscous nature of the red mud at that stage. Such methods reduce dust created by red mud waste (RMW) after they have dried out by atmospheric exposure, and do so only after considerable damage has been done by dust to the surrounding areas. Nevertheless, the strength of this indurated layer exceeds the maximum for penetration by plant roots (Harris 2009). Plant growth is therefore difficult on gypsum-indurated red muds. In an effort to decrease the production of fine wind-blown dust while concurrently enhancing plant root growth conditions, a study using decomposable phyto-organics to bind finely crushed particles (<1 mm diameter) of gypsum-treated red muds is hereby proposed. The objective of this study was therefore to bind dispersible clay particles into larger quasi soil aggregates to resist wind-blown dust.

1.3 Dust Definitions

According to the International Standardization Organization (ISO 4225—ISO 1994a, b), "Dust: small solid particles, conventionally taken as those particles below 75 μm in diameter, which settle out under their own weight but which may remain suspended for some time". The Glossary of Atmospheric Chemistry Terms (IUPAC 1990), characterizes dust as: "Small, dry, solid particles projected into the air by natural forces, such as wind, volcanic eruption, and by mechanical or man-made processes such as crushing, grinding, milling, drilling, demolition, shovelling, conveying, screening, bagging, and sweeping. Dust particles are usually in the size range from about 1–100 μm in diameter, and they settle slowly under the influence of gravity." Based on these definitions, decreasing air-borne dust entails the decrease of sub-100 μm particle sizes. It is at the micro-aggregate size level that excess sodium from the Bayer process produces separation of clay particles. Since micro-aggregates bind <5 μm particles into larger entities at the <250 μm level, and wind-blown dust particles do not exceed 100 μm in diameter, it follows that production of stable micro-aggregates from <5 μm particles should reduce the rate of wind-blown dust.

1.4 Chemistry of Red Mud Dust

Bauxite red mud consists primarily of the insoluble fraction of the bauxite ore that remains after extraction of the aluminium-containing components. Iron oxides (10–30 %), titanium dioxide (2–15 %), silicon oxide (5–20 %) and undissolved alumina (0–20 %) make up the residue, together with a wide range of other oxides which will vary according to the initial bauxite source (Jones and Haynes 2011). The high concentration of iron compounds in the bauxite gives the by-product its characteristic red colour (Fig. 1), and hence its common name "Red Mud" (AAC 2012).

The particle dimension of red mud is usually less than 1 mm (Hai et al. 2014) (actually usually lesser by $<10^{-1}$ of 1 mm), and dust generated from the residue drying area mostly consists of fine clay particles and a small fraction of sodium carbonate crystals (Alcoa 2007). The sodium carbonate is precipitated on the surface of residue as entrained moisture evaporates (Alcoa 2007). Therefore, dry red mud easily spreads into the air and causes dust pollution (Hai et al. 2014).

1.5 Red Mud Dust Impacts

Wind speeds in excess of 6.5 m s^{-1} (23 km h^{-1}) can pick up and transport the fine particles of bauxite mine residue if the dry residue surfaces in toxic red muds are not carefully managed (Alcoa 2007). This was discovered in the Weipa Mines of Western Australia (Alcoa 2007). The distance over which these particles are transported depends on a variety of factors including atmospheric conditions, the size, shape and mass of the particles and the surrounding infrastructure such as roads, embankments and drains (Alcoa 2007). Rolling topography and irregular terrain can increase wind speeds (Jackson and Hunt 1975) and such terrain exists in close proximity to all Jamaica bauxite mine locations (Fig. 1), thereby exacerbating the

(a)

(b)

Fig. 1 **a** Surface crusts in dried red mud bauxite waste form after several years of atmospheric carbonation, producing dusts. Even after decades, vegetation is inhibited. Na^+ prevents particle agglomeration by occupying available attachment sites with its weak, hydrated, single charge. **b** Detailed view of carbonated, dried red mud waste. Note the whitish efflorescence of soluble salts close to, and on, the upper surfaces. Seemingly comminuted particles can be seen between the hardened masses (after Alcoa 2007)

volume and unpredictability of wind-blown dusts. Furthermore, as stated above, Jamaican bauxites contain a far larger clay fraction (75 %) than that of all other major bauxite locations (Wehr et al. 2006).

Though particle size analysis shows that Jamaican Terra Rossa bauxites contain 20 % sand (O'Callaghan et al. 1998), it is the secondary (clay) fraction which contains the bauxite. Similar processing techniques in Jamaica (Bayer process) to that of the Weipa (West Australia) kaolinites indicate similar post-beneficiation clay particle sizes. Above 11 m s^{-1} (40 km h^{-1}), such dust emissions from the residue area can increase rapidly and wind speeds above 14.5 m s^{-1} (50 km h^{-1}) are predicted to be the largest source of dust. Attaining speeds often exceeding 13 m s^{-1} during winter months (Macpherson 1991), unobstructed trade winds in Jamaica and the northern Caribbean should, on the above basis, normally carry fugitive dust released from the dried surface of red mud dumps.

1.6 Effects on Health

Friesen et al. (2009) examined the associations between alumina and bauxite dust exposure and cancer incidence and circulatory and respiratory disease mortality among bauxite miners and alumina refinery workers. Their preliminary findings of the very few cases in the limited population study suggest that cumulative inhalable bauxite exposure may be associated with an excess risk of death from non-malignant respiratory disease and that cumulative inhalable alumina dust exposure may be associated with an excess risk of death from cerebrovascular disease. Nevertheless, neither exposure appeared to increase the risk of incident cancers. Further, there was no apparent danger and no association between every bauxite exposure and any of the outcomes (Friesen et al. 2009), though there was a borderline significant association between every alumina exposure and cerebrovascular disease mortality.

Friesen et al. (2009) also found that exposures to bauxite dust, alumina dust, and caustic mist in contemporary best-practice bauxite mining and alumina refining operations have not been demonstrated to be associated with clinically significant decrements in lung function. Exposures to bauxite dust and alumina dust at such operations were also not associated with the incidence of cancer (Friesen et al. 2009).

Despite these favourable findings, adverse health effects of bauxite dust are possible. This is because red mud residue and sand consist primarily of alumina, silica and iron oxides (USEPA 2004) and silicosis is an irreversible condition with no cure (Wagner 1997). Silicosis is the most common form of pneumoconiosis (lung-related diseases), which is caused by occupational exposure of free silica dust. Treatment options currently focus on alleviating the symptoms and preventing complications. According to Pattajoshi (2006), dust is inevitable in mineral industries, hence the risk factor leading to pneumoconiosis is assessed on the basis of equation:

$R = f\,(d, f, s, i, t,)$

where R = Health Risk, d = particle size, f = concentration of finer dust, s = specific noxiousness, i = individual factor (susceptibility) t = time of exposure. Also, Nouh (1989) reports that a non-occupational form of silicosis has been described that is caused by long-term exposure to sand dust in desert areas, with cases reported from the Sahara, Libyan desert and the Negev, and that the disease is caused by deposition of this dust in the lung. It is thus suggested here, that long, non-occupational exposure to silicates in red mud dust could produce similar outcomes.

1.7 St. Croix—Virgin Islands Red Mud Dust Hazards

Located within the North East Trade Wind Belt, the major concern for northern Caribbean aluminium refining locations is the transportation and deposition of such caustic dust into residential and commercial areas. According to the USGS (2014), St. Croix has a total area of about 80 square miles and is dominated in the north-western (Northside Range) and eastern (East End Range) parts by highland areas that rise to altitudes of 1088 and 866 ft, respectively, and that are underlain chiefly by poorly permeable intrusive, volcanic, volcani-clastic, and sedimentary rocks. The Kingshill aquifer, which lies between these two ranges, is in a low-land that slopes from the south flank of the Northside Range southward to the sea (USGS 2014). The surface of the limestone and marl that compose the aquifer has been deeply eroded, but about one-fourth of it is covered by a blanket of alluvium, alluvial fan, debris flow, and slope wash deposits as much as 80 ft thick, which moderates the dissected topography and forms a broad rolling plain through which low, rounded limestone hills are exposed. St. Croix has four major streams that flow intermittently. All four rise in the Northside Range, and River Gut, the largest, flows southward across the Kingshill aquifer to the sea. The other three streams flow mostly across volcanic rocks to the west coast. The remainder of the streams on the island are very short and flow only after heavy rains (USGS 2014).

The following documentation from the USEPA Report (2012), in September 1998 depicts the Virgin Islands red mud dust problem: "Wind-blown fugitive dust, identified as bauxite by Virgin Islands Department of Planning and Natural Resources (VIDPNR), was deposited onto residential properties located 750 ft north of the red mud site as a result of Hurricane Georges. Observations by VIDPNR indicated that reddish material had been deposited onto the residential structures and their drinking water cisterns. EPA sampling confirmed that bauxite was in the cisterns. VIDPNR issued a Notice of Violation (NOV) to St. Croix Alumina (SCA) for the fugitive dust release, determining that there were no precautionary measures in place to secure the stockpiled bauxite from hurricane winds. Residents located north of the Site have reported red dust deposition on rooftops, inside cisterns, in outdoor gardens, and on indoor furniture associated with hurricanes and other high wind events."

The USEPA Report (2012) states that one of the side effects of gypsum applications is gradual lithifying through atmospheric carbonation. Thus, less than 10 years after treatment, such additives routinely transform red mud ponds into masses of hard, clinker-like cobble-stones of low individual permeability, low porosity and high strength with water movements unsuitable for plant growth. Such physical changes may have increased agglomeration of dust-sized particles, thereby increasing the mass and resistance to wind transport of each entity.

The Kingshill aquifer is located in the central to southwestern parts of the island of St. Croix in the Virgin Islands of the United States. The aquifer consists of limestone and marl that has a maximum saturated thickness of about 200 ft. Although the aquifer produces only small volumes of water of marginal chemical

Fig. 2 The typical rolling landscape of Jamaican mines. Such curved topographies increase the speed of air currents compared to flat terrain. Source: Jamaica Bauxite Institute

quality, it is the only significant aquifer in the U.S. Virgin Islands and supplies a large proportion of the water needed for public supply and industry on St. Croix. Water is scarce on St. Croix. Streamflow is meager and not reliable; aquifers are small and yield mineralized water, much of which is unfit for human use (USEPA 2012).

The USEPA (2012) reports that the contamination poses a potential threat to receptor areas including mangrove wetland areas, water-bird and endangered/threatened species habitats, fisheries, and fresh ground water in the Kingshill aquifer (Fig. 2). The "site conditions and analytical results provide evidence of hazardous substance migration in both the surface water and ground water pathways from on-site disposal areas." (Fig. 3)

According to the USEPA (2012): "VIDPNR reviewed and approved a plan for the pre-design preparation work in October. It calls for St. Croix Alumina to conduct a series of studies, especially to find the boundaries of bauxite residue through test pits and shovel holes and to look into how the stuff behaves in the ground. A hydrology study is needed to evaluate storm water drainage requirements too. Then comes a greenhouse study looking at suitable plant species to cover the site. And SCA must acquire all the permits and approvals for drilling, test pits and the remediation work, according to VIDPNR."

1.8 Particle Cementation

Cohesiveness between clay particles can be significantly increased by inorganic cementing agents such as $CaCO_3$, and Fe and Al oxides (Zhang et al. 2015). In a study of African Ultisols and Oxisols, Ahn (1979) observed highly stable microaggregates not dependent upon organic matter. Therefore, binding agents need not be from organic sources. After adding gypsum to two different red-brown earths, Shanmuganathan and Oades (1984) noted a reduction in the amount of dispersible

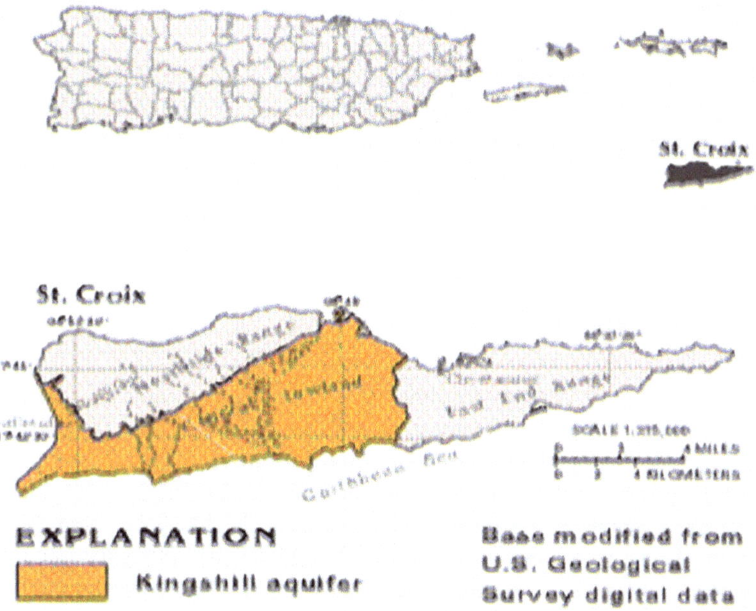

Fig. 3 The Kingshill aquifer is in central and southwestern parts of the island of St. Croix in the U.S. Virgin Islands. *Source* USGS

clay, an increase in the proportion of water-stable aggregates sized 50–250 μm diameter, and an increase in soil friability. Nevertheless, before carbonation occurred, mechanical strength was reduced by gypsum addition.

Due to its divalency, Ca^{2+} binds strongly to clay particles, therefore is not readily lost from the soil via leaching. Ca^{2+} therefore promotes aggregation and is non-toxic. Rengasamy et al. (1984) reported that dialyzed Ca-clays did not disperse in water even after a week, and it has been found that dry aggregates of montmorillonite or illite saturated with Ca^{2+} did not disperse when immersed in water (Rengasamy et al. 1984). Thus, replacement of the sodium on the clay colloids with a non-toxic cation which produces better structure is achieved, as the Ca^{2+} replaces the leached Na^{+} cations (Rengasamy et al. 1984). Applying the above principles, Harris (2009) infused finely grounded phytogenic materials into gypsum-treated remolded red mud aggregates during 6-month incubation. He then detected marginal to significant increases in aggregate stability under wet-sieving as compared to the controls. However, it is hereby postulated that his method would have achieved greater aggregation had he applied a smaller particle size than 2-mm to increase the surface area and the number of reactive surfaces during incubation.

1.9 Aim

Without dislocation there is no dust. The treatment of this study therefore aimed to "fix" the clay particles into soil aggregates. These were to bind the clay particles against dislocation by the wind. This procedure was also aimed at resisting natural field impacts such as rain and running water which dislocate fine clay particles prior to them becoming airborne dust.

2 Materials and Methods

Twenty kilogram of (1) gypsum-lithified red mud (2) red mud beneath the gypsum-treated layer were randomly collected from the Kirkvine Pond 6 Bauxite residue storage area in Jamaica. Whereas the gypsum-lithified red mud (G) exists in rock-hard form in the 0–30 cm zone, the non-gypsum fraction (R) exists adjacently below 30 cm as a semi-viscous constituent, having been seemingly unaffected by the gypsum which O'Callaghan et al. (1998) had ploughed into the zone (G) directly above, in 1996. Both materials are of the same age but very different physico-chemically. The top layer is non-dispersive while the 15–30 cm layer is highly dispersive (Fig. 4). Prior to air-dry, this 15–30 cm depth layer had the physical consistency of untreated red mud waste that supplies wind-blown dust. Samples were crushed at air-dry in a mortar and pestle and passed through a 1-mm aperture sieve. Decomposable organic material as finely chopped (<2 mm) leaves from *Acacia senensis* ("Kasha", an invasive legume of southern Jamaica) was thoroughly mixed in with either above-mentioned bauxite waste in dry, grinded homogenous (<1 mm) form and incubated at room temperature for 42 days at a water content of 60 % of the field capacity of each soil. Of the two crushed red muds, field capacity (water-holding capacity) was substantially higher for the gypsum-treated samples (Table 1).

There were six treatments, each having three replicates: (1) G, (2) G25 (i.e., gypsum-treated red mud + 25 % phyto-organics), (3) G50 (i.e., gypsum-treated red mud + 50 % phyto-organics) (4) R (red mud), (5) R25 (red mud + 25 % phyto-organics), (6) R50 (red mud + 25 % phyto-organics). Phyto-organic treatments are referred to hereafter as GP or RP.

2.1 Measuring Potential Wind-Blown Dust

Clay dislocation: 1st submergence

After incubation for six weeks, the samples at air dry were subjected to submergence in (1) de-ionized water which simulated rain water and (2) ionized water from a tap carrying ground water. Deionized water was included because Khattab

(a)

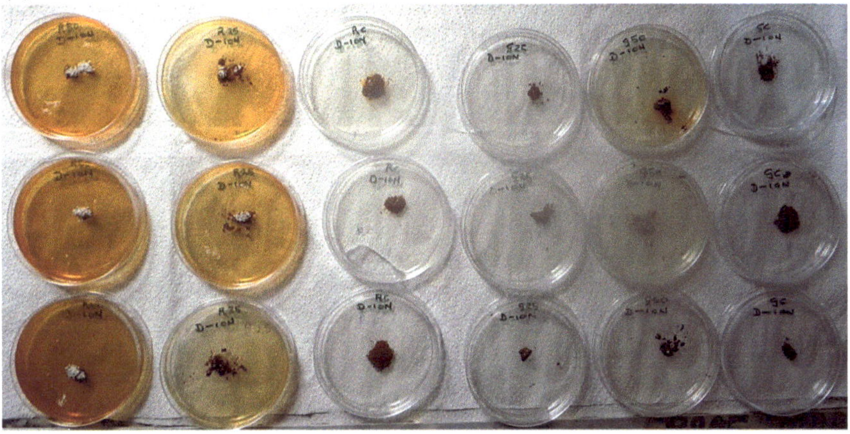

(b)

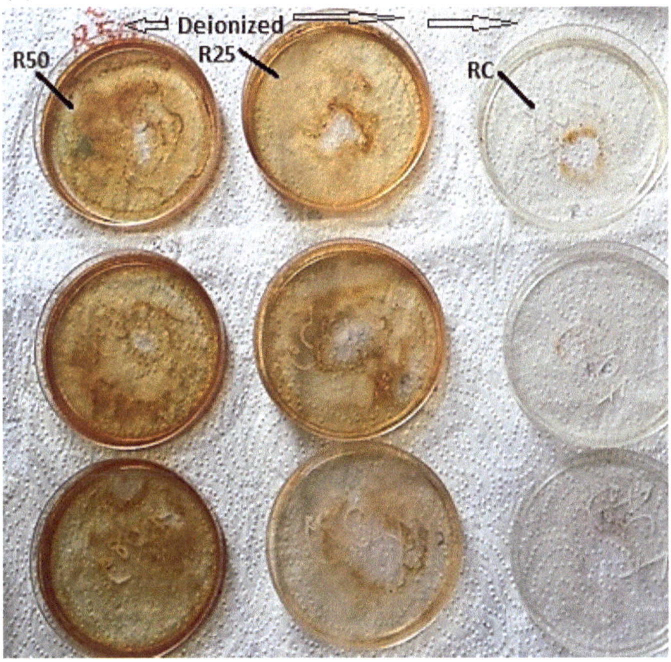

Fig. 4 **a** Before drying, clay dislocation in deionized water for remolded aggregate samples from the 15–30 cm depth in carbonated red mud. At extreme left are the R50 samples; second from left are the R25 samples with less clay dispersion depicted by the *lighter colour*. **b** Dried samples. Clay dislocation in deionized water for remolded aggregate samples from the 15–30 cm depth in carbonated red mud. Samples at left had 50 % decomposable phyto-organics (R50), centre had 25 % (R25), and the 3rd column had no phyto-organics (RC)

Table 1 Properties of red mud subjacent to 0–15 cm depth gypsum-treated red muds

Properties	Values
Red mud pH (saturated extract)	12
EC (saturated extract: dS. m^{-1})	0.25
Organic carbon (%)	0.3
CEC (c molc kg^{-1})	40
Al_2O_3 (%)	16
CaO (% w/w)	7
Fe_2O_3 (%)	47
Na_2O (%)	3
P_2O_5 (%)	2
SiO_2 (%)	3
TiO_2 (%)	6
Bound H_2O (%)	14
Particle size	0.001-mm

and Othman (2013) noted a general reduction of strength of rocks with an increase the number of wet-dry cycles when using distilled water. Replicates of treated aggregates having a diameter of approximately one cm were each placed in a watch glass, after which water was slowly added because the weak structure of bauxite waste is subjected to breakdown by electrolytes. The relative rates of clay dislocation from the aggregates signified the amount of dust particles which would have emerged on drying. Without clay dislocating forces on the aggregates there would be no potential dust. The extent of clay particle dislocation from the aggregates was determined by using weight differences. The clay dislocation process was monitored for the ensuing days until no further change was observed. At air dry, aggregates were removed from the watch glasses and the masses of air-dry residues of fine dust weighed by difference between watch glasses before usage post-usage containing the samples.

Clay dislocation: 2nd submergence
After clay removal from watch glasses at air-dry, the clay dislocation procedure was repeated. Each aggregate was then mechanically agitated in an end-over-end shaker for a pre-determined time period. At air-dry, all particles were again weighed. The separated fractions were air-dried. They were weighed after sieving on a nest of three scales: 2 mm, 500, and 100 μm.

2.2 Flocculation Measurements by Wet-Sieving

For the following reasons, some incubated aggregates were subjected to wet-sieving at air-dry: (1) Aggregate breakdown under rain water (and running water) destroys unstable soil aggregates (2) because wet sieving simulates the action of the electrical and mechanical disintegrating forces inflicted by the high di-electric constant of rain water, (3) the energy inflicted on the red mud waste by wet-sieving, being greater than that imposed by expected daily winds, indirectly indicates resistance of the aggregates to the production of wind-blown dust at high wind velocities. Residues of aggregates disintegrated by wet-sieving were air dried until no further loss in weight. The ratio of <100 μm:>100 μm sized floccules was used to determine the effectiveness of each treatment.

3 Results

3.1 Clay Dislocation: 1st Submergence

3.1.1 In De-ionized Water

While evaporating after submergence samples were observed daily until there was no further change (due clay dislocation) in appearance. After just the first day, and onwards, the inner surfaces of petri dishes containing the 15–30 cm subjacent depth R-phyto-samples became covered with what seemed like a layer of fine <100 μm sized dislocated particles (Fig. 4). On the assumption that this colouration was due to dislocated clay particles, Table 2 shows that after incubation plus air-dry and 5 days of submergence in deionized water, relatively little observable clay particle dislocation occurred from any of the non-phyto-organic treatments as compared to that of the phyto-organic treatments (Table 2; Fig. 4). The R50 dislocated 33 % of the mass of the sample as clay, and the R25 dislocated 29 % of the sample mass for an average of 29 % dislocation by the subjacent phyto-treated samples. The R50 dislocated more clay than did the R25. This was unexpected because these soil particles were considered bound by organic matter.

Further, it can be seen that the quantity of clay dislocated was much greater from the 15 to 25 cm depth compared with the 0–15 cm depth, i.e., in a ratio approximating 15:1 (Table 2). For the 0–15 cm depth the samples containing 25 % phyto-organics dislocated no clay, while the samples containing 50 % phyto-organics produced a 4 % w/w rate of clay dislocation. Thus for the 0–15 cm depth, rate of phyto-organics determined effectiveness to suppress clay dislocation in deionized water. The reverse is true at the 15–30 cm depth (Table 2), where the 25 % phyto-organics treatment dislocated approximately 25 % more clay than the 50 % phyto-organic treatment. This was surprising. Clearly, the factor which produced the initial dislocation in deionized water existed in the phyto-samples. Yet, the greater concentration of phyto-material the smaller was the clay dislocation. This anomaly is examined later.

Table 2 Clay (% of mass) dislocated after incubation treatments and submergence in de-ionized water for 8 days

Red mud (depth in cm)	0 % phyto-organics	25 % phyto-	50 % phyto-	Average
0–15	0	0	4	2
15–30	0.01	33	25	29

After 8 days of submergence in deionized water, non-phyto-organic treatments from all samples dislocated very little clay above the level of detection (Table 2; Fig. 5).

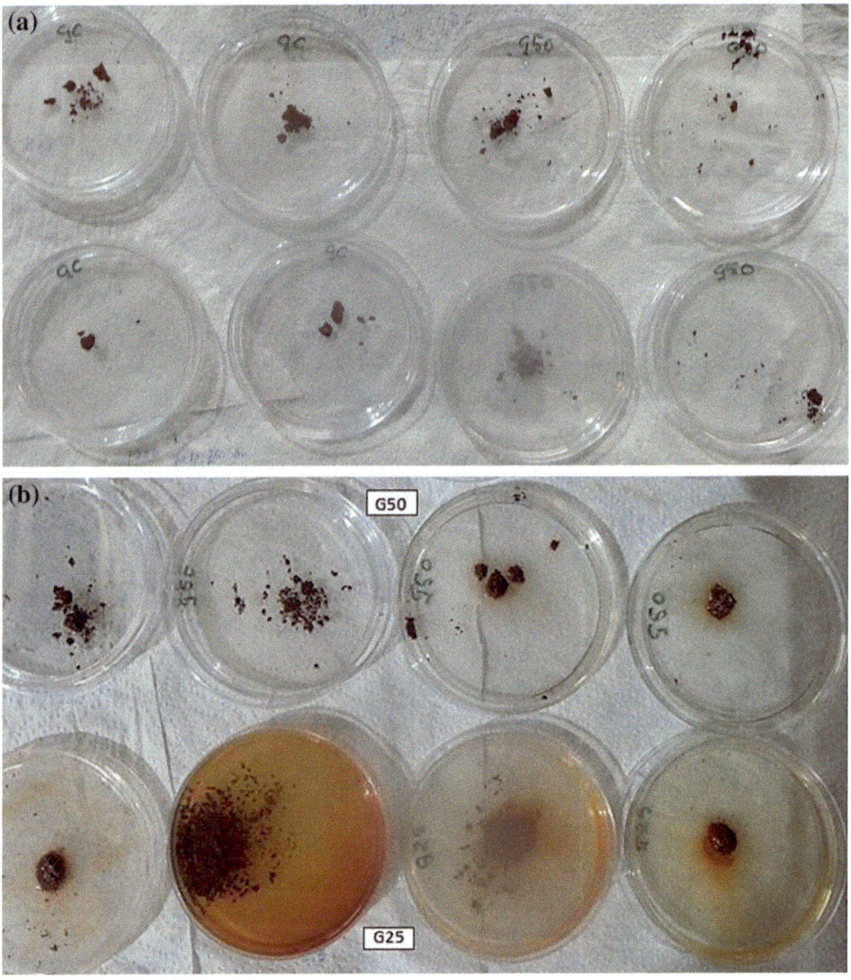

Fig. 5 **a** Post-incubation gypsum-treated samples in distilled water after 1st submergence. The *colourless water* indicates that even after 8 days of submergence, no dislocated clay particles were observed at the <50 μm diameter size. The four samples at right which were treated with 50 % w/w phyto-organics show only a slight discoloration. Effective flocculation of red mud individual particles therefore occurred in all the above samples. **b** Post-incubation gypsum-treated samples in non-deionized water after 1st submergence. In contrast to the reaction with deionized water (Fig. 5a) where the colourless water indicated no clay dislocation, here, samples from within the same treatment dislocate clay when added to non-deionized water, especially the samples with the lower rate (G25) of added phyto-organics. The possible explanation could lie in an increased number of negative charges in the system (see text)

3.1.2 Non-deionized Water

Non-de-ionized water also dislocated clay from the samples but there were minor but significant differences. Firstly, the total mass of clay dislocated was higher for the deionized water (Table 3), where the 15–30 cm level dislocated almost twice as much clay under the deionized water compared to the non-deionized water. On the other hand, for the 0–15 cm level, the non-deionized water caused a 4-fold increase in clay dislocation compared to the deionized water (Fig. 6). Again, only the phyto-treatments caused substantial clay dislocations.

For the G samples (no phyto-), submergence in non-deionized water produced small, barely visible levels of clay dislocation (Table 3). This nevertheless was greater than that for de-ionized water, where no dispersion appeared from any of the non-phyto samples. Similarly, from the 0 to 15 cm depth, only those GP samples treated with 25 % phyto-organics increased clay dislocation up to half the level of the phyto-organically treated R samples. Of all the phyto-organic treatments, GP samples treated with 50 % phyto-organics exhibited the lowest levels of clay dislocation from the aggregates (Table 3). The possible reasons are discussed later.

The clay dislocations in deionized water were less severe for the G samples. Only 2 % of the GP mine-waste dispersed, whereas 29 % of the R samples dislocated in the form of clay particles. This is almost a 15-fold difference. As electrical stresses on soil aggregates vary directly with purity of water, a high concentration of electrolytes (i.e., >EC) in a soil should increase clay dislocation in deionized water. Thus the R samples, with a higher electrical conductivity than the G samples, released more clay particles in deionized water than in non-deionized water.

The relative rates of clay dislocation from the aggregates signified the amount of dust particles which would have emerged on drying. After incubation for 6 weeks, the result for either the G or R samples was that little or no observable clay dislocation occurred in deionized or non-deionized water.

In deionized water, gypsum-treated bauxite waste without phyto-organics (GC) showed no dispersion of clay-particles at the microstructural level (<100 μm diameter) for the 0–15 cm-depth. For non-phyto samples taken from the subjacent level, a very slight clay dislocation effect was visible. This was below the detection level of the weighing equipment. However for the RP samples a large amount of dust averaging 25 % w/w would have been generated by phyto-organic treatments. This was unexpected, and possible reasons are explored later. Based on these results, the GP samples were overwhelmingly less dispersive than the RP samples.

Table 3 Clay (% of mass) dislocated by non-deionized water after incubation treatments

Red mud (depth in cm)	0 % phyto-organics (controls)	25 % phyto-organics	50 % phyto-organics	Average
0–15 (G)	0	8	8	8
15–30 (R)	0	16	16	16

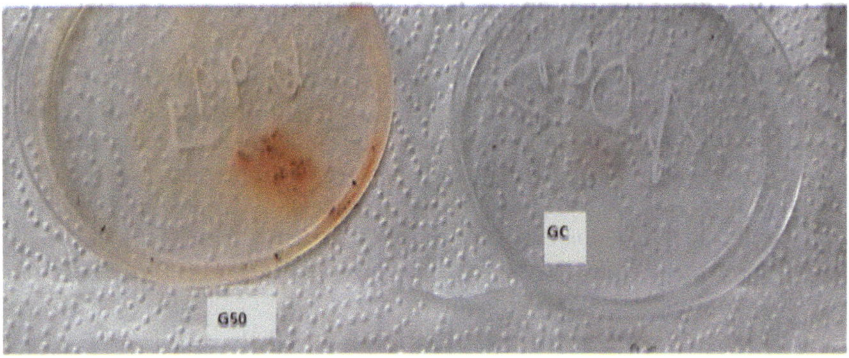

Fig. 6 Post-incubation gypsum-treated samples in non-distilled water after the 1st submergence. Though it can be seen that the non-phyto-treatment at right dislocated very little clay at this stage (hence almost colourless water), dislocated clay particles were observed from the 50 % phyto-treatment at *left* (G50). This strange reaction by the G50 (discussed later in the text) contrasts starkly with that which occurred in de-ionized water. Stippled background contains 1-mm diameter circles

3.1.3 Clay Dislocation: 2nd Submergence

After removal from Petri dishes at air-dry, and after the second episode of clay dislocation in water the remolded aggregate masses were weighed and measured. Even after the 2nd submergence, the phyto-organically treated samples in general retained stability in water to a far greater degree than that of the 1st submergence. Clay dislocation procedure was repeated on the aggregates and the results show that no dislocation occurred at all from any sample. It can be seen that the 2nd submergence produced no dislocation of clay particles (Table 4) as the total mass for each sample submerged was zero.

In comparison to the low levels of dislocation observed from the G, R, and G50 treatments, the level of clay dislocation exhibited by all RP treatments was more than 10 times as high. However, for all phyto-organic treatments, especially in the case of the RP, the second submergence produced far less clay dislocation (>10-fold decrease) than the first submergence (Table 4). On the established fact that totally untreated Bayer red muds are (1) totally dispersive (AAC 2012) and (2) hence totally disaggregated (Bardossy 1982), clay dispersion would therefore have been far greater in such totally untreated red muds than had occurred with the

Table 4 Mass of clay (grams) dislocated from aggregates after shaking in water (wet-sieving) and 2nd submergence for 8 days

Red mud (depth in cm)	0 % phyto-organics (controls)	All phyto-organics	Total mass	Difference
0–15	0	0	0	0
15–30	0	0	0	0

treatments of this study. Overall, greater clay dislocation was more evident from the de-ionized samples, especially the R and RP samples.

The lack of dislodged clay particles from the 2nd submergence could have meant that all the dispersible clay had been released at the first submergence (Table 4). To test this assumption, the aggregates were again air-dried and subjected to dry sieving in a shaker where the aggregates were rolled and bounce on a sieve having a 100 μm aperture. Each aggregate was then mechanically agitated in an end-over-end shaker for a pre-determined time period. At air-dry, all particles were again weighed. The separated fractions were viewed and measured under magnification using a microscope with a graticuled lens. In the rating scheme for this study, the clay dislocation ratio for the RP as compared with the GP samples was >10:2.6.

Therefore, as (1) the 2nd submergence of this study caused no observable clay dislocation, the maximum total potential dust for any treatment remained at 29 % w/w. After the 2nd submergence, samples were air-dried and shaken in a sieve having apertures 100 μm in diameter. Results are shown in Table 5. It can be seen that, whereas the phyto-treatments continued to dislocate very little clay even during dry sieving, the non-phyto-treatments dislocated at least twice as much clay as the least efficient phyto-treatment during this procedure. In fact, R samples (non-phytogenic-) dropped 60 % of their mass from the dry-sieve as dislocated <100 μm clay particles (Table 5). As noted in Sect. 1.0 above, among the world's bauxite ores Jamaican bauxite texturally contains by far the largest clay fraction at 75 % (Alcoa 2007). Therefore, had this been a field study, it is reasonable to assume that the R (non-phyto-organic) samples would have potentially released 42 % of its mass as wind-blown clay. Again, as stated above, several variables determine the actual concentration of wind-blown dust at any given location. Therefore no attempt is made here to quantify the potential rate of wind-blown dust. Nevertheless, it can be seen that despite being by far the least effective treatment of this study, the R (non-phyto-) samples reduced potential wind-blown dust by 33 % (i.e., 75 − 42 = 33). Using Tables 2, 3, 4 and 5, similar calculations estimate the potential dust reduction by other treatments as follows:

Table 5 Proportion of <125 μm particles after dry-sieving of aggregates following two submergences in water for a total of 15 days after incubation

Red mud (depth in cm)	0 % phyto-organics (controls)	25 % phyto-organics	50 % phyto-organics	Total
0–15 (G)	4	2	1	7
15–30 (R)	60	2	0	62

Table 6 Impact of large falling water drops on remolded aggregates (days) after incubation and submergence in water for 15 days as depicted by the counted number of drops (CND) before shattering

Red mud (depth in cm)	0 % phyto-organics	25 % phyto-organics	50 % phyto-organics	Total
0–15	150	30	430	670
15–30	50	70	330	450

$$
\begin{aligned}
G50 &= 75 - (4 + 8 + 0 + 1) = 67\,\% \\
G25 &= 75 - (8 + 0 + 2 + 3) = 67\,\% \\
R50 &= 75 - (16 + 0 + 0 + 13) = 56\,\% \\
R25 &= 75 - (16 + 0 + 2 + 3) = 54\,\% \\
G &= 75 - (0 + 0 + 0 + 4) = 71\,\% \\
R &= 75 - (.01 + 0 + 0 + 60) = 14.99\,\%
\end{aligned}
$$

Prior to the test of aggregate stability, the leading treatment for dust reduction here is still gypsum (71 %), with gypsum + phytogenic being second at 67 %. Predictably, without gypsum, incubation yielded only 15 % dust reduction for non-gypsum treatments.

3.1.4 Aggregate Stability

In a physical sense, resistance to shattering impacts reduces dust production. After 15 days, the samples were subjected to falling drops of water which simulated raindrops from a pre-determined height. The results showed that the phyto-organically treated samples at 50 % from either the 0–15 cm depth or the subjacent red mud withstood more than twice as many impacts before shattering, as each of the non-phyto-organically treated samples (Table 6). Additionally, the G50-proved substantially more resilient in this respect than the R50 samples (Table 6). Without organics, the G aggregates withstood three times as many raindrop impacts as the R aggregates. The overall G treatment produced 50 % greater resiliency under simulated raindrop impacts than the R treatments (Table 6). However, the counted number of drops (CND) resistance of the R25 exceeded that of the G25 samples by 50 %. This apparent anomaly is discussed later.

3.1.5 Stability in Stagnant Water

See Table 7.

Table 7 Longevity of remolded aggregates (days) after incubation and submergence in water for 15 days

Red mud (depth in cm)	0 % phyto-organics (controls)	25 % phyto-organics	50 % phyto-organics	Total
0–15	2	3	13	21
15–30	12	3	13	28

Table 8 Aggregate longevity in minutes of stability during wet-sieving after incubation treatments

Red mud (depth in cm)	0 % phyto-organics	25 % phyto-organics	50 % phyto-organics	Total
0–15	11	2	2	15
15–30	12	3	13	28

3.1.6 Stability in Moving Water

A vast difference of water-stable longevity amongst treated samples can be seen, because samples from the 15–30 cm depth with either 50 % phyto-organics (R50) or without phyto-organics (R only) are by at least a 5-fold margin the most resilient under wet sieving (Tables 7 and 8). Aggregates subjected to wet-sieving maintained their integrity in the following order of decreasing longevity: R50 = G50 >R = G25 = R25 = G25 (Tables 7 and 8).

4 Discussion

4.1 Fine-Particle Dislocation

As dust results from dislocated clay, and contact with water weakens soil aggregates and dislocates clay (Rengasamy et al. 1984), the most resilient aggregates are the lowest potential producers of dust. It can be seen that despite being submerged in still water for 15 days, the phyto-organic (P) samples outlasted all others. In fact, before the end of day 1, several non-P samples had already disintegrated (Table 6), some having done so within a few minutes of submergence.

For aggregates without phyto-organics from the 0–15 cm previously gypsum-treated layer, relatively few measurable clay particles (potential dust) were dislocated during submergence in water for several days. At the end of day 2 of submergence, those samples incubated from the 15–30 cm layer (i.e., under the gypsum-treated layer, without phyto-organics) released no visible clay particles. Wind power increases exponentially with wind speed such that the power of the wind varies as the cube of its speed (Kovarik et al. 1979). Thus an increase of even 2 or 3 km h^{-1} is a significant increase in terms of energy expended and hence the amount of dust removable from a surface. As Na^{+} causes mutual repulsion among clay particles, and the 15–30 cm depth contained Na^+ far in excess of the surface layer, the low level of clay dislocation from that depth was unexpected. This is further discussed later.

In the first submergence, the gypsum plus phyto-organics treatment dislocated >5-times the amount of fine clay particles (potential dust) compared to the amount from the gypsum-only treatments. This also was against expectations because applying similar treatments to sodic (Na^+-rich or high ESP) subsoils, Harris and Rengasamy (2004) used sub-micron particle analysis to show that phyto-organics with gypsum substantially flocculated (agglutinated) clay particles. Thus by increasing the mass sizes, they decreased the release of individual clay particles. Further, for submergence #1, the greatest reduction of potential dust in this study occurred as follows in descending order of efficiency:

G >R >G50 >G25 >R50 = R25.

Decomposable phyto-organics have been shown to bind sodic clay particles (Harris and Rengasamy 2004) and inorganic particles in bauxite waste (Harris 2009), and as discussed later, wet-sieving results showed that strong binding of inorganic particles also occurred in this study. Yet, for the ultra-fine (highest dust-potential) inorganic colloids, such a binding force from phyto-organics proved, particularly at the lower phyto-organic rates, ineffective initially in this study. Isomorphic atomic substitution causes the negative charge exhibited by all clay particles. Muneer and Oades (1989) noted an increased number of negatively charged particles formed as decomposition and breakdown of phyto-organics advances in remolded aggregates. They concluded that small negative charges from organic colloids which were the same approximate length as clay particles, were dispersing the clay particles. In this study, it can be seen that, compared with non-phyto-organic-treatments, it was the phyto-organic samples that produced more clay dislocation in the 1st submergence, especially the phyto-organics at the lower rate (R25, and to a lesser extent, the G25). On the other hand, though they readily broke down even in still water, non-phyto-treated macro-aggregate samples from the 0–15 cm depth (G) did not dislocate ultra-fine clay particles when those macro-aggregates were submerged in water (Table 1). Furthermore, the following observations were made:

- No clay dislocation from the G50 or G25 samples occurred in deionized water
- Between G50 and G25 samples, clay dislocation in non-deionized water was greater from G25

Clay dislocation by the G25 and G50 in the non-deionized water could have been due to a greater number of negatively charged colloids (organic and/or inorganic) intrinsic to that water as compared to the deionized (less contaminated) type. On this assumption, the total mass of negative colloids would be greater in the non-deionized system. The G25, with less microbial substrates than the G50 and hence with more competition among microbial decomposers, would have been expected to have undergone more advanced breakdown of organic matter, thereby presenting smaller colloids to the system. As noted earlier, highly comminuted negative colloids of similar sizes mutually repel. It is thus more likely that of the two treatments, the G25 would have been more likely to contain organic colloids fine enough to repel clay particles of similar sizes. Such electronic forces could have dislocated the clay particles at a higher rate from the G25 than from the G50 aggregates.

For similar reasons, in a study incubating gypsum with sodic sub-soils, Harris and Rengasamy (2004) observed no stable macro-aggregates (i.e., >250 μm) at any stage of incubation. They concluded that gypsum alone cannot produce stable macro-aggregates, though stabilization occurred at the <250 μm size level. They also reported that phyto-organic additions caused a slight increasing trend.

The ability of negatively charged clay colloids to disperse clay particles is illustrated by the immediate settling out by flocculation of clay colloids or organic colloids whenever small amounts of multivalent cations from $Al_2(SO_4)_3$ (alum) are added to such suspensions. It is very likely that after the first submergence, most or all of the organic colloids became trapped in the dried residue. This may explain why, after dislocating such high levels of clay in the first submergence, at submergence #2, the phyto treatments failed to dislocate any clay at all.

4.2 *Natural Flushing*

The natural flushing by rainfall for more than a decade (Harris 2009) would have removed much sodium sulphate in the 0–15 cm layer. This relative increase in the ratio of Ca^{2+} to Na^+ ions promoted the binding action among clay particles. However, annual rainfall also subsequently depleted the unprotected Ca^{2+} ions in the upper (0–15 cm) layer. It was expected that phyto-organic gums from the decomposing phyto-organics, would have induced stable aggregation among the G25 and G50 samples. This did not occur. On the other hand, with the same levels of gypsum in the same bauxite waste, Harris (2009) observed stabilization of macro-aggregates. The explanation may be that Harris (2009) incubated the samples for a much longer period of time (more than 24 weeks) compared to the 6-week incubation of the present study. With more advanced levels of decomposition, stronger binding between inorganic particles may have occurred in that earlier study.

Aggregate stability

Yet, paradoxically, in this same study of just six weeks, underneath the surface, the 15–30 cm layer, despite having massively accumulated Na^+ ions from the prolonged elluviation from the 0–15 cm layer plus residual Na^+ from the Bayer Process, produced stable macro-aggregates of 6-fold the longevity under wet-sieving than those produced by the N^+-deficient 0–15 (surface) layer. Moreover, as stated above, Harris and Rengasamy (2004) concluded that with or without gypsum applications, phyto-organics cannot macro-aggregate sodic clays to a level of water-stability, though stable micro-aggregation (<250 μm) can be achieved. The object of flushing is to remove the Na^+ in the form of Na_2SO_4 from the exchange reaction with $CaSO_4$ to provide greater opportunities for the adsorption of Ca^{2+} onto clay particles. Yet despite the obvious absence of flushing at the 15–30 cm level (indicated by high Na^+ levels), strong Ca^{2+} adsorption as proven by high aggregate stability, had clearly occurred subsequently (i.e., after gypsum addition to the superjacent layer). The only samples to exhibit efflorescence were the phyto-treated R samples (R25 & R50). Therefore, this white efflorescence on the outside of the aggregates (Fig 4a: the six samples on the left) may more clearly explain the substantial resilience of the R50 aggregates during wet-sieving. In this particular system, the white efflorescence could be drawn only from the following major group of substances—Na_2SO_4, NaCl, Na_2CO_3, $CaSO_4$, $CaSO_4.2H_2O$, $CaCO_3$, $Ca(OH)_2$. All forms of sodium sulphate, sodium chloride, and sodium carbonate are white solids that are highly soluble in pure water. The efflorescence exhibited relatively low solubility in water at <0.0030 kg L^{-1} in deionized water at 20 °C. As this is substantially lower than that of all the above substances except gypsum and calcium carbonate, the efflorescence is comprised of either calcium carbonate or gypsum, or a combination of both. The explanation may be that with such unusually high levels of Ca^{2+} in the 15–30 cm level, exchange sites became dominated by the more strongly bound Ca^{2+}. Purely from the standpoint of stability, this nullified the need for Na_2SO_4 flushing. Interestingly, as stated above, the R25 samples dislocated more clay than all other treatments. It is therefore surprising that the R25, despite the above-mentioned statements, exhibited the least resistance to wet sieving. However the relatively low level of phyto-organics, compared to the R50, may explain this condition (Tables 7 and 8).

On the other hand, the lower resistance to wet-sieving of the 0–15-cm gypsum-treated phytogenic macro-aggregates suggests an inactivation by carbonation to $CaCO_3$ of a substantial quantity of the Ca^{2+} released in the early stages of the gypsum treatment applied several years before this study. Being exposed sub-aerially, greater atmospheric CO^2 would have been available in the 0–15 cm depth compared to the subjacent layer.

Another possible explanation for the longevity of the R samples to wet sieving entails thixotropically affected entities, referred to by Coughlan et al. (1973) as “fortuitous agglomerates.” These fortuitous agglomerates survive wet-sieving but are not true aggregates, being case-hardened during storage possibly in this case by oxidation of iron oxides.

As stated above (Sect. 1.2), based on Wehr et al. (2006), the total clay fraction for the Jamaica bauxite waste is 75 %. On the assumption that all clay particles are <100 μm in diameter, potential dust in Jamaica bauxite waste, according to his table, is at least 84 % w/w. As the best treatments of this study increased more than 95 % of the particle size of the bauxite waste to well over the <100 μm-diameter threshold after submergence in de-ionized water, it is concluded that the phyto-organic treatments substantially decreased potential dust not just from Jamaica bauxite wastes, but potentially for all other wastes listed above. This includes red muds from St. Croix (Virgin Islands), Guyana, and Haiti.

The relatively high level of aggregate stability under wet-sieving observed for the crushed G (no phyto-organics) samples in this study was not expected because an opposing result occurred in a study of the same lithified red mud waste by Harris (2009). In yet another study, Harris and Rengasamy (2004) also found that no macro-aggregate stability in a sodic sub-soil when treated with gypsum.

However, in both of those studies, they used <2 mm particle sizes, whereas in the current study the particle size used was <1 mm. Particles must be fine enough to provide a sufficient reactive surface area for the solid-state chemical reactions (Mehta and Monteiro 1993). It is thus postulated that the larger surface area (more than twice as large) of the smaller particles of this study afforded far greater opportunities for cohesive forces of the Ca^{2+} ions among inorganic particles.

It is the binding action of organic cements and electronic attraction of cations that produce aggregation and, by extension, can reduce dust formation in sodic soils and spoil heaps. It is interesting that both the mine tailings of the present study and the soil studied by Harris and Rengasamy (2004) reacted similarly to phyto-organic additives. Neither sodic entity was influenced by phyto-organics acting alone. In other words, without gypsum, sodic clays do not react significantly with phyto-organics to from stable macro-aggregates. This is primarily because organic colloids and sodic clays are mutually repulsive, being both negatively charged. Without multi-valent cations there are few positive charges to link the clay particles. Harris and Rengasamy (2004) found that the role of gypsum is to flocculate the clay particles, as has been found by Baldock et al. (1994); and Muneer and Oades (1989), where Ca^{2+} in soil solution did not improve macro-aggregation. In this study, however, gypsum acting alone stabilized macro-aggregates from finer inorganic particles.

Applying similar treatments to sodic subsoils, Harris and Rengasamy (2004) found that clay particles had been aggregated by decomposing phyto-organics into a size range not exceeding 30 μm. In this study, agglomerates from the RM + phyto-organics treatment remained stable under physical pressure, and did not release dust particles. Yet, these agglomerates of varying sizes up to 4 mm from the RM + phyto-organics treatments were readily unstable under wet-sieving (Table 2), as was the case for those produced in an experiment conducted by Harris and Rengasamy (2004). The implications here are that under field conditions, such stability would be destroyed during the first rain shower, and that dust would be produced on drying of the crumbled masses. However, destruction of unstable macro-aggregates produce progressively smaller aggregates with an inverse change in soil strength (Dexter 1988) caused by stronger inter-particle binding mechanisms (Seguel and Horne 2006). Thus, below a particular wind speed, micro-aggregates above dust-size produce less dust per volume than macro-aggregates. As reported above, dust particles of up to 100 μm can be airborne. As soil micro-aggregates range up to 250 μm (diameter), an appreciable proportion can be airborne. Though micro-aggregate size was not measured in this study, Harris and Rengasamy (2004) found that micro-aggregates which formed under similar conditions had an average size of 30 μm (diameter). It is therefore reasonable to conclude that a large proportion of micro-aggregates in this study would be below 100 μm, and hence susceptible to creating dust hazards.

Based on the above, it is clear that (1) the water-stability of red mud waste with or without added phyto-organics varies directly with addition of gypsum (2) water-stability of red mud waste determines dust production levels. Under specific conditions of this study, water-stability was achieved not only during wet-sieving trials, but prior to that, during two episodes of several days of submergence.

5 Conclusions

The potential effect of gypsum treatments and phyto-organics on quantities of wind-blown dust from red mud wastes was examined. Gypsum treatment decreased dust-sized clay particles by >80 %. Average dust particle sizes of <100 μm in original samples were transformed in the gypsum-treated samples in soil crumbs. The average dust particle size showed a >4-fold increase over that of the controls. The size of dislocated particles was <100 μm. Under the influence of simulated de-ionized water, decomposable phyto-organic additives initially increased small quantities of the finest clay particles. It is likely that this minor dislocation was due to an increase in negatively charged organic colloids formed by phyto-organic decomposition during incubation. Nevertheless, increasing the size of particle clusters in this study potentially produced greater resistance against wind saltation and airborne movements.

6 Geotechnical Applications

6.1. One successful approach to dust reduction is to spray bitumen emulsions onto red mud embankment walls.

6.2. Unfortunately, the waste oil used has a short life-span in the environment due to its chemical reaction with caustic (NaOH and its derivatives), and natural biodegradation (Alcoa 2007). Studies have shown that the oil is undetectable below 2 m (Alcoa 2007). However the protection of the ground water below and around the tailings dump can be guaranteed only with an effective waste dump seal. Since such protective seals were never installed at the base of Caribbean red mud waste ponds, the bitumen emulsion could seep into those underlying aquifers. Moreover, as added gypsum removes the caustic in the following reaction:

$$Ca(SO_4) \cdot 2H_2O + 2NaOH \rightarrow Na_2SO_4 + Ca(OH)_2 + 2H_2O, \quad (1)$$

The bitumen emulsion would not degrade as easily wherever gypsum had been previously applied. In Jamaica, gypsum had been applied at over 10 t/ha, as has been the case for Kirkvine Pond 6 (O'Callaghan et al. 1998). Potentially, therefore, excess hydrocarbon oils could remain in gypsum-treated red muds after application of bitumen emulsions. Such excess oils in the environment create non-wetting soils (Ward et al. 2015). Non-wetting soils are often prone to dispersion and structural breakdown (Ward et al. 2015).

6.3. A non-toxic regime is preferred as follows:
Cover dry solid red mud waste with approximately 30 cm of incubated dried red mud mixed with grinded phytogenic material. Roughen the material before adding grass seeds.

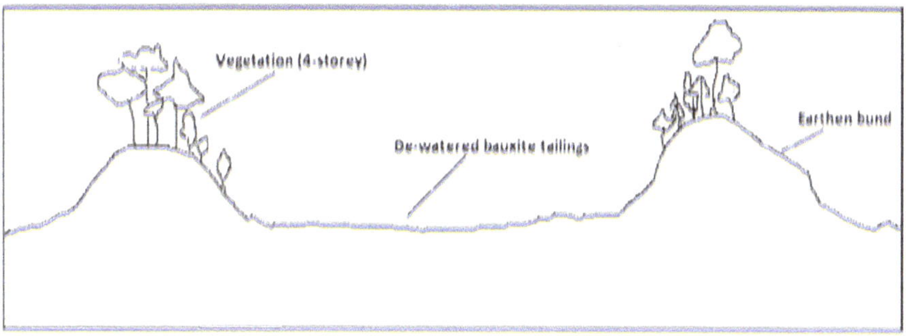

Fig. 7 Earthen bunds planted with native vegetation comprised of several storeys for trapping wind-blown dust

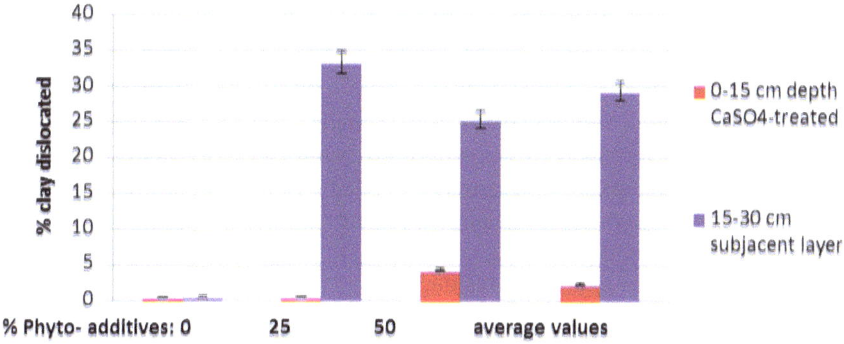

Fig. 8 Mass of clay dislocated after incubation and soaking in deionized water for 8 days

To conserve water, cover with thin polythene to simulate a horizontally extended glass house. After several weeks, add grass seeds.

6.4. Other field methods:

Raised banks (bunds) can be planted with wind-breaking vegetation to trap dust (Figs. 7, 8), being non-toxic compared to bitumen emulsions (Fig. 9).

6.5. Phosphorus depletion

P-fixing in such acidic soils prevents the successful growth of most plants except *Cajanus cajan*, which has the ability to extract P from P-fixing soils. *C. cajan* exudes piscidic acid which enhances the availability of phosphate from iron-phosphate and rock-phosphate (Ae et al. 1990). Thus, unlike other legumes, pigeon pea is one of the few crop species that can utilize iron bound P efficiently making it capable of binding clay particles under P-limiting conditions.

6.6. On the other hand, growth media, by producing aggregation of individual soil particles, can minimize dust from bauxite mine tailings (residues).

Fig. 9 Bitumen emulsion being sprayed to reduce dust from embankment walls. Adapted from Alcoa World Alumina Australia, March 07.

6.7. As bauxite is a residual product containing extremely fine particles, suspensions of red mud are ubiquitous wherever bauxite is mined. Hence mud reduction strategies are trans-continentally applicable. In Queensland, Australia (Alcoa 2007), specific dust controls employed by Alcoa to minimize dust generation include:

6.8. Turning over the mud in the residue area;
The use of sprinklers and water carts;
Spraying exposed banks with bitumen;
Applying rock aggregate to large areas that are not required in the short term;
Road management through application of emulsified waste oil as a dust suppressant and
Restricting access and; planting grasses or native vegetation.

6.9. However, most of the above efforts each have major drawbacks ranging from high cost (sprinklers and water carts, applying aggregate) to environmental unsustainability (waste oil, bitumen). Hence, for the developing Caribbean region, the most sustainable from the standpoint of cost and environmental damage are turning over the mud in the residue area (a short-term corrective) and planting grasses or native vegetation (a long-term strategy). For the latter strategy, the major problem is the edaphically hostile growth media of red muds including pH >11, and lack of aeration.

References

AAC (Australian Aluminium Council) (2012) Bauxite residue in Australia www.alluminium.org.aubauxite.world-aluminium.org/refining/bauxite-residue-management.html

Ae N, Arihara J, Okada K, Yoshihara T, Johansen C (1990) Phosphorus uptake by pigeonpea and its role in cropping systems of the Indian subcontinent. Science 248:477–480

Ahn PM (1979) The optimum length of planned fallows. In: Mongi HO, Huxley PA (eds) Soils research in agro-forestry. Nairobi, pp 15–39

Alcoa (2007) Pinjarra Alumina refinery efficiency upgrade. Alcoa World Alumina Australia. Dust management plan for the Alcoa Pinjarra bauxite residue disposal area

Baldock JA, Aoyama M, Oades JM, Susanto Grant CD (1994) Structural amelioration of a South Australian red brown earth using calcium and organic amendments. Aust J Soil Res 32:571–594

Bardossy G (1982) Karst bauxites. Elsevier Scientific Publishing Co, USA

Coughlan KJ, Fox WE, Hughes JD (1973) A study of the mechanisms of aggregation in a krasnozem soil. Aust J Soil Res 11:65–73

Dexter AR (1988) Advances in characterization of soil structure. Soil Till Res 11:199–238

Friesen MC, Fritschi L, Del Monaco A, Benke G, Dennekamp M, de Klerk N, Hoving JL, MacFarlane E, Sim MR (2009) Relationships between alumina and bauxite dust exposure and cancer, respiratory and circulatory disease. Occup Environ Med 66:615–618. doi:10.1136/oem.2008.043992

Graham GA, Fawkes R (1992) Red muds disposal management at QAL. Proceedings of an international bauxite tailings workshop. Australian Bauxite and Alumina Producers, Pereth, Western Australia

Hai LD, Khai NM, Quy TV, Huan NX (2014) Material composition and properties of red mud coming from alumina processing plant Tanrai, Lamdong, Vietnam. Int J Res Earth Environ Sci. ISSN 2311–2484 http://www.ijsk.org/ijrees.html

Harris MA (2009) Structural improvement of age-hardened gypsum-treated bauxite red mud waste using readily decomposable phyto-organics. Environ Geol 56:1517–1522. doi:10.1007/s00254-008-1249-5

Harris MA, Megharaj M (2001) The effects of sludge and green manure on hydraulic conductivity and aggregation in pyritic mine tailings materials. Environ Geol 41:285–296. doi:10.1007/s002540100434

Harris MA, Rengasamy P (2004) Sodium affected subsoils, gypsum, and green-manure: Inter- actions and implications for amelioration of toxic red mud wastes. Environ Geol 45(8):1118–1130

IUPAC (1990) Glossary of Atmospheric Chemistry Terms (1990). http://old.iupac.org/reports/1990/6211calvert/glossary.html. Accessed 8 June 2015

Jackson PS, Hunt JCR (1975) Turbulent wind flow over a low hill. Quart J R Met SOC (1975). 101, pp 929–955

Jones BEH, Haynes RJ (2011) Bauxite processing residue: a critical review of its formation, properties, storage, and revegetation (2011). Crit Rev Environ Sci Tech 41:271–315

Khattab S, Othman H (2013) Durability & strength of limestone used in building. Al Rafidain Eng 21(3)

Kovarik T, Pipher C, Hurst J (1979) Wind Energy. Quality Books Inc, NSW

Li LY (2001) A study of iron mineral transformation to reduce red mud tailings. Waste Manage 21(6):525–534. doi:10.1016/S0956-053X(00)00107-0

MacPherson J (1991) The making of the West Indies. Longmans

Mehta PK, Monteiro OJM (1993) Concrete: Microstructure, Properties and Materials. McGraw Hill, USA

Muneer M, Oades JM (1989) The role of Ca_2 organic interactions in soil aggregate stability. II. Field studies with 14C-labelled straw, $CaCO_3$, and $CaSO_4\_2H_2O$. Aust J Soil Res 27:401–409

Nouh MS (1989) Is the desert lung syndrome (nonoccupational dust pneumoconiosis) a variant of pulmonary alveolar microlithiasis? Report of 4 cases with review of the literature. Respiration 55(2):122–126. doi:10.1159/000195715. PMID 2549601

O'Callaghan WB, McDonald SC Richards DM, Reid RE (1998) Development of a topsoil-free vegetative cover on a former red mud disposal site. Alcan Jamaica rehabilitation project paper

Pattajoshi PK (2006) Assessment of airborne dust associated with chemical plant: A case study. Indian J Occup Environ Med 2006 10:32–34. doi: 10.4103/0019-5278.22893

Rengasamy P, Greene RSB, Ford GW, Mehanni AH (1984) Identification of dispersive behaviour and the management of Red-brown earths. Aust J Soil Res 22:413–431

Seguel O, Horne RF (2006) Structure properties and pore dynamics in aggregate beds due to wetting-drying cycles. J Plant Nutr Soil Sci 169(2). doi: 10.1002/jpln.200521854

USEPA (2004) Air quality criteria for particulate matter (Final Report Oct. 2004). U.S. Environmental Protection Agency. Washington DC EPA 600/P-99/002aF-bF, 2004

USGS GROUND WATER ATLAS of the UNITED STATES Alaska, Hawaii, Puerto Rico and the U. S. Virgin Islands HA 730-N *pubs.usgs.gov/ha/ha730/ch_n/N-PR_VItext5.html*

Wagner GR (1997) Asbestosis and silicosis. Lancet 349(9061):1311–1315

Ward P, Margaret MR, Ramona J, Shayne M, Stephen D (2015) On-row seeding as a tool for management of water repellent sands. 2015. Building productive, diverse and sustainable landscapes. Proceedings of the 17th ASA conference, 20–24 September 2015, Hobart, Australia. Web site www.agronomy2015.com.au

Wehr JB, Fulton I, Menzies NW (2006) Revegetation strategies for bauxite refinery residue: a case study of Alcan Gove in Northern Territory, Australia. Environ Manag 37:297–306

Zhang XW, Kong LW, Cui XL, Yin S (2015) Occurrence characteristics of free iron oxides in soil microstructure: evidence from XRD, SEM and EDS. Bull Eng Geol Environ pp 1–11. doi:10.1007/s10064-015-0781-2)

Fragility and Bearing Capacity of Gray Shales in North-Eastern Jamaica After Landslips: Origins, and Implications for Open-Cut Mines

Mark Anglin Harris

Abstract Steepened open-cast mines with shortened slopes are prone to landslipping during prolonged rainfall accompanied by sudden increases in intensity. This paper concerns (1) The role of antecedent rainfall in the causation of landslips in deeply weathered rock formations, and (2) the effects of wet-dry cycles on the erodibility of landslip-exposed shales, and causes of the landslips. The only landslip episode in living memory of relatively deep landslide correlated with the most active hurricane season on record. For the period 1923–2015, a span of 92 years (living memory of some residents) the probability of a landslip occurring in any particular year is 1/92, or approximately 1.1 %. Being less than 5 %, this rejects the landslip as a chance event in 2005. It is therefore concluded that antecedent rainfall from the hurricanes of 2005 provided the saturation levels required for the landslip after subsequent high-intensity bursts of rainfall. Rapid weathering of strong, hitherto un-weathered shale rocks was fast enough to have been measurable weekly (4 mm depth per week) after exposure by a landslip. Regular early morning rain showers followed by hot, dry afternoons caused frequent wet/dry cycles which facilitated rapid slaking of the shale. Several landslips occurred concurrently on the south bank of the Wild Cane River, where the rock beds dip towards lower ground. Resistance to impacts in the rock fabric was equally low after decades, or weeks of exposure. It was concluded that the combination of high frequency wet/dry cycles on steeply dipping shale beds, and rock attitude caused very rapid weathering. Rapid slaking of the illitic shales could cause high turbidity levels in contiguous aquatic environments. Building foundations and roads could be rapidly de-stabilized even on such freshly exposed, apparently consolidated sites.

Keywords Antecedent rainfall • Mass movement • Mobilization function • Slaking • Tropical hazards • Tropical weathering • Weathered shales

Original article: **Fragility of a dark gray shale in north-eastern Jamaica: effects and implications of landslip exposure.** *Environmental Earth Sciences* July 2010, Volume 61, Issue 2, pp. 369–377.

M.A. Harris, *Geobiotechnological Solutions to Anthropogenic Disturbances*, Environmental Earth Sciences, DOI 10.1007/978-3-319-30465-6_3

Highlights

- High frequency wet-dry episodes caused rapid weathering of strong, hitherto un-weathered shale rocks at 4 mm depth per week eroded after exposure by a landslip.
- Building foundations and roads could be rapidly de-stabilized even on such freshly exposed, apparently consolidated and initially indurated sites.
- High hurricane activity with antecedent rainfall can cause landslips in open-cast mines.

1 Introduction

Landslips are a common occurrence in Jamaica (Ahmad 1995) and a widespread cause of disaster in the humid tropics (Ahmad 1995). It is widely recognized that soil slips and debris flows due to shallow failures are triggered by short intense storms (Aleotti 2004; Flentje et al. 2000; Paronuzzi et al. 2002), with failures usually occurring in regions of the world where steep slopes consisting of residual soils are subjected to periods of prolonged and heavy rainfall (Garcia-Urquia and Axelsson 2015). The influence of rainfall on landslides differs substantially, depending upon landslide dimensions, kine-matics, material involved, while most deep-seated landslides are affected by long-term variation of annual rainfall which has to last several years (Bonnard and Noverraz 2001). However the landslip itself is not the only danger. The strength of the newly sub-aerially exposed rock can be deceptive. Botts (1986) reported strength reduction in clay shales of 80 % after just one wet/dry slaking cycle under confinement. Using such exposed rocks as building foundations or as building stone is unsafe.

The storm that hit north-eastern Jamaica on Friday 25 November 2005 was notable even for a wet-season event. Rainfall episodes of high intensity are frequent in the area: in 2004, more than 180 mm fell in 12 h; but in November 2005 165 mm of rain fell in 6 h, clearly of a substantially higher intensity, triggering one of the largest and most destructive series of mass movements in more than 60 years. Four landslips (Fig. 1) occurred within 24 h, and a permeability unconformity clearly exposed by one of the landslips (Fig. 2). The direct and immediate impacts of the landslips were obvious. Ten houses were buried beneath 10,000 m^3 of rock and soil, and 25 local inhabitants made homeless. Fortunately no lives were lost. Damage was estimated at more than J$400,000,000 (US$6,500,000). As population pressure forces the urban boundary lines onto the hills of many major cities of the world (Schuster and Highland 2007), human intervention is significantly contributing to the occurrence of many urban landslides. In densely populated areas, even the smallest landslides demand significant attention due to the disturbances and losses (Chatterjea 2011). However, in many cities of Latin America, landslides usually affect the poorest sector of the population and lead to numerous deaths that could be prevented if urban planning codes were rigorously

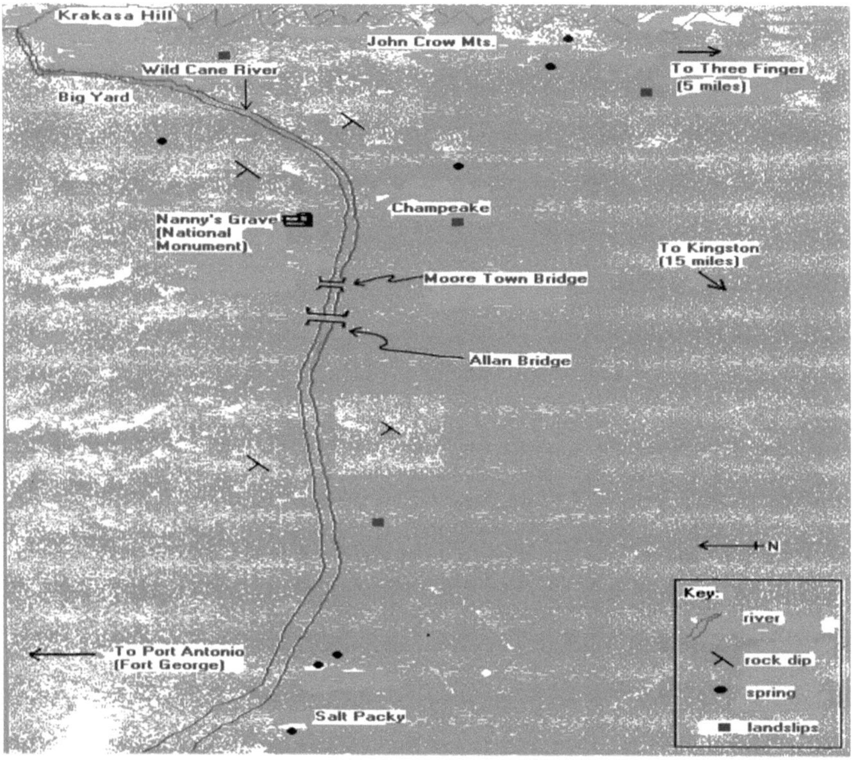

Fig. 1 Location of four landslips in north-eastern Jamaica on 25/11/05. As shown, rock beds strike at azimuth 300 degrees while dipping at 30 degrees (Scale: 1 cm = 50 m)

enforced to prevent development on slopes highly susceptible to landslides (Alexander 2005). Garcia-Urquia and Axelsson (2015) Despite the regularity of high intensity rain, this study region had never in living memory produced landslides of the scale of November 2005. What caused the mass movements? Why were they restricted to one bank of the river, and what are the implications for this and other freshly exposed shale permeability unconformities under similar climatic conditions?

1.1 Antecedent Rainfall

Numerous case studies have recognized the importance of *antecedent rainfall* (*AR*) as the controlling agent of the soil moisture in slopes (Jemec and Komac 2012; Khan et al. 2012). The influence of antecedent and triggering rainfall have been demonstrated by Garcia-Urquia and Axelsson (2015), who constructed an intensity-antecedent duration scale, allowing *landslide days* (*LDs*) to exhibit their *critical rainfall intensities* (*CRIs*) within the time window dictated by the threshold

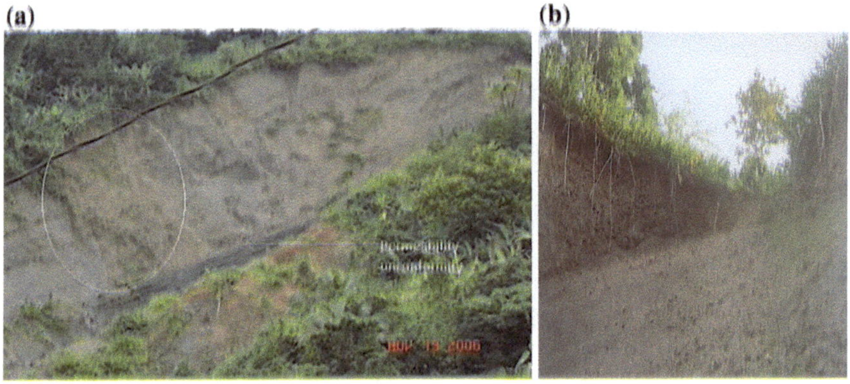

Fig. 2 **a** Permeability unconformity (*dark gray* bedding) still visible 12 months after landslip at Champeake, in north-eastern Jamaica. It can be seen that the upper weathered block is saprolitic, as compared to the darker gray fissile beds lower down. Banana trees (3 m high) at top left can be used to indicate the depth of the weathered section above the darker more resistant shale. The area in the *white circle* depicts an advanced stage of weathering. It is highly permeable. From this weak spot a tension crack can initiate a future landslip. **b** Tension crack at head of landslip exposes permeability unconformity (*dark grey* bedding). Here, the brown permeable unconsolidated saprolitic layer appears more advanced than at the lower slope levels (Fig. 2a) on the subjacent relatively impermeable base

duration. Shallow landslips usually follow a long saturation period that is punctuated by an intense burst of precipitation over several hours or a few days. If a slope is artificially over-steepened (such as in road cuts and excavations such as occur in open-cut mines), the material will become unstable and begin to move downslope (Murray 2007). Thus Fourie (1996) discussed a technique for predicting whether a particular rainfall event (defined in terms of intensity, duration and return period) will cause ingress of a wetting front to this critical depth, with particular reference to the failure of a number of road embankments.

One of the most common approaches to address AR is that of establishing the frequency of high-magnitude rainfall events responsible for landslide occurrence (Floris and Bozzano 2008; Frattini et al. 2009; Petrucci and Pasqua 2009; Polemio and Sdao 1999). Bui et al. (2013), Dahal and Hasegawa (2008), have attempted to determine the AR duration that allows the best discrimination between landslide and non-landslide events. Sengupta et al. (2010) attest to the contribution of both antecedent and triggering rainfall by showing that landslides in Sikkim, India do not occur on days of extreme rainfall and have established a rainfall threshold based on a 15-day cumulative rainfall of 250 mm. Ibsen and Casagli (2004) show that in the Porretta-Vergato region of Italy, AR as far back as 6 months prepares the terrain for failure, while an intense storm is needed to trigger landslides. In the Indian Himalayas, Mathew et al. (2013) combined an intensity–duration threshold with a probabilistic assessment of AR to evaluate landslide occurrence. Similarly, Gabet et al. (2004) concluded that while regolith thickness conditions seasonal rainfall accumulation in the Nepalese Himalayas, slope angle controls the daily rainfall required to initiate landslides (Garcia-Urquia and Axelsson 2015). Figure 2 above shows that both these predisposing factors—great regolith thickness and slope angle—preceded the landslips in north-eastern Jamaica. On the other hand, many researchers have concluded that deep-seated landslides are strongly tied to relatively long periods of rainfall (Martelloni et al. 2012; Petrucci and Pasqua 2009; Rizzo et al. 2001).

1.2 Weathering and Mass Movements in Humid Tropics

The chemical weathering process has produced several ores such as bauxite that are valuable, and soil. However, prolonged and deep weathering, with the added abundance of water and luxuriant vegetation, create thick, weathered mantles of regolith, even on steep slopes. Normally, the availability of water and consistently high temperatures increase the efficiency of chemical reactions such that very rapid chemical weathering of rocks occur in hot humid regions. The weathering processes are almost continuous, and more active than those in temperate areas. Deeply weathered rocks are common in tropical areas (Robinson and Thagesen 2004). In Madagascar for example, micas (one of the more stable minerals) in granites have rotted after just decades of exposure (Twidale and Campbell 2005). In Puerto Rico, the igneous rocks in the Upland Province weather rapidly to form a deep, predominantly coarse-grained soil mantle called saprolite. When saturated, saprolite produces debris slides, debris flows, and slumps ranging from a few to several hundred meters long. In May 1985, severe storms in west-central Puerto Rico triggered hundreds of such debris slides and debris flows, which choked streams, blocked roads, and destroyed homes and other structures. Hurricanes David and Frederic in 1979 produced extreme rainfalls in north-eastern Puerto Rico that triggered several debris slides as long as 750 m and as deep as 25 m on slopes of deeply weathered intrusive igneous rock. Rock falls from natural slopes and particularly from the ubiquitous steep road cuts also are common in the igneous and sedimentary rocks of the Upland Province. Thus consistently heavy rainfall (Twidale and Campbell 2005) and permeability unconformities (Simonnet 1967) are most important as major factors of rapid mass movements in the humid tropics.

1.3 Weathering of Shale (Review)

Shale is a detrital sedimentary rock (Fig. 3) composed of very fine clay-sized particles. Detrital sedimentary rocks are composed of the weathered and eroded particles of larger pieces of rock.

Clay forms from the decomposition of the silicate mineral feldspar. Other minerals present in shale are quartz, mica, pyrite, and organic matter. Hydration

Fig. 3 Shale parent materials which produces secondary clay minerals. Source: Adapted from Department of Geology & Planetary Science, University of Pittsburgh

occurs when a mineral reacts with either the H^+ or the OH^- from water to form a new mineral. Hydration affects the clay mineral montmorillonite; if it is saturated with H^+ it decomposes. Protons (H^+) will react with the Al in the clay structure, the clay collapses, and the Al becomes exchangeable.

As most rock-forming minerals were formed under high temperature and high pressure conditions, they are un-stable at low temperatures and pressures which exist at the earth's surface, especially in the presence of H_2O, O_2, CO_2, and micro-organisms. Weathering thus is a process in which rock minerals (or primary minerals) are transformed into soil minerals (or secondary minerals) according to the process:

$$\Delta\ \text{weathering}/\Delta\ \text{time} = \text{f(climate, topography, parent material \& biosphere)}. \quad (1)$$

Thus weathering is faster on steep slopes than on flat areas (Twidale and Campbell 2005). Clay minerals form in the presence of water and are an important fraction of soils and hence useful to agriculture. Among the weathering forces affecting freshly exposed minerals in shale are dissolution, hydration, and oxidation. Some silicates, such as pyroxene, will dissolve in carbonic acid (rain water):

$$MgSiO_3 + H_2O + 2H_2CO_3 \rightarrow Mg^{2+} + 2HCO_3 + HSiO_4\ \text{(silicic acid)} \quad (2)$$

Some types of minerals shrink and swell as they are wetted and dried. For example, montmorillonite clay expands when wet and shrinks when dry. The resulting soil instability can lead to mass wasting or downslope movement of soil. They are prone to mass movements either en bloc or with flowage. Clay and water accumulating at the base of the regolith at the interface between regolith and bedrock facilitate slippage. Masses of material sliding along structural surfaces such as bedding planes are known as landslips. Copious rainfall, highly weathered surface, and long steep slopes often characterizing young and active locations are major ingredients of rapid mass movements in the humid tropics. Water tends to move because it is incompressible. This is termed "pore pressure". Sudden saturation of the ground loosens the soil and triggers the slide.

1.4 Shallow Landslips, Trade Winds and Pore Pressure

Pore pressure increases under moisture-laden Trade Winds of the northern Caribbean. For example, rock falls in Puerto Rico from steep limestone slopes in the southern flank of the Upland Province and in the Northern Karst province, have repeatedly blocked major and minor roads. As in north-eastern Jamaica, in the same latitude, annual rainfall in northeast Puerto Rico ranges from 2500 to more than 4000 mm, with rainfall intensities of up to 65 mm h^{-1} having been recorded (Larsen 1990). Most of these landslides are shallow, with failure depths of 0.5–7 m, and are associated with periods of intense, prolonged rainfall. Southward into the higher northern Caribbean mountains there is an orographic increase causing heavy runoffs which produce steep, deeply dissected surfaces (Balgin and Coleman 1965).

Most of the limestone in Puerto Rico is porous and fractured (Larsen and Parks 1997), so infiltration of rainfall can build up pore pressure within the rock and trigger rock falls. Though observing considerable surface runoff at the study sites in north-eastern Puerto Rico, the data of Larsen (1990) showed moderate to rapid increases in pore pressure in response to short duration storm events. They found that pore-pressure increases are greatest in the lower sections of concave slopes apparently due to convergent flow. Limestone rock falls have triggered debris flows on steep colluvial slopes below near-vertical bedrock faces (Fig. 4). At one site, large limestone boulders fell from a bedrock face and impacted the head of the colluvial slope; this either disrupted the saturated colluvium enough to cause it to flow downslope, or rapidly increased the pore pressure in the colluvium and caused it to mobilize (Larsen (1990).

Fig. 4 Karst topography, Puerto Rico. The steep topography in fractured limestone exacerbates pore pressure plus frequency and danger of landslips. Steep-sided inselbergs of Tertiary and Cretaceous sedimentary projecting through young sediments also produce landslips. Note suspended sediments in swollen river. Adapted from: (http://pubs.usgs.gov)

1.5 Implications of Slope Cuts

According to Ahmad (1995), most of the landslips in the humid tropics are shallow, occurring on unconformities (permeability unconformities for example) in weathered shales. All the landslips of the present study occurred on a permeability unconformity on the southern slope of the Upper Wild Cane River Valley (Fig. 2), where shale beds dipped at >20° towards lower ground. Alexander (2005) observed that for shallow landslips, man's activities represented an indirect contributory cause of failure in almost 30 % of the recorded cases; due, particularly, to road cuts/embankment on hill slopes, inadequate drainage of surface water, and overloading. Further, most of the material in the landslides consisted of incoherent deposits, and to a lesser extent, of debris and loose soil. Almost everywhere, the most consistent damage caused by slope failures were along primary and secondary roads (70 %) and, secondly, in buildings (16 %) (Alexander 2005).

As evident above, the problems are not unique to shales. Landslips increase in frequency when slopes are cut, and the mountainous topography of most of the Caribbean Archipalego often necessitates the cutting of highways through slopes. On a spatial data base of 1609 landslides, Larsen and Parks (1997) demonstrated a simple method to assess the spatial association of mass-wasting with highways. Using a geographic information system, they analyzed landslide frequency in relation to highways. A 126 km long transportation network in a 201 km^2 area of humid-tropical, mountainous, forested terrain in Puerto Rico in conjunction with a series of 20 buffer (disturbance) zones varying from 5 to 400 m in length, was measured perpendicular to the highways. Average landslide frequency in the study area at distances greater than 85 m from roads was about six landslides per square kilometre. At distances of 85 m or less on either side of a highway, landslide frequency was about 30 landslides per square kilometre. On average, this elevated disturbance rate affected 330 $m^2\ km^{-2}\ a^{-1}$ within the 170 m swath. The mass-wasting rate outside of the disturbance zone affected 40 $m^2\ km^{-2}\ a^{-1}$. These results indicate that the rate of mass-wasting disturbance is increased from five to eight times in a 170 m wide swath along road corridors (Larsen and Parks 1997). They found that the lateral extent of the environmental impact of roads in their study area is greater than is commonly perceived. This result is similar to the landslip locations of the current study, all four of which occurred within 12 hours, and within road cuts and a river valley (Fig. 1).

1.6 Slip Surfaces

Geological unconformities produce slip surfaces. Larsen and Parks (1997) note that the sedimentary rocks flanking the igneous interior of the island of Puerto Rico produce rock falls from steep cliffs and road cuts, large rock and debris slumps, and rock block slides. The deeper landslides form where limestone or sandstone overlay weaker silts and clays that act as slip surfaces. The colluvium and residuum on steep slopes composed of mudstone and particularly limestone are susceptible to failure as debris slides and debris flows; such landslides generally fail at the interface between the weathered surface material, commonly 0.5–2 m deep, and the un-weathered bedrock. Even deeper is that of the study area of north-eastern Jamaica, where depth to the un-weathered permeability unconformity ranged between 4 and 10 m (Fig. 2).

1.7 Study Area: North-Eastern Jamaica

The study area is located in north-eastern Jamaica, where relative humidity regularly exceeds 70 % with an average daily temperature of 28–32 °C and a seasonal range below 10 °C (Finchan and Draper 1998). The diurnal range often exceeds 10 °C only in the "winter" season (McPherson 1965). However the high relative humidity facilitates frequent early morning rain showers throughout the year, with greater intensity and frequency in winter. Rainfall in the study area is mainly orographic, often exceeding 5000 mm (Finchan and Draper 1998). Near here at Mill Bank, 6 km away, world records for rainfall intensity have been recorded (Mcpherson 1965). Located in the interior foothills of the John Crow Mountains (Fig. 5), the terrain is deeply incised by river gorges. The John Crow Mountain

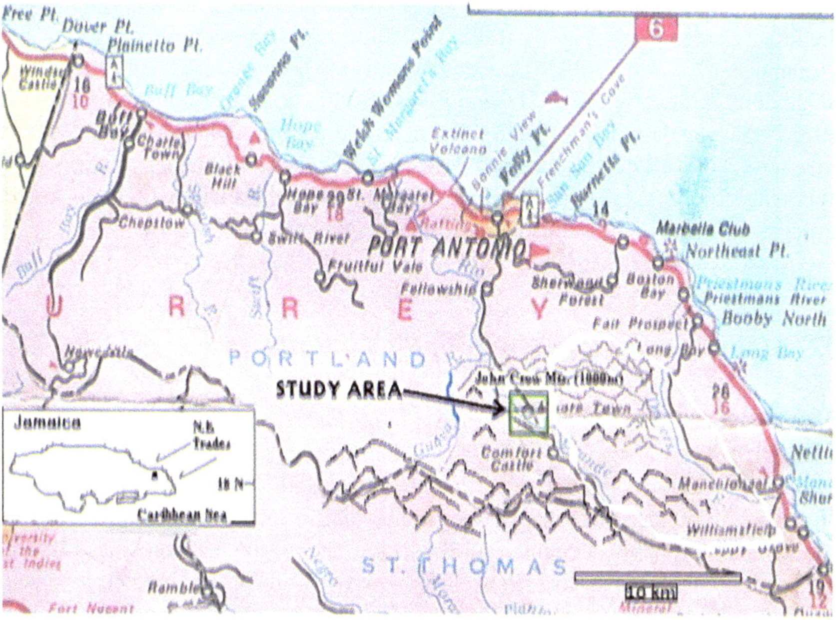

Fig. 5 Study area in the John Crow Mountain range of eastern Jamaica (Scale: 1 cm = 3.5 km)

Fig. 6 A typical gorge in the John Crow Mountains, north-eastern Jamaica. Here, world record intensity rainfalls have not only removed the tertiary limestone capping, but have produced several perennial springs (see Fig. 1 for location guide)

Range is comprised of a capping of white tertiary limestone, the only remnants of its kind in eastern Jamaica (Finchan and Draper 1998). In the upper section of the Wild Cane River Valley, highly jointed fissile low quartz dark grey shales (Jackson 1998) of early Paleocene age (Jackson et al. 1998) have been exposed by the intense weathering (Fig. 6) of the crystalline white limestone of the John Crow Mountain series, which tilt downwards towards the northeast (Jackson et al. 1998). Shale units incised by the Wild Cane River also dip down towards the northeast at 20–35°. The shale is 20 m thick, without visible organic remains. Deep chemical weathering of >15 m has occurred in the shale, whose secondary mineralogy is dominated by illitic clays (Jackson et al. 1998). Here the shattered, highly jointed structure of the rocks is indicative of low cohesion and high porosity.

1.8 Mobility and Morphology of the Landslip

The author estimated the areal extent of the landslip at 0.08 ha, with average thickness of 1 m at the top and 10 m in the middle, tapering at the end. Being comprised largely of brittle regolith rather than soil, low particle cohesiveness produced a flattened, fan-shaped toe, rather than a lobate appearance. The low cohesiveness probably accounted for the lack of any visible tension cracks beyond the head, though the up-slope limit of the mass movement was marked by a scar.

This scar, as the slip surface (Fig. 2b), reveals that the impermeable layer (permeability unconformity) tapered asymptotically towards the head, such that the mass of movable unconsolidated material resting on the unconformity (Figs. 2a, b) increased rapidly in thickness towards the lower altitude.

Iritano et al. (1998) introduced a so-called mobilization function,

$$Y(t), \tag{3}$$

that indirectly describes all the factors that contribute to trigger a landslide and that is dependent, at every time t, on the amount of water infiltrated in the soil before the time t. The common imperative in all landslips is intense rain, and rainfall intensity and duration can increase the rate of landslips. In addition, of all landslip episodes observed during the past fifty years in the study area, all were fewer in number and of lower magnitude (though heavy rain preceded each) than those of November 25, 2005 (personal observations and individual interviews). Therefore it is reasonable to conclude that in the absence of other factors, the unusually high rate of landslips in the study area occurred in response to bursts of rainfall intensity with an unusually long background duration.

Yet, though AR is a determinant of landslips, one uncertainty is based on the fact that the Champeake landslip (the largest) of this study occurred at the permeability unconformity of bedding planes (Fig. 2a). In permeable material, the above-mentioned mobilization function operates when further infiltration ceases, because the depth of a slip surface determines cessation of infiltration. In other words, if the permeability unconformity (Fig. 2a) of this study had been located at a shallower depth, landslips might have occurred at Champeake in the years preceding 2005. Implications for mine safety are that unconformities often mark the depth of open-cut mines. For example, in Jamaican bauxite mines it is often the crystalline limestone unconformity on which the ore rests. The similarities between unstable land in shallow landslips and that of open-cut mines suggest the potential for landslips in open-cut mines.

It seems therefore that intrinsic geologic factors can dominate the mobilization function. As stated in Eq. 1 above, the variables for weathering rate include parent material, and all variables in the equation are common to both landslips in this study. Both are (a) of weathered shales and (b) are on the same side with slopes shortened by a road cut. Yet, both landslips in the locality have shown vastly different frequencies of occurrence. At Salt Packy (Fig. 1), landslips occurred in the rainy season of every year until 1951, when the installation of a deep-drain system prevented any further episodes. On the other hand, no Champeake landslip had ever occurred in living memory before that of November 2005. Between both locations, the only uncertainty is the depth to the permeability unconformity. Whereas this is known for Champeake, it has not been ascertained for Salt Packy.

The 1st hypothesis for this study therefore is stated thus: there is a factor which at any time dominates and supersedes the mobilization function.

The 2nd hypothesis: Exposed shales will slake due to unloading.

1.9 Aim

The purpose of this study was to:

1. To establish the main cause(s) of the deep landslip of the study area and determine appropriate coping strategies for similarly induced landslips.
2. Determine the structural integrity of the shale after sudden exposure at the permeability unconformity.

2 Method

2.1 Rock Tests

According to Gemici (2001), the mechanical and physical properties of rocks help to predict the behaviour of the extended rock mass. He further affirmed that weathering processes, by affecting the physical and mechanical properties of the rock mass, cause a reduction in strength and an increase in the deformation and permeability of the rock mass. To determine their mechanical properties, (1) a drop impact resistance test, (2) a unit weight determination (UWD), (3) a rate of slaking test, and (4) a temperature stress test was carried out on rock samples from the disturbed area.

2.2 Drop Impact Test

Readily disintegrated rocks increase the chances of burial of life and property by the rubble thereby created. Toughness is related to the momentum required for breakage on collision. For example, normal glass, but not lead, shatters readily on collision, though lead is softer. The final energy release of a rock mass when it comes to rest changes spatially if it disintegrates while falling or rolling downhill. As the force is spread over a wider area, the kinetic energy is reduced at each point of collision. Collisions cause fracturing, thus for each event, the more collisions during the landslip, the smaller the final impact. Hence, changes with time in the destructive parameters of a landslip may be predicted by determining the ease of disintegration of its rocks. To simulate the impact of a directed force as rocks collide during a landslip, rocks in this study were subjected to the contact of a heavy, falling mass. From a pre-determined height, a flat-faced concrete block (4.5 kg) mass was dropped onto each rock sample. A sample was deemed "shattered" when at least one crack appeared in its middle region. The number of impacts required to shatter the sample was recorded as the counted number of impacts (CNI).

2.3 Unit Weight Determination

In situ rock alteration indicates removal and/or substitution of original minerals with smaller molecular units in the fabric of a rock. When prolonged, this process inevitably provides entry points for air and water, as indicated by a decrease in rock mass (Craig 1994). Unit weight differences within an originally homogenous rock type are therefore significant because in humid areas this indicates levels of absorption of water, solution, and mineralogical alteration by exposure to water (hydrolysis and hydration). The purpose of this test was to determine the extent to which density of the weathered shale differed from that of the apparently un-weathered variety. Density differences would be expected to vary directly with extent of weathering, and hence ability of the rock to contain water. Similar sized samples were weighed after saturation, oven-dried and reweighed.

To determine the volume of each sample, samples were waterproofed with a thin layer of lacquer after oven-dry, and the volume of water displaced by each sample ascertained. The density was then calculated. The mass of each was determined by crushing the rocks and passing the particles through a 1-mm sieve. The unit weight of the soil was determined by placing 100 ml of the sieved material in a pre-weighed, graduated volumetric cylinder, and the mass found by difference.

2.4 Rate of Slaking

As slaking is defined as a crumbling of (brittle) material (Webster 1972), and by Gemici (2001) as a crumbling as of quicklime when mixed with water, the rate of slaking was evaluated in this study as the rate of crumbling. The depth of the crumbling material was ascertained with a narrow-gauge soil auger—a graduated depth gauge comprised of a conically tapered graduated cylindrical steel rod (2 cm diameter) pushed with a helical action into the crust of newly weathered material at the permeability unconformity. During measurement the auger was kept in a perpendicular position to the surface using a Bosch DWM4OL Digital Angle Finder Protractor. The soil augur was used because the thinness and texture of crumbled layers at this study site seemed unlikely to have facilitated the operation of the conventional cone of a conventional penetrometer (Craig 1994). Two readings were taken after the landslip, on days 2 and 42 at randomly selected points within squares on a grid. The recorded slaking depth (0 mm) of the un-weathered freshly exposed solid rock on the day 2 after the landslip was used as a reference. To determine the rate of slaking of the shale, the depth of crumbling was divided by number of weeks of exposure and the values interpolated.

2.5 Expansion and Contraction

To determine the role of temperature extremes alone (without added moisture) in the slaking of the shales, samples of varying sizes were subjected to alternate episodes of high and low temperatures by autoclaving forty times for 4 min at 200 °C, followed by sudden cooling in a freezer set at 0 °C. After each cycle, samples were observed for cracks using a 129 magnification lens.

Three series of 40 cycles were conducted over a period of 3 days.

2.6 Particle Settling Velocity

As turbidity varies inversely with sediment settling velocity (Middleton 1996), the particle size analysis method of Bouyoucos (Kaddah 1974) was used to compare particle settling velocity of particles from the dark grey shale and its regolith with that of a reference kaolinite dominated soil (O'Callaghan et al. 1998) well-known from prior tests for exhibiting a rapid settling velocity (Harris and Omoregie 2007). A kaolinitic soil was chosen as a reference because in the humid tropics the most unstable slopes occur on rocks in which weathering has progressed far enough to produce illite and montmorillonite as distinct from the more mature kaolinite which does not expand and contract so easily and needs more moisture to reach its liquid limit (Gibbs 1967). Samples from the consolidated dark grey shale beds and the regolith therefrom were to be compared with the reference soil. Both Champeake samples were crushed using a pestle and mortar and all samples passed through a 0.5-mm aperture sieve prior to further treatment.

Statistics

The Student's *t*-test was applied to determine significant differences between two landslips within the study area.

3 Results and Discussion

Interviews with older residents of the study area revealed that during the period 1923–1955, twenty landslips (one in the rainy season of each year) occurred at Salt Packy, a location less than 2 miles from Champeake (Fig. 1). Before 1955 when deep drains stopped further landslips, landslips occurred every year during the rainy months which coincided with the Christmas season. In that era, valued dishes such as meats and fish-kind were preserved mainly with salt in a gourd called a Packy. As locals were guaranteed jobs clearing the landslips, their meats were assured, and the name Salt Packy was given to that location. During the annual rainy season, high intensity rainfall normally occurs amid long-duration rainfall. Yet prior to 2005, a landslip had not occurred in living memory at Champeake. As cited earlier, the characteristics of a deep landslip must include AR. It is hereby concluded that Salt Packy was not a deep landslip, otherwise landslips there would not have occurred at such a high frequency of once per year, i.e., consistently without AR. During the rainy season of October–November, annual high intensity rainfall is expected, and, most often occur in the study area. But no landslips ever occurred at Champeake. However, long before late November of 2005, a build-up of AR began in the most active hurricane season in recorded history. This is discussed later.

For several days after the landslip at Champeake, i.e., after the cessation of rain, accumulated water of clear colour continued to gush from the junction between the decayed shale and the newly exposed un-weathered shale, i.e., at the permeability unconformity. Being of similar decomposition and permeability to the Salt Packy lithology, Champeake should have similarly produced landslips in response to high rainfall intensities each year over the same decades. It did not. Despite the absence of borehole samples for Salt Packy, the only marked substantial dissimilarity between both locations seems to be a difference in depth to the permeability unconformity. Champeake is deep, hence requires AR to initiate a landslip; Salt Packy = depth unknown, yet responded each year to intense rainfall of long duration with a landslip. Therefore, based on the above-mentioned workers (Sect. 1), i.e., Aleotti (2004), Flentje et al. (2000), Paronuzzi et al. (2002), Garcia-Urquia and Axelsson (2015), who found that soil slips and debris flows due to shallow failures are triggered by short intense storms where steep slopes consisting of residual soils are subjected to periods of prolonged and heavy rainfall, Salt Packy is a shallow landslip location with a permeability unconformity closer to the surface than that of Champeake.

Statistics

The Student's *t*-test or paired difference test of the null hypothesis that the difference between two responses measured on the same statistical unit has a mean value of zero was applied. Results show that the difference between the two locations was extreme enough to have occurred from a different population of landslips. Thus the null hypothesis was rejected at the 0.01 level of significance. On the assumption of flat land having a depth of 10 m to the impermeable layer, and taking into account the volume occupied by the rock fabric, hence assuming porosity of 20 %, the rainfall required for full saturation would be considerably less 10,000 mm, i.e., specifically, 2000 mm. Within dipping beds, the required volume of rainfall for saturation would however increase. But thick layers of clayey regolith can withhold antecedent water for a considerable time, as demonstrated by the sustained (for several days) water leakage at Champeake following the landslip. In other words the rainfall level required to saturate a flat-lying highly porous rock is substantially less than that required to fully occupy the same depth of, say, a lake. Thus, depending on the rock type, the AR required for a landslip varies. Yet accounting for the differences in landslip occurrences between Champeake and Salt Packy requires identification of substantial AR in the weeks and months prior to November 25, 2005. The following section presents the evidence.

Source of antecedent rainfall (AR)
Owing to the extensive area affected by a single hurricane, torrential rains often occur within hundreds of km from the eye. In 2005 no hurricanes hit Jamaica directly, but according to NOAA (2006), the 2005 Atlantic hurricane season was the most active Atlantic hurricane season in recorded history, shattering numerous records with twenty-eight tropical and subtropical storms formed, of which a record fifteen became hurricanes. A record seven of these strengthened into major hurricanes, a record-tying five reached Category 4 hurricanes, and a record four acquired the highest categorization for hurricanes on the Saffir-Simpson Hurricane Scale: that of Category 5 strength. Among these Category 5 storms were the two most intense and costliest Atlantic hurricanes on record, namely, Katrina and Wilma.

Recorded information by the Meteorological Office of Kingston Jamaica (2006) contains the following reports of the storms which brought rainfall to Jamaica in 2005.

(1) June 7–11: Arlene. Although Arlene did not affect the island's weather directly, other weather conditions "resulted in more episodes of flash-flooding through to the end of the month."
(2) July 5–8: Hurricane Dennis. Its closest pass was about 50 km northeast of Port Antonio in the early afternoon. Dennis became a Category 3 hurricane at 4:00 p.m. and reached Category 4 by 10:00 the same evening. The highest rainfall total over the three-day period was detected at Mavis Bank in St. Andrew and measured 623.2 mm. The most in one day was 497.6 on July 7, which exceeded the 50-year return period. The rate of rainfall at that station reached 65 mm h^{-1} between 5:00 and 6:00 p.m.
(3) July 11–17 Hurricane Emily. July 16 with moderate to heavy showers spreading from eastern to central and then to western parishes. The highest one-day rainfall total was 297 mm on July 17, while the highest two-day total for the event was 392 mm; both in southwestern Jamaica.
(4) October 13–20: Hurricane Wilma. The island was severely impacted by the heavy and persistent rainfall that occurred during the period October 13–20, 2005. An elevated ground water table, the result of rainfall of the days preceding these events, exacerbated the widespread damage. The highest one-day rainfall total occurred in south-central Jamaica and amounted to 278 mm on October 17.

In open-cut mines, where deep tropical weathering produced thick overburdens exceeding 8 m, substantial AR can accumulate, particularly during years of prolific hurricanes. Prolonged infiltration can saturate regolith so as to require extended periods for complete drainage. Such conditions maintain continuous lubrication at permeability unconformities. This promotes conditions for landslips.

3.1 Drop Impact Test (Counted Number of Impacts: CNI)

The CNI's of the rock samples taken from the freshly exposed un-weathered permeability unconformity stratum were more than 70 times (150 impacts to shatter) that of the weathered samples (Fig. 5). There were no significant differences in the CNI of rock samples of all sizes (3–8 cm diameter) from the surface down to the junction (Fig. 2) of un-weathered rock. All samples shattered after just one or two impacts from the 4.5 kg mass. All the rock samples from the surface down to a level of 6–8 m therefore responded equally to the impact of a heavy, falling mass, and were relatively weak compared to the reference shale from the deeper un-weathered zone, which required 150 impacts to shatter.

According to the test results, rocks which were <1 m below the surface did not exhibit greater strength than those at a more advanced stage of weathering several meters deeper. Such a level of weakness was therefore achieved early in the weathering cycle of the shales. In other words, maximum weakness is achieved long before maximum breakdown of the shale fabric occurs, and under conditions of the study area, such shales quickly acquire their maximum weakness long before the weathering cycle is completed. This result was unexpected. Rocks which suddenly became sub-aerially exposed at the permeability unconformity were observed to begin to crumble within 20 days of sub-aerial exposure. In a laboratory slaking test, Botts (1986) observed nearly a 75 % drop in the shear strength of Pierre Shale samples, or a 6-degree drop in the internal friction angles and a reduction of the cohesion from 848 to 0 k Pa, after one wetting and drying cycle (Schaefer and Birchmier 2013). The wide variation in strength shows the engineering complications in classifying the dark grey shale of this study as either a rock or soil. (Knight 1963), using similar shale material, found that extensive problems were encountered during the construction of a dam. The relatively high shatter resistance of the un-weathered shale of this study is indicative of a relatively good bearing ratio as a building stone for foundations or as substrata for roads. However, in the Niger Delta Sedimentary Basin, Emujakporue (2011) found that high initial bearing capacities of shales decreased in time with exposure to slaking. Similarly, high shatter resistance of the dark grey shale in this study suggests a high bearing capacity. Yet, as shown above, such lithology, when exposed to, or near, the earth's surface, can catastrophically deteriorate. Such entities potentially possess insufficient strength as foundations, foundation stones, or road sub-strata.

The newly sub-aerially exposed shale layers softened and swelled to a hummocky appearance creating uneven air spaces under the thin surface (<2 cm) at the upper bedding planes. These brittle surface layers readily shattered when walked on, as the behavior of weathered clay minerals in the shale can make it very unstable (Anderson and Cobb 2008). They found that a lens of smectite or expanding clays in shale caused the shale to move through compression and expansion. This may have been the major problem with structural stability such as cracking of internal walls in buildings they assessed. When this smectite comes into contact with water it will also expand and heave the foundation; when the shale dries out, the site will then shrink as could be the case in this study where a clay lens can be seen (Fig. 2). Though this would be a perpetual problem, based on their analysis of the clay minerals in the shale, the smectite problem seems minor compared to the weathering of pyrite zones in the clay (Anderson and Cobb 2008). There were

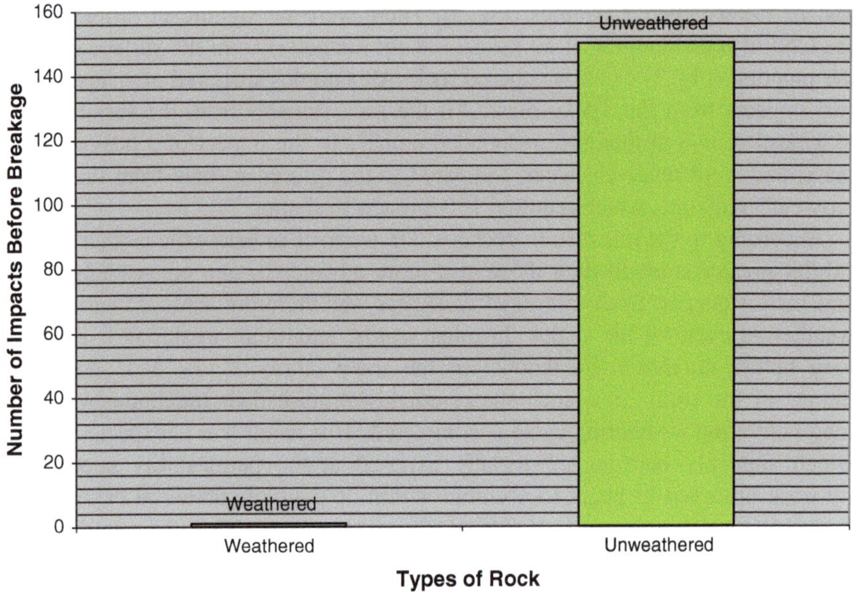

Fig. 7 Shatter resistance of weathered shale at the Champeake landslip, Moore Town, north-eastern Jamaica. Note comparison with un-weathered shale layer as depicted by repeated impacts from a dropped 4.5 kg mass. Shatter resistance indicates the internal cohesiveness and toughness of the rock fabric, and hence the saprolitic status of the weathered rock

no pyrite zones detected in the grey shale of this study. However, it may be borne in mind that core and outcrop descriptions could possibly reveal details producing similar effects for the study area. Botts (1986) found that such clay-shales are typically characterized as a stiff, fissured clay or shale that is susceptible to significant deterioration as a result of interaction with water. They are usually highly overconsolidated clays with high plasticity, a high content of montmorillonite clay, and a tendency to undergo extreme slaking with wetting and drying cycles (Fig. 7).

For several days after the landslip occurred (with no further rain), spring water still flowed on the exposed un-weathered slip surface from under previously weathered, intact material upslope. All samples from the permeability unconformity which were under continuously running water remained un-weathered for several days after the landslip, this contrast thereby indicating the devastating weathering role of wet-dry cycles at this location. At day 4 after the landslip, surface water still flowed on the permeability unconformity, but had abated to a trickle of 50 $cm^3 s^{-1}$ from its high on day 2 of >3000 $cm^3 s^{-1}$. This indicated not only the sustained high intensity of the rains in the 24-h period immediately preceding the landslip, but the high water retention characteristic of the clay-like regolith layer due to its secondary mineralization. All rock fragments in the rubble had colours ranging from greyish brown to medium brown. Thus there were several weathering stages from initial to advanced. Weathering was observed to cause the durable, brittle, rock-like shale to turn into a weathered, soil-like material at the site. Yet there was little difference in strength between the recently weathered rocks and those exhibiting signs of advanced weathering.

The staining of the rock particles was most often on sides perpendicular to the bedding plane. Reading et al. (1995) found that in the humid tropics the zone of weathered rock typically is highly permeable; in the lower sections discoloration and alteration of minerals are more pronounced along joints and other planes of weakness. It is in these locations along joints perpendicular to the bedding plane that drying frequency would be expected to be higher compared to the more water-logged bedding surfaces.

3.2 *Rate of Slaking*

The depth of friability was 32 mm at 42 days after the landslip. This indicated a rate of slaking estimated at (4 mm per week in the first 6 weeks after the landslip (Table 1). Were these comminuted particles excipiently or incipiently formed? If excipiently deposited, they must have been water-borne, due to the large (5 mm diameter) sizes and would have been transported from the head of the landslip. However, such a cause seemed unlikely due to the high angularity, low hardness, and poor sorting of the particles. Nevertheless, as these particles were observed at just 0–16 m downslope of the landslip head, there would not have been extended processes on the landslip within a mere 16 m from the head, such as saltation, entrainment, or mechanical abrading in flowing water to have rounded the edges. However, had they been formed at the head of the landslip and washed down quickly during heavy rains the angularity could have been accounted for. This question was further investigated with stratified random sampling along transects 2 m apart along the surface of the landslip towards its summit (0–5 cm depth) to detect any horizontal flow indicated by sorting of particles in the downslope direction. The lack of any grading (Table 2) suggests that the particles were incipiently produced.

This left only two possible incipient causes of comminution: (a) un-loading, and (b) in situ atmospheric stresses. It was noted that the Champeake Landslip rock blisters were observed only where the rocks were sub-aerially placed—never under continuous water even as shallow as 10 cm, where no evidence of weathering was observed or detected. Being a mechanical reaction to relieved overlying stress, the swelling and expansion caused by unloading cannot be restricted

Table 1 Depth of slaked shale particles after 42-day exposure to wet/dry cycles on permeability un-conformity exposed by the Champeake landslip

Days exposed	Random locations of 1 m^2	Average slaking depth (mm)
2	1, 2, 3, 4, 5, 6, 7, 8	0
42	1, 2, 3, 4, 5, 6, 7, 8	34

Table 2 Particle sizes of comminuted particles on the Champeake landslip in four surface transects from the top towards the toe

Distance from head stress fracture (m)	Transects # at 4 m intervals	Average particle size (mm)
0–4	1, 2, 3, 4	8
4–8	1, 2, 3, 4,	9
8–12	–, 2, 3, 4	8
12–16	–, 2, 3, 4	7

(Strahler and Strahler 1973), even by shallow overlying water bodies, yet no comminution of particles, or loose slabs of rock were observed in the submerged locations on the landslip. It is therefore concluded that unloading action was not responsible for the comminution of the shale. On the other hand, it is swelling accompanying water absorption, and shrinking of fine silt and clay in alternate periods of rain and drought that causes slaking of shales (Strahler and Strahler 1973), and during winter in the study area, morning precipitation was observed to occur on 2 out of 3 days in the rainy season, alternating with dry, sunny conditions in the late mornings and afternoons.

Further, Strahler and Strahler (1973) stated that the efficacy of temperature changes alone as an agent of rock disintegration has not been confirmed by laboratory experiments. This view was corroborated in this study, as repeated episodes of extreme heating and cooling did not substantially slake, or produce a single visible crack in shale blocks of various sizes (Table 3). Heat capacity C_p, also known as specific heat, is the ability of a material to store heat. The tendency for clay shales to weather, and to often slake upon drying and rewetting has been well documented. Stress relief allows for opening of existing discontinuities and may lead to formation of new discontinuities such as stress relief fracturing in shale rock types (Price and De Freitas 2009). The degradation causes the material to soften and lose strength, possibly leading to slope failures (Schaefer and Birchmier 2013). It can be seen that small rates of flaking was achieved in some samples in the proportions 0.06, 0.006, and 0.003 % of sample mass after several cycles of heat/cold, but only on the smaller masses of shale (Table 3). Flaking varied inversely with mass.

Table 3 Response of suddenly exposed shale to three sets of 40 °C cycles of rapid, moisture-devoid temperature changes (10–100 °C)

Total minutes of cooling + heating	Visible cracks	Mass flaked off (mg)
600, 600	0	30
1350, 1350	0	8.8
2250, 2250	0	5.0
3015, 3015	0	0
5400, 5400	0	0

Thus, despite being subjected to longer periods of extreme temperatures, the larger masses seemed totally unresponsive to the extreme temperature treatment (Table 3). It is commonly known that even a thin coverage by rock particles will severely restrict evaporation from trapped ground moisture. Though thermometers placed on the surface of the exposed shale rocks showed diurnal ranges exceeding 30 °C, daytime temperatures below the surface in the slaked material would have been much lower than at the surface and thus a substantially lower temperature range would have existed in the slaked material below the surface. Had extreme temperature changes per se accounted for the high slaking rate seen in this field study, subsequent slaking of deeper layers in the field would not have occurred due to the protection from the heat provided by the surface particles and, additionally, by the resultant insulating air spaces below the surface. Gemici (2001) concluded that rocks begin to deteriorate and disintegrate immediately on exposure to atmospheric conditions, and shales, when exposed to the weather (for example after landslips) tended to slake. Shale degrades into various clay and sulphate minerals and has a lack of shear strength when subjected to loads (Anderson and Cobb 2008). Shales have fissility, which causes them to split after swelling, into thin layers by moisture absorption (Fig. 6). The author found that the swollen layers of Champeake (Fig. 6) cracked underfoot when walked on and were <3 cm in thickness. It is therefore of interest that the shale samples which flaked minutely after the heat/cold treatments of this study were all <3 cm thick. Also, the larger masses required almost ten times longer to acquire the same number of cooling/heating cycles (Table 3). As in the case of the treated samples, thinner layers in the field (formed by swelling at the permeability unconformity) would have similarly undergone faster cooling, heating and drying during a wet/dry cycle, thereby increasing the frequency of their wet/dry episodes. Continuing erosion and loosening is exacerbated by the moisture changes caused by wetting and drying (Gemici 2001). In providing the base level of interflow prior to the landslip, the permeability unconformity was for decades continuously covered with water for longer periods than the layers above it. The landslip revealed it as being completely un-weathered (Fig. 2). This indicates the protective role of continuous water coverage in the weathering of shales, in contrast to the rapidity of weathering caused by alternate wet/dry cycles. On day 1 after the landslide, three slickensides, each approximately 950-mm long by 10-mm deep, and 3 to 10-mm wide were observed on the surface of the relatively hard, newly-exposed grey permeability unconformity (Fig. 2). Before day 40 the slickensides ceased to exist. It was concluded therefore that the slickensides created by the landslip of this study and quickly removed from view were the victims of rapid weathering which caused in situ slaking under diurnal wet/dry conditions prevailing in the study area. Barry et al. (2009) cited hydraulic gradient as a major factor influencing the speed of groundwater interflow. Thus the steep hydraulic gradient in the study area would have helped to eliminate water stagnation, thereby enhancing the wet/dry process. The presence of moisture cycles, rapid acquisition of low hardness levels, in situ rock blisters (Fig. 8), and poor sorting of particles at the study site therefore suggest multiple wet/dry episodes with temperature changes as the main cause of slaking in this study.

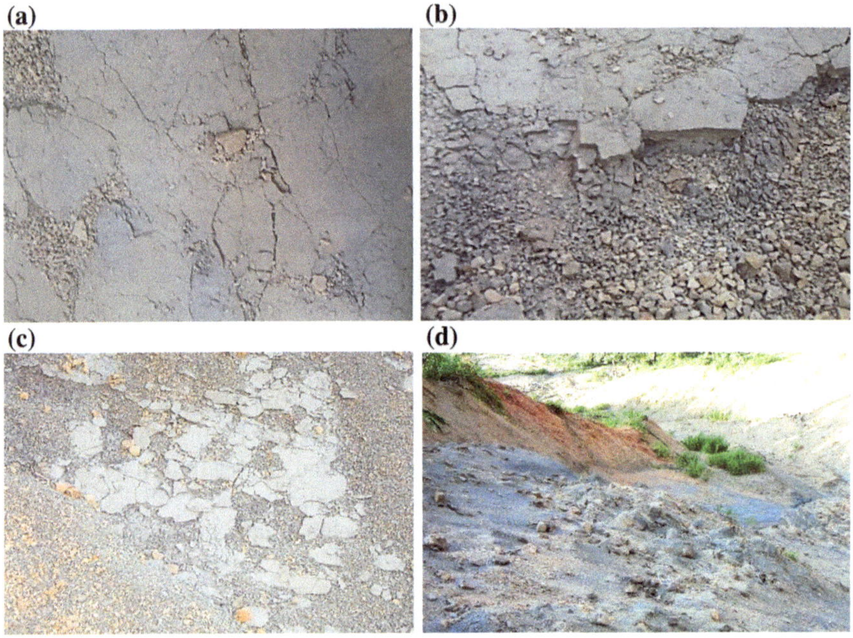

Fig. 8 **a** Week 1. The slaking process. Rapid swelling, blistering, cracking and slaking of shale observed five weeks after exposure by landslip. On the day of the landslip the exposed dark grey totally un-weathered and consolidated shale appeared to be resilient and had no cracks (Fig. 2). Yet, unloading and daily wet/dry cycles eroded observed slickensides within 20 days. **b** Week 3. The slaking process continued. **c** Week 6. Increased slaking and comminution within 40 days of landslip. **d** Week 9. Continued slaking at day 60, prodicing the finest particles

3.3 Unit Weight Determination (UWD)

Results of the UWD showed that the unit weight of the un-weathered rocks was >33 % higher than that of the weathered rock (Fig. 9). Subsequent oven-drying of the crushed rock at 105 °C indicated water content for the weathered samples of 21 %, compared to <2 % for the un-weathered material. Thus even while carrying a more than 10-fold water content as that of the un-weathered shale, the weathered shale was much less dense than the former. This is an indication of the intensity of the weathering process which had occurred prior to the landslip. One explanation could be that weathering is faster on steep slopes than on flat areas (Twidale and Campbell 2005), and the dip angle of the shale in this study exceeded 40°. Secondly, Twidale and Campbell (2005) cited the role of surface tension in impeding water movement, such that highly porous rocklike shales are of low permeability, partly because the pore spaces are so small. Such impedance, along with a high, evenly distributed rainfall regime, has in this study caused the entrapment of water for long periods in the pores, thereby facilitating reactions leading to deep weathering.

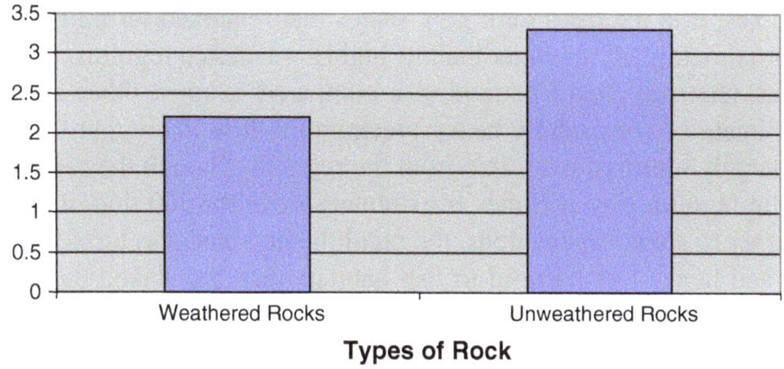

Fig. 9 Unit weight determination (relative density: g cc^{-1}) of dark grey early paleocene shale rocks

3.4 Particle Settling Velocity

Compared with the reference soil, the samples from the study area, particularly in the clay fraction sizes of <0.002 mm, exhibited much lower settling velocities:

Reference soil = <0.5 min; Brown regolith = 110 min; Dark grey shale = 160 min.

These results confirm the relatively high potential of the rapidly slaking illitic dark grey shales to increase the turbidity of water. There was insignificant turbidity from the kaolinitic reference clay (Fig. 10). Koo et al. (2010) found that applying large amounts of illite clays to natural aquatic habitats increased turbidity at 2 m downstream from the application point. They noted that though turbidity returned to levels of controls after 4 h, it continued to fluctuate due to fish movements. In this study, settling of both shale-derived samples occurred in <4 h. However, of the

Fig. 10 Settling velocitiy for <5 μm particles from a landslip in north-east Jamaica compared to that of a reference kaolinite

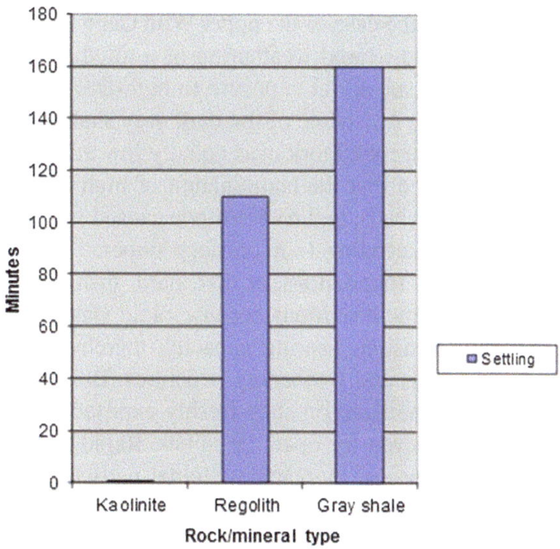

three samples, it is the fresh dark grey shales that remained turbid for the longest time (approximately 30 % longer than its highly weathered regolith).

With its relatively small particle size compared to most other clay minerals, under a climate of consistently heavy precipitation it is likely that the illitic fraction was largely removed over time from the regolith. Though the settling speed of the landslip regolith clay particles is extremely slow at >200 that of the reference clay, contrary to some expectations, the regolith contained less turbidity-producing material, and hence less harmful to fish habitats than the slaked dark grey shale. Nevertheless, as shown in this study, the rapid breakdown on subaerial exposure of the un-weathered Paleocene dark grey shale raises environmental concerns, not least being the danger to aquatic life.

4 Conclusions

The only landslip episode in living memory of relatively deep landslide correlated with the most active hurricane season on record. For the period 1923–2015, a span of 92 years (living memory of some residents) the probability of a landslip occurring in any particular year is 1/92, or approximately 1.1 %. Being less than 5 %, this repudiates the acceptance of the landslip as a chance event in 2005. It is therefore concluded that antecedent rainfall from the hurricanes of 2005 provided the saturation levels required for the landslip after subsequent high-intensity bursts of rainfall.

Though over-steepening by man—such as in road cuts and excavations causes material to become unstable and begin to move downslope, in this study, the short distances from the river of all the landslides suggest that a similar series of landslips is enhanced even by natural steepening by a river over relatively long periods of time as in the case of the upper Wild Cane River valley.

In this study, rapid weathering of a tilted fissile dark grey shale was observed after sudden sub-aerial exposure to high-frequency wet/dry cycles. All tests done showed a low resistance of the dark grey shales to physical stresses. Resistance to impacts in the rock fabric was equally low after decades, or weeks of exposure. It was concluded that the combination of high frequency wet/dry cycles on steeply dipping shale beds, and rock attitude caused very rapid weathering. This led to rotting depths exceeding 15 m on steep slopes.

Such rock foundations, at first, hard, firm, and hence seemingly secure, could quickly become hazardous because rapid slaking quickly decreases rock strength, and, by extension, bearing capacity, thereby potentially unsettling foundations, cracking walls and weakening structures. Building and road foundations could be rapidly de-stabilized on such freshly exposed, apparently consolidated sites, with implications even for open-cut mines. Rapid slaking of fragile illitic shales could increase release of inorganic colloidal particles, hence exacerbating turbidity levels in contiguous aquatic environments. This could jeopardize aquatic ecosystems. Moreover, under a combination of frequent wet/dry cycles, downward dipping beds, a predicted increased frequency of extreme weather, and the rapidly slaking propensity of low-quartz shales, the current hazards to life, property and aquatic environments in the humid tropics can be exacerbated if methods to prevent landslips in such locations are not adopted.

5 Geotechnical Applications

5.1 Tegucigalpa, Honduras: Role of Antecedent Rainfall and Human Interference

Landslides are the main hazard in the city of Tegucigalpa, Honduras, a city which suffers from the occurrence of destructive landslips on a yearly basis (Garcia-Urquia and Axelsson 2015). Analyses of monthly and annual rainfall means have showed how extreme rainfall events as experienced during hurricane Mitch in October 1998 have significantly contributed to the initiation of landslides (Garcia-Urquia and Axelsson 2015). The so-called mobilization function, $Y(t)$, (Irritano et al. 1998) cited above has been applied in Tegùcigalpa. Thus, in Tegucigalpa, a critical region where minimum triggering and antecedent rainfall amounts are necessary for landslide occurrence has been established to reduce the number of false alarms. Garcia-Urquia and Axelsson 2015 noted the critical rainfall intensity for every landslide day, denoting the antecedent rainfall amount yielding the highest return period divided by its antecedent duration. This they used for the construction of four rainfall thresholds with durations of 7, 15, 30 and 60 days. The accuracy of each threshold in discriminating between landside and non-landslide days was determined by comparing the antecedent rainfall amounts of each day in the study period with the threshold's rainfall requirements. Nonetheless, the threshold's low rainfall requirements suggest that human intervention has significantly predisposed Tegucigalpa's slopes to failure (Garcia-Urquia and Axelsson 2015). Such human interventions include cutting into slopes and building on slopes.

5.2 The Potentially Rapid Weathering and Slaking of Shales Require Safety Precautions Where the Cutting of Slopes Increases the Frequency of Landslips

- A foundation is the part of a structure which transmits the weight of the structure to the ground. All structures constructed on land are supported on foundations. A foundation is, therefore, a connecting link between the structure proper and the ground which supports it. On grey shale rocks, interior floors and non-load-bearing concrete-block walls have cracked and heaved considerably, bending and breaking copper water lines and radiators and breaking ceiling plaster

(Anderson and Cobb 2008). Building foundations should therefore not be made in such apparently strong, recently exposed, un-weathered dark grey shale rocks.

- Other construction activities such as road building require that the structural design of the paved area is dependent on the bearing capacity of the sub-grade and the bearing strength of the paving materials (Robinson and Thagesen 2004). The properties of rocks and soils in and near the road line is the basis for appropriate road design. The strength of the grey shale rocks indicated in this study is insufficient as a substratum for vehicular traffic.
- Excavation of a slope causes a change in stress regime in the rock mass, often denoted as stress relief. This may be a change in magnitude or orientation (Tating et al. 2013 Ingeokring Newspaper). The dominant deterioration processes that affect the durability of a man-made rock slope after excavation are stress relief and weathering (Sheoran et al. 2010). Thus the stability of cuts and fills is dependent on the properties of the rocks and soils concerned (Robinson and Thagesen 2004).
- Sub-surface drains are installed to relieve pore pressure during heavy rains. These should be placed some distance above, and not directly on, the permeability unconformity, because ground-water follows paths curved concavely upward (Strahler and Strahler 1973; Montgomery 1989).
- After installation of a deep drain system (Fig. 11) 1957, landslips ceased at Salt Packy. Further, by decreasing the incidence of landslides (and hence shale exposure), the above-mentioned drainage procedure could also reduce the amount of illitic sediment in streams from potentially slaking dark grey shales.
- On the south bank of the Wild Cane River in Moore Town, the deep weathered shale lithology extends westwards (Fig. 1) from the western end of the 2005 Champeake landslip (north-eastern Jamaica). As there are no devices constructed in this location to relieve hydrostatic pressure, other landslips are possible (west of the Champeake landslip) during conditions similar to those which caused the landslips of 2005. Pre-emptive devices include either underground conduits which relieve pressure partly by pulling up stored underground water upwards (Fig. 12), or deep, concrete-lined surface drains such as those at Salt Packy (Fig. 11).
- There are dangers of falling rock in hilly areas where the excavations have been cut into the hillside. Currently, several houses constructed of concrete and masonry are located on the deeply weathered shale (regolith) on slopes adjacent to the landslip zone in this study. Residents under such climates should not return and rebuild dwellings on similarly exposed shale unconformities.
- In the past, expansion and contraction due to temperature variations was thought to be a major factor of weathering, but it has been proven that this process, to be effective, requires moisture (McLean and Gibble 1992). Thus materials subjected to decades of temperature changes under high humidity can be deeply weathered. According to McLean and Gibble (1992), this may not be readily discerned in hand specimens. Therefore use of rocks in the humid tropics for road surfaces should be based on mineralogical investigations.

(a)

(b)

Fig. 11 **a** A section of the Salt Packy deep drain system, north-eastern Jamaica. Seen at centre, each drain is 80 cm × 200 cm deep. Concrete-lined, it completely stopped landslides, which previously had been an annual event. After its installation, no landslide occurred here (i.e., within the last 60 years). **b** Water flowing in the Salt Packy deep drain system, northeastern Jamaica

Fig. 12 Hydrostatic pressure, as shown here, makes ground water move "uphill," thereby relieving the pore pressure

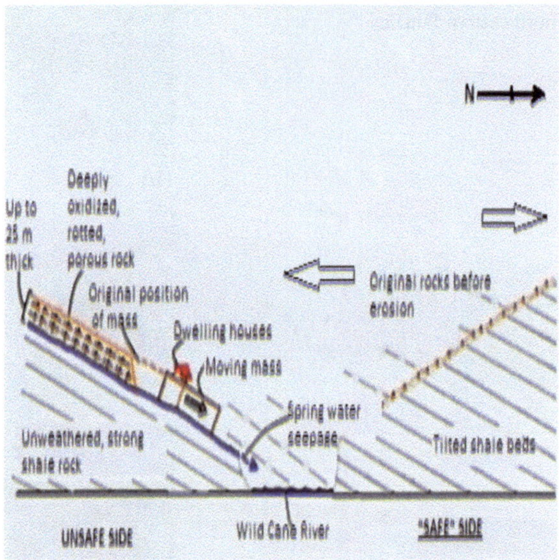

Fig. 13 For dipping beds, slope cuts or river valleys create a greater likelihood of landslips on down-dip side of the cut

- Regularly inspect the substratum for tension cracks. Tension cracks running roughly parallel to the out slope indicate that the area is unstable and likely to slump (Sheoran et al. 2010). Only while the gravitational forces are unable to overcome the frictional forces keeping the material in place will soil and regolith remain on a hillslope. The following factors increase frictional resistance relative to downslope forces. Their effects can be reduced with the following actions:
- Cease increased overburden from structures (construct smaller houses; cease such construction).
- Reduce soil moisture (construct deep trenches parallel to the slope and along the length of it).
- Increase roots holding the soil to bedrock (plant quick-growing deep-rooted drought resistant trees).

- Cease all excavations; install retaining walls when undercutting of the slope occurs by excavation or erosion.
- For weathering by wet/dry cycles: retain vegetative ground cover to retain soil moisture.
- In each high-frequency landslip province, establish a critical region where minimum triggering and antecedent rainfall amounts are applied to landslide occurrence to reduce the number of false alarms, after Garcia-Urquia and Axelsson (2015).
- The safer side of road cuts or river valleys can be determined by evaluating the dip of the rock layers, as shown (Fig. 13).

Acknowledgments The author wishes to acknowledge the assistance of Dr. Rowl Twidale of the School of Earth and Environmental Sciences, the University of Adelaide for his suggestions for this paper. Many thanks also to the encouragement and keen interest shown by Dr. Paul Gyles and Dr. Vincent Wright, of Northern Caribbean University in research of this nature.

References

Ahmad R (1995) Landslides in Jamaica: extent, significance and geologic zonation. In: Barker D, McGregor DFM (eds) Environment and development in the Caribbean. Geographical perspectives. The Press, University of the West Indies, Barbados, pp 147–169

Anderson JH, Cobb JC (2008) Foundation problems and pyrite oxidation in Chattanooga shale, Estill County, Kentucky. Report of investigations 18, Series XII, 2008. www.uky.edu/kys. ISSN: 0075–5591.

Aleotti P (2004) A warning system for rainfall-induced shallow failures. Eng Geol 73(3–4):247–265. doi:10.1016/j.enggeo.2004.01.007

Alexander D (2005) Vulnerability to landslides. In: Glade T, Anderson M, Crozier MJ (eds) Landslide hazard and risk. Wiley, New York, pp 175–198

Anderson MG, Holcombe E, Blake JR, Ghesquire F, Holm-Nielsen N, Fisseha T (2011) Reducing landslide risk in communities: evidence from the Eastern Caribbean. Appl Geogr 31(2):590–599. doi:10.1016/j.apgeog.2010.11.001

Balgin WGV, Coleman A (1965) Puerto Rico geography 50(3):274–286. Published by: Geographical Association Stable URL: http://www.jstor.org/stable/40567047

Barry F, Ophori D, Hoffman J, Canace R (2009) Groundwater flow and capture zone analysis of the Central Passaic River Basin, New Jersey. Environ Geol 56:1593–1603

Bonnard CH, Noverraz F (2001) Influence of climate change on large landslides: assessment of long term movements and trends. In: Proceedings of the international conference on landslides: causes impact and countermeasures, Gluckauf, Essen, Davos, pp 121–138

Botts (1986) Mechanisms of strength loss during wetting and drying of pierre shale. Botts (1986 www.geotech-fr.org/sites/default/files/congres/cimsg/1183.pdf)

Bui DT, Pradhan B, Lofman O, Revhaug I, Dick AVB (2013) Regional prediction of landslide hazard using probability analysis of intense rainfall in the Hoa Binh province, Vietnam. Nat Hazards 66(2):707–730. doi:10.1007/s11069-012-0510-0

Chatterjea K (2011) Severe wet spells and vulnerability of urban slopes: case of Singapore. Nat Hazards 56(1):1–18. doi:10.1007/s11069-009-9362-7

Craig RF (1994) Soil mechanics. Chapman & Hall, London 316

Dahal RK, Hasegawa S (2008) Representative rainfall thresholds for landslides in the Nepal Himalaya. Geomorphology 100(3–4):429–443. doi:10.1016/j.geomorph.2008.01.014

Emujakporue OG (2011) Determination of the bearing capacities of sandstones and shale lithologies from shear wave velocity in the Niger Delta Sedimentary Basin. Asian Journal of Earth Sciences 4(4):223–227. doi:10.3923/ajes.2011.223-227

Finchan AG, Draper G (1998) Jamaica underground: the caves, sinkholes and underground rivers (2nd edn). Press, University of the West Indies, p 35

Flentje P, Chowdhury RN, Tobin P (2000) Management of landslides triggered by a major storm event in Wollongong, Australia. Proceedings of the II

Floris M, Bozzano F (2008) Evaluation of landslide reactivation: a modified rainfall threshold model based on historical records of rainfall and landslides. Geomorphology 94(1):40–57. doi:10.1016/j.geomorph.2007.04.009

Fourie AB (1996) Predicting rainfall-induced slope instability. Proc Inst Civil Eng Geotech Eng 119:211–218

Frattini P, Crosta G, Sosio R (2009) Approaches for defining thresholds and return periods for rainfall-triggered shallow landslides. Hydrol Process 23(10):1444–1460. doi:10.1002/hyp.7269

Gabet EJ, Burbank DW, Putkonen JK, Pratt-Sitaula BA, Ojha T (2004) Rainfall thresholds for landsliding in the Himalayas of Nepal. Geomorphology 63(3–4):131–143. doi:10.1016/j.geomorph.2004.03.011

Garcia-Urquia E, Axelsson K (2015) Rainfall thresholds for the occurrence of urban landslides in Tegucigalpa, Honduras: an application of the critical rainfall intensity. Geografiska Annaler Series A Phys Geogr 97(1), 61–83, March 2015. doi:10.1111/geoa.12092

Gemici U (2001) Durability of shales in Nrider Izmir, Turkey, with an emphasis on the impact of water on slaking behaviour. Environ Geol 41:430–439

Gibbs RJ (1967) The Amazon River: environmental factors that control its dissolved and suspended load. Science 156:1734–1736

Harris MA, Omoregie SN (2007) Post-mining deterioration of bauxite overburdens in Jamaica: storage methods or subsoil dilution? Environ Geol 54(1):111–115

Ibsen ML, Casagali N (2004) Rainfall patterns and related landslide incidence in the Porretta-Vergato region, Italy. Landslides 1(2):143–150

Irritano G, Versace P, Sirangelo B (1998) Real time estimation of hazard for landslides triggered by rainfall. Environ Geol 35(2–3):175–183

Jackson T (1998) Caribbean geology: into the third millenneum: transactions of the fifteenth Caribbean geological conference. University of the West Indies Press, p 248

Jackson T, Draper G, Robinson E, Fluegeman RH (1998) Field guide to the geology of eastern Jamaica. Contrib Geol 3:10–21

Jemec M, Komac M (2012) Rainfall patterns for shallow landsliding in perialpine Slovenia. Nat Hazards 67(3):1011–1023. doi:10.1007/s11069-011-9882-9

Kaddah MT (1974) The hydrometer method for detailed particle size analysis: 1. Graphical interpretation of hydrometer readings and test of method. Soil Sci 118(2):102

Khan YA, Lateh H, Baten MA, Kamil AA (2012) Critical antecedent rainfall conditions for shallow landslides in Chittagong City of Bangladesh. Environ Earth Sci 67(1):97–106. doi:10.1007/s12665-011-1483-0

Knight DK (1963) On dam, geology, embankments and cut slopes. J Soil Mech Found Div ASCE 89(2):99–125

Koo B-J, Fletcher DE, Hinton TG, Barton CD (2010) Assessment of stream fish mortality from short-term exposure to illite clays used as an in situ method for remediating 137cs contaminated wetlands. Int J Soil Sediment Water 3(2). http://scholarworks.umass.edu/intljssw/vol3/iss2/2

Larsen MC (1990) Rainfall-soil moisture relations in landslide-prone areas of a tropical rain forest, Puerto RicoConference Title: tropical hydrology and Caribbean water resources. In: Proceedings of the international symposium on tropical hydrology and fourth caribbean islands water resources congress, San Juan, Puerto Rico. https://water.usgs.gov/.../larsen_torres-sanchez_abstract1990.html. Accessed 24 Feb 2016

Larsen Matthew C, Parks John E (1997) Natural hazards and disasters: landslides in Puerto Rico. How wide is a road? The association of roads and mass-wasting in a forested montane environment. Unit for disaster studies UWI Mona. Wiley, New York http://www.mona.uwi.edu/uds/Land_Puerto_Rico.html. Accessed 29 April 2015

Martelloni G, Segoni S, Fanti R, Catani F (2012) Rainfall thresholds for the forecasting of landslide occurrence at regional scale. Landslides 9(4):485–495. doi:10.1007/s10346-011-0308-2

Mathew J, Babu DG, Kundu S, Kumar KV, Pant CC (2013) Integrating intensity-duration-based rainfall threshold and antecedent rainfall-based probability estimate towards generating early warning for rainfall-induced landslides in parts of the Garhwal Himalaya, India. Landslides, 11(4). doi:10.1007/s10346-013-0408-2

McLean AC, Gibble CD (1992) Geology for civil engineers. Chapman and Hall, London

Mcpherson J (1965) The making of the West Indies. Longmans, London, pp 130–132

Meteorological Office of Kingston Jamaica (2006) The hurricanes of 2005. www.metservice.gov.jm. Accessed 3 April 2016

Middleton GV (1996) Small scale models of turbidity currents and the criterion for auto-suspension. J Sedimentol Res 36(1):202–208

Montgomery Carla W (1989) Environmental geology. William C. Brown (Publisher), Dubuque, pp 301–373

Murray T (2007) The challenge of Sustainability. Washington forest protection association—annual report 2007. ar2007.wfpa.org. Accessed 6 July 2015

NOAA (2006) NOAA reviews record-setting 2005 atlantic hurricane season. National Oceanic and Atmospheric Administration. www.noaanews.noaa.gov/stories2005/s2540.htm. Accessed 3 April 2016

O'Callaghan WB, Mcdonald SC, Richards DM, Reid RE (1998) Development of a topsoil-free vegetative cover on a former red mud disposal site. Alcan Jamaica rehabilitation project paper

Paronuzzi P, Del Fabbro M, Maddaleni P (2002) Frane superficial litipo slide debris flow causate dal nubifragio del 21/22 giugno 1996 nella Val Chiarso' (Alpi Carniche, Friuli). Memorie della Societa Geologica Italiana 57:443–452

Petrucci O, Pasqua A (2009) A methodological approach to characterise landslide periods based on historical series of rainfall and landslide damage. Nat Hazards Earth Syst Sci 9(5):1655–1670. doi:10.5194/nhess-9-1655-2009

Polemio M, Sdao F (1999) The role of rainfall in the landslide hazard: the case of the Avigliano urban area (Southern Apennines, Italy). Eng Geol 53(3–4):297–309. doi:10.1016/S0013-7952(98)00083-0

Price DG, De Freitas MH (eds) (2009) Engineering geology-principles and practice. Springer, Berlin

Reading Alison J, Thompson Russell D, Millington Andrew C (1995) Humid tropical environments. Blackwell, Cambridge, pp 188–199

Rizzo V, Fragale F, Terranova O (2001) Widespread landslide and flood events in the Catanzaro's Isthmus (Calabria, Italy): relationships with rainfall data. In: Glade T, Albini P, Francés F (eds) The use of historical data in natural hazards assessments. Kluwer Academic Publishers, The Netherlands, pp 169–183

Robinson R, Thagesen B (2004) Road engineering for development, 2nd edn. Spon Press. Taylor & Francis group, London and New York

Schaefer VR, Birchmier MA (2013) Mechanisms of strength loss during wetting and drying of pierre shale. In: Proceedings of the 18th international conference on soil mechanics and geotechnical engineering, Paris 2013

Schuster R, Highland L (2007) The Third Hans Cloos Lecture. Urban landslides: socioeconomic impacts and overview of mitigative strategies. Bull Eng Geol Environ 66(1):1–27. doi:10.1007/s10064-006-0080-z

Sengupta A, Gupta S, Anbarasu K (2010) Rainfall thresholds for the initiation of landslide at Lanta Khola in north Sikkim, India. Nat Hazards 52(1):31–42. doi:10.1007/s11069-009-9352-9

Sheoran V, Sheoran AS, Poonia P (2010) Soil reclamation of abandoned mine land by revegetation: a review international journal of soil, sediment and water. Int J Soil Sediment Water 3(2):13 Iss. 2, Art. 13 ISSN: 1940-3259
Simonnet DS (1967) Landslide distribution and earthquakes in the Bewani and Toricelli Mountains, New Guinea. In: Jennings JN
Strahler AN, Strahler AH (1973) Environmental geoscience: interaction between natural systems and man. Hamilton Publishing Company, Santa Barbara, pp 268–269, 306–307
Tating F, Hack R, Jetten V (2013) Weathering and deterioration as quantitative factors in slope design in humid tropical areas; case study northern Kota Kinabalu, Sabah, Malaysia. Ingeokring Newsletter, vol 33, pp 22–28. ISSN: 1384-1351
Twidale CR, Campbell EM (2005) Australian landforms. Rosenburg Pty. Ltd, NSW, pp 84–132
Webster N (1972) The webster encyclopedic dictionary of the english language. Consolidated Book Publishers, Chicago, p 908 (Environ Earth Sci (2010) 61:369–377 377)

Part II
Mine Waste: Utilization and Recycling

Hydrocarbon-Contamination of Soils Around a Trinidadian Pitch Lake: Red Muds as a Resistance to Potential Soil Toxicants

Mark Anglin Harris

Abstract Hydrocarbon-contamination can change hydraulic conductivity (HC) in soils, and hence affect the spread of aqueous toxicants in the ground. A constant head permeameter used in the laboratory to measure HC of soils taken from near the Pitch Lake in Trinidad determined that the HC exceeded that of a reference soil having "normal HC" for a loam. Although water moved rapidly through it, the Pitch Lake soil (PLS) remained dry due to water repellence. Treatment consisted of either of two red mud bauxite wastes mixed at 25 and 50 % w/w with PLS at air dry. One of the bauxite wastes had been treated with gypsum several years before. At 25 % w/w the non-gypsum-treated red mud waste decreased HC of the PLS by 50 %, and at 50 % w/w inflicted a 10-fold decrease of HC on the PLS. The gypsum-had treated red mud waste no effect on the HC of the PLS. The drastic decrease in HC of the hydrocarbon-contaminated soil implies blocking of hydraulic channels by inorganic particles. The high levels of Na^+ released in the Bayer beneficiation process dispersed and released fine <5 mμ clay particles from the non-gypsum-treated red muds. This suggests that the rapid movement of aqueous pollutants in such hydrocarbon polluted soils could be similarly curtailed under field conditions.

Keywords Saturated hydraulic conductivity · Sodicity · Water drop method · Water repellence

Highlights

- Water repellence caused high hydraulic conductivities in hydrocarbon-contaminated soils.
- Field soils from around a pitch lake were hydrocarbon-contaminated and non-wetting.
- Hydrocarbon contamination increased ($p < 0.05$) saturated hydraulic conductivity.
- Laboratory applied clay-dislocating soil potentially reduced spread of aqueous toxicants.

M.A. Harris, *Geobiotechnological Solutions to Anthropogenic Disturbances*, Environmental Earth Sciences, DOI 10.1007/978-3-319-30465-6_4

1 Introduction

Water repellent soils normally exhibit low infiltration rates (Roberts 1966; Blackwell 2000). Hydrophobicity can reduce the affinity for soils to water such that infiltration or wetting may be delayed for periods ranging from as little as a few seconds to in excess of weeks (Hall 2009). Soil hydrophobicity is thought to be caused primarily by a coating of long-chained hydrophobic organic molecules on individual soil particles. Hydrophobicity influences soil hydrological and ecological functions (Takawira et al. 2014). A study by Lourenco et al. (2015) confirmed the hypothesis that hydrocarbon contamination induces water repellence and reduces soil moisture retention at low suction (<100 kPa) for laboratory contaminated soils. Water is not easily absorbed by such non-wetting soils. If aqueous contaminants are released in them, unrestricted rates of spread (Fig. 1) could rapidly contaminate large areas of soil.

Surface seepages of petroleum hydrocarbons cause hydrophobicity (personal observation) in soils adjacent to the Pitch Lake in southern Trinidad (Fig. 2). In other fractured fields where surface seepages occur, there may be potentially similar results. Non-wetting soils are often prone to dispersion and structural breakdown (Ward et al. 2015). Water repellence poses challenges to stability of building foundations (Pietruszczak and Oulapour 1999), pesticide concentration, and leaching (Blackwell 2000), and for crop production in terms of crop establishment, nutrition, and weed control (Ward et al. 2015). The exact chemical composition of substances responsible for the development of water repellence in soils is difficult to identify, but they are generally organic compounds that accumulate on and between soil particles (Dekker et al. 2009). Two approaches to the modification of soil hydrophobicity, and hence water movements, are (1) inorganic and (2) organic.

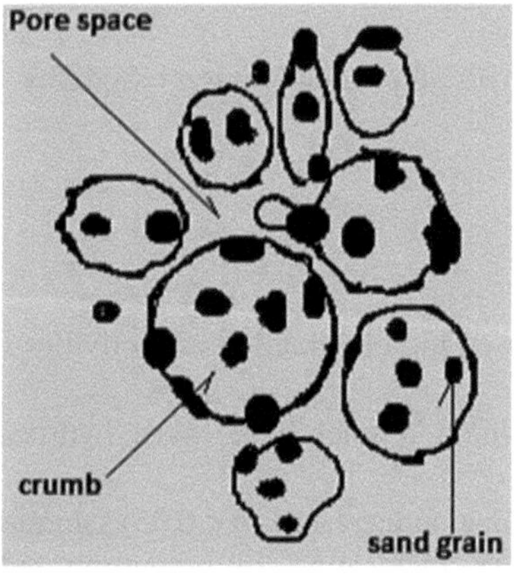

Fig. 1 Typical soil aggregates without excess hydrocarbon oils, hence with no barriers to internal wetting within each crumb

Fig. 2 Part of the Pitch Lake at LaBrea, Trinidad. The dark gray area is the central section. Note the limitations to vegetation luxuriance on southern section. This is due largely to water-repellent soils. Adapted from Geological Society of Trindad & Tobago (GSTT)

1.1 Inorganic Approach

Water retention in soil is critical for remediation of contaminated sites (Lourenco et al. 2015). According to Roy and McGill (2002), soil water repellence is a function of soil surface chemistry. For example, abrasion of sand particles during light sieving had only small effects on repellence, but more vigorous abrasion through rotational movement of the sand reduced repellence markedly (King 1981). Alternatively, soil hydraulic conductivity (HC) is related to the composition and concentration of exchangeable cations and soluble electrolytes (Frenkel et al. 1992). Thus Quirk and Schofield (1955) showed that the HC of a soil decreased with increasing exchangeable sodium percentage (ESP) and decreasing electrolyte concentration in the soil solution. Essington (2004) described dispersive conditions as mutual repulsion of tactoids fully surrounded by associated Na^+ and waters of hydration (ESP < 15). McNeal and Coleman et al. (1966) found a linear relationship between reduction of HC and the degree of macroscopic swelling of the extracted soil clay. Thus de-flocculation and movement of clay into the conducting pores was proposed as the second mechanism for explaining the reduction in HC. The decrease in soil HC due to swelling and dispersion of clay is explained by the presence of monovalent exchangeable cations. Such changes were responsible for the loss of nearly all drainable macro-pores in the soil, which was evidenced by the considerable reduction in cumulative infiltration amounts and rates (Lourenco et al. 2015).

Clay mineralogy has also been shown to have a large influence on reductions in HC (Churchman et al. 1993; McNeal and Coleman 1966). Due to their 2:1 layer structure which accommodates a high amount of exchangeable Na^+ within its interlayer space (Arienzo et al. 2012; Churchman et al. 1993), smectites show extensive swelling and dispersion (Fig. 3). Na^+ is a large monovalent ion

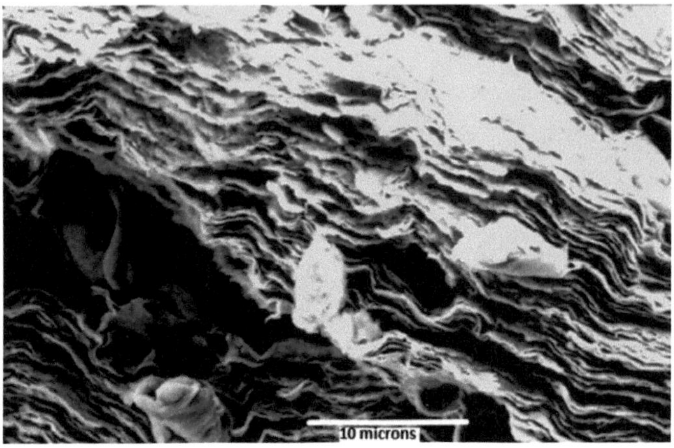

Fig. 3 A 2:1 clay displays cavities containing hydrated Na^+ ions. The presence of Na^+ exacerbates very active shrink/swell behavior which releases clay particles into the soil solution. These particles can block soil pores to reduce hydraulic conductivity. *Source* https://pubs.er.usgs.gov/publication

and more effectively forces clay tactoids (i.e., particles) apart than Ca^{2+} or Mg^{2+} (Quirk 1986). Swelling occurs with increasing Na^+ concentration, as hydration of Na^+ leads to the expansion of the interlayer (ESP > 15).

1.2 Organic Approach

Rainfall infiltration into water repellent soils is characteristically patchy, resulting in patchy and staggered crop germination of crops and of weeds (Ward et al. 2015). They speculated that improvements in soil water availability, and increases in organic matter, might encourage microbial activity, leading to increased degradation of the compounds causing water repellence.

Roper et al. (2013) showed that soil water contents in water repellent sand were greater in crop rows where crop residue was retained. Based on observations, Ward et al. (2015) showed that soils in crop rows showed lower levels of water repellence when crop rows were established on or close to previous crop rows, compared with crop rows established on the inter-row spaces. This was attributed to the likelihood of greater organic matter and nutrient accumulation, the former increasing water availability, which could have encouraged microbial activity. Such microbial activity could have led to increased degradation of the compounds causing water repellence. Nevertheless, compared to naturally-occurring and fire-induced hydrophobicity, limited information is available on the impacts of hydrocarbon contamination on water repellence and hydraulic properties (Lourenco et al. 2015). Increased saturated hydraulic conductivity associated with laboratory contaminated soils contradicted their original hypothesis (Lourenco et al. 2015) such that their findings imply that storms falling on initially dry recently contaminated soils may trigger contaminant transport and erosion via enhanced surface runoff, and rapid spreading of contaminants once they reach the groundwater systems.

1.3 Aim and Hypothesis

The aim of this study is to determine, for a hydrocarbon-contaminated non-wetting soil, (1) the hydraulic conductivity and hence (2) the likelihood of rapid spread of aqueous pollutants, and (3) a potential method to reduce the spreading rate of such contaminants in that hydrocarbon-contaminated soil.

Sources of Na^+ bentonite, which can reduce infiltration rates are rare in the Caribbean. Therefore it is hypothesized that abundant Na^+-rich dried red muds from bauxite waste (of relatively low toxicity) could release reduce very high infiltration rates in hydrophobic hydrocarbon-contaminated soils.

2 Materials and Method

2.1 Site Details

The La Brea Pitch Lake occupies 36 ha in southwestern Trinidad at 62° W and 10.5° N (Fig. 4). Average annual rainfall at La Brea is 1250–1750 mm, most of which falls between May and December. Temperature is lowest in January when it averages 25 °C, rarely rising higher than 33 °C in June through September, which are the hottest months.

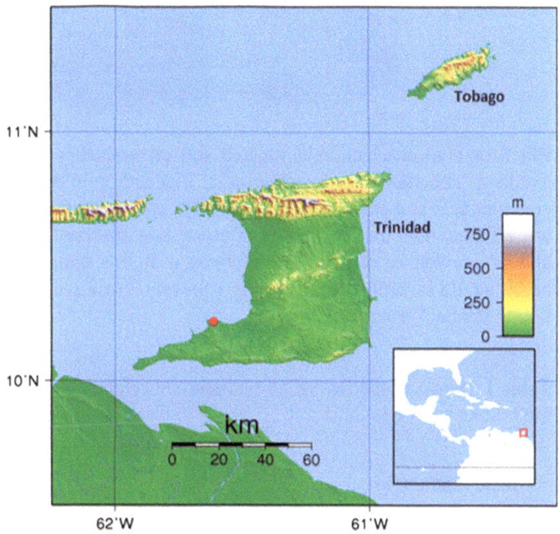

Fig. 4 Location of the Pitch Lake at La Brea (*circular dot*) in Trinidad and Tobago

2.2 Soil Sampling

Soil cores were collected on the periphery of the Pitch Lake along a transect perpendicular to the centre in April 2012, to a depth of 20 cm and brought to a laboratory on the same day. The soil was brought to air dry after exposure for three days. Black petroleum tar contaminated some soil aggregates (Fig. 5) at the site (Table 1).

(a)

(b)

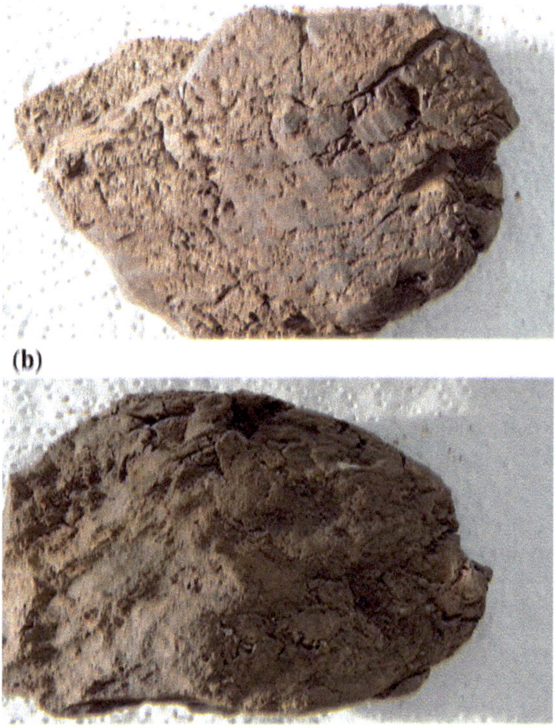

Fig. 5 **a** Hydrocarbon-contaminated soil on periphery of Pitch Lake in LaBrea, southwestern Trinidad. The diameter of each circular indentation in the background is 1 mm. **b** Hydrocarbon-contaminated soil sampled 3 m further out on the periphery of Pitch Lake in LaBrea, southwestern Trinidad than that in **a** above. Despite less observed evidence of hydrocarbon contamination and more evidence of good soil structure in **b**, this sample was more dispersive than that of **a**. It hence would be probably less stable after rain. For a scale, the diameter of each circular indentation in the background is 1 mm

Table 1 Some properties of an undisturbed reference soil from Mount Nelson, Jamaica, and a Pitch Lake soil (Trinidad)

Property	Reference soil	Pitch Lake soil
Organic matter (%)	4.2	56.6
pH	5.3	6.9

2.3 Water Repellence Determination

Water repellence was measured by counting the number of seconds required for water to be absorbed into the soil (WPDT) after the method of DeBano (1981). Fifty replicates were used, where water repellence is classified as follows: lower than 3 s = highly wetting (0); 3–6 s = wetting (1); 6–60 s = slightly water repellent (2); 60–600 s moderately water repellent (3); >600 s = extremely water repellent (4). Samples were measured at air-dry were crushed and sieved to less than 2.0 mm to remove large organic matter and gravel.

2.4 Dispersion and Hydraulic Conductivity

Measurement of spontaneous dispersion and hydraulic conductivity were conducted respectively using the Modified Emerson Water Dispersion Test (Emerson 1967). Based on the objective of potentially using the treatment as a barrier for aqueous contaminants in field soils, potentially large volumes of soil additive could be required to affect interflow under field conditions. Therefore, the constant head method for hydraulic conductivity Klute (1983) used was modified to include soil columns of thin cross-sections, i.e., of 2.5 cm thickness.

2.5 Bauxite Waste Treatments

Samples of gypsum-treated bauxite waste (O'Callaghan et al. 1998) collected from Kirkvine Pond 6 during 2007 from two depths: 0–15 and 15–30 and stored at air-dry, were grinded and passed through a 1.0 mm diameter sieve, and thoroughly mixed in with each of the pitch lake soils (PLS) from La Brea, Trinidad. Two depths were selected as potential treatment additives because only the 0–15 cm layer was gypsum treated. It has been shown (Harris 2008) that the gypsum applied to a depth of 15 cm had little or no physical effect on soil below 15 cm depth. Therefore the original chemistry of the RMW at 15–30 cm was not as altered as that of the 0–15 depth (GRMW). It was therefore postulated that the greater ESP and SAR of material from 15–30 cm may enhance displacement of inorganic ions in an hydraulic column. Removal of divalent ions from the exchange complex can contribute to reductions in aggregate stability and decreases in HC (Kopittke et al. 2006).

2.6 Red Mud Safety Concerns

Red mud can contain elevated levels of metals, and the pH of red mud characteristically ranges from 10 to 12 due to the use of caustic soda during the extraction process. Red mud is also known to contain technologically enhanced naturally occurring radioactive material (TENORM), including thorium and uranium. TENORM is naturally occurring radioactive material (NORM) that has been processed in such a manner that its concentration has increased (EPA 2012).

However, concerns about the radioactivity in red mud waste were allayed because (Pinnock 1991) in World Aluminium (2015) showed that using 100 % bauxite residue gave a dose equivalent to just over 2 m Sv y^{-1} and was judged to be acceptable. However, other work on Hungarian bauxites has recommended a maximum addition of 15 % bauxite residue to avoid exceeding a level of 0.3 m Sv y^{-1} (Pinnock 1991 in World Aluminium 2015).

2.7 Statistics

Sample means compared using Tukey's Honestly Significant Difference ($P \leq 0.05$) to examine differences in HC between treatments.

3 Results

3.1 Water Repellence

Water repellence of the PLS was found to be at the highest value, i.e., extremely water repellent (4), compared with the reference soil, which had a value of highly wetting (0), i.e., showing very rapid water absorbance. Rapid water absorbance is unusual for hydrophobic soils. Further, the fact that soil aggregates after several days in water, and even aggregates as small as <2 mm remained dry after several hours of water infiltration showed the extremely hydrophobic nature of the Pitch Lake soils (Fig. 6, Table 2).

Fig. 6 Pitch Lake soil agglomerates after being subjected to through-flow in a water column. Soil particles remained dry even after water flowed for several (>30) minutes through the mass. The agglomerates are floating in the water, as proven by the lack of shadows under the soil particles at left. Each circular depression in background cloth is 1 mm in diameter

Table 2 Water repellence rating[a] of normal soils and bauxite waste as additives

Soil/bauxite waste	Water repellence
Reference soil	0
Pitch Lake soil	4
Dried red mud	0
Gypsum-treated red mud	0

[a]Water repellence is classified as follows: lower than 3 s = highly wetting (0); 3–6 s = wetting (1); 6–60 s = slightly water repellent (2); 60–600 s moderately water repellent (3); >600 s = extremely water repellent (4)

Table 3 Hydraulic conductivity (cm min^{-1}) of Pitch Lake soils after treatment with dried red muds

Reference soil	55
Pitch Lake soil + 25 % dried red mud	47
Pitch Lake soil + 50 % dried red mud	20
Pitch Lake soil + 25 % gypsum-treated dried red mud	202
Pitch Lake soil + 50 % gypsum-treated dried red mud	202

3.2 Changes in Hydraulic Conductivity

There was little, if any, restriction to water-flow through the hydrocarbon contaminated samples. The hydrocarbon-contaminated soils (HCS) all exhibited high values for hydraulic conductivity, exceeding 150 cm min^{-1} and exceeding that of the reference soil by >35% (Table 3). This flow rate is similar to that through a fine-to-medium (<2 mm) quartz sand.

When RMW was added at 25 % of the volume of the PLS in the column, the RMW effect on the HC of the PLS was expected to be proportional to its volume (25 %). Yet the 25 % treatment of RMW changed PLS hydraulic conductivity, not, as was expected, by 25 %, but by a far larger proportion, i.e., by 400 % (decrease) in HC (Table 3).

Even more dramatically, RMW added at the rate of 50 % to the PLS reduced HC in the column not by 50 %, but by a 10-fold margin (1000 %). Conversely, the GRMW in similarly added proportions of 25 and 50 % had no effect on the HC of the PLS in the column. There was no observed or detected reaction. Though the reduction in porosity of the soil column was dependent on concentration of the additive, it must be noted that the particle diameters of all treatments, including the non-effective (gypsum-based) ones, were identical.

3.3 Hydrophobicity and Dispersion

Though all PLS soil samples along the transect taken across the periphery of the La Brea Pitch-Lake were water repellent, there was no observed pattern of soil hydrophobicity increasing inversely with distance from the centre of the Pitch

Table 4 Modified emerson dispersion test levels of dried bauxite waste and Pitch Lake soils

Reference soil	0
Pitch Lake soil (low tar)	4
Pitch Lake soil (high tar)	2
Dried red mud (no gypsum)	4
Dried red mud (gypsum-treated)	0

Emerson dispersion levels: *1* slight milkiness *2* obvious milkiness, less than 50 % of the aggregate affected *3* obvious milkiness, greater than 50 % of the aggregate affected *4* total dispersion leaving only sand grains. *Source* Emerson Aggregate Test (1967)

Fig. 7 Dispersion of a Pitch Lake soil. A single soil crumb totally disintegrates in water after 2 h. Yet, HC remained high due to inherent hydrophobicity. Samples closer to the centre of the Pitch Lake exhibited less dispersion. *Scale* each circular depression in container is 1 mm in diameter

Lake. This suggests a low threshold of hydrocarbon contamination required to produce hydrophobicity in a soil. The Modified Emerson Water Dispersion Test showed that the PLS exhibited slaking and dispersion at level 4 (the second highest level) after just 2 h (Table 4, Fig. 7). Contrastingly, no dispersive reaction was detected for the reference soil (Table 4), even after several weeks of contact with water. Though all the aggregates were water repellent, and, to varying degrees dispersive, those aggregates which visibly contained large amounts of high density black tar were the least dispersive, while the PLS with the lowest levels of hydrocarbon contamination were the most dispersive.

4 Discussion

4.1 Infiltration

On the assumption of high levels of hydrocarbon contamination in the PLS, the inherently high level of saturated hydraulic conductivity correlating with hydrocarbon-contamination of the soil in this study is corroborated by Lourenco et al. (2015), who, in a laboratory simulated hydrocarbon contamination, induced soil water repellence. Saturated hydraulic conductivity (K_S) increased linearly with level of (hydrocarbon) contamination ($p < 0.05$; r2 $\approx$ 0.8), indicating rapid flow of water attributed to a reduction of the dielectric constant, and hence water–soil matrix interactions (Lourenco et al. 2015). A humic soil is defined as soil in which organic carbon exceeds 50 %. Yet, the PLS, with 56 % organic carbon (Table 1) is not a humic soil, there being no visible humus (Fig. 6). Therefore, it is reasonable to conclude that hydrocarbons represent most of the organic C in PLS. On close microscopic inspection of the mineral soil, there was no humus visible. In this study, the soil with lower observable levels of hydrocarbons (black tar) exhibited greater dispersion. Hence the less dispersive samples occurred closest to the centre of the Pitch Lake PLS. This was against expectations.

In the samples closer to the centre of the Pitch Lake, dark, tarry petroleum occupied a greater proportion of the soil (Fig. 5). Asphalt generally contains some of the heaviest and least volatile fractions of petroleum distillates. In the aquatic environment, asphalt will sink to bottom as a dark tarry substance. Petroleum distillates in order of decreasing volatility include:

1. Petroleum ether or benzene
2. Gasoline
3. Naphtha
4. Mineral spirits
5. Kerosene
6. Fuel oil
7. Lubricating oils
8. Paraffin wax
9. Asphalt or tar.

As gravitational forces act on all objects including those with horizontal movement, lighter, more mobile short-molecular chain hydrocarbon fractions would have been expected to have travelled furthest out from the centre of the Pitch Lake. It is plausible then, that the smaller hydrocarbon chains of the lighter fractions would have more easily penetrated the micro-pores and microstructure of those soil aggregates in direct variation with distance from the centre of the Pitch Lake. Thus the greater the distance, the greater the sorting of, and concentration of, contaminants, in such soil samples. This leads thereby to (1) greater effective hydrophobicity and/or (2) greater dispersion amongst such (more distant) samples.

The results of this study show that a dispersive clay synergistically decreased hydraulic conductivity in a hydrophobic hydrocarbon-contaminated soil. This outcome supports the hypothesis of this study which was based on the activity of sodium-affected soils in an aqueous medium. Even as little as 1 % of the total clay when dispersed affects the hydraulic conductivity by blocking micro-channels in the soil mass (Goldberg and Glaubig 1987). Tuffour et al. (2015) found that finer sediments were highly effective in altering soil properties even at low concentrations. They showed that after mixing with 10 % dry Na^+ Bentonite, wetting resulted in complete filling of skeletal pore space by a jell-like clay fabric. With a 3 % level of sodium oxides in Jamaican red muds, the Na^+ levels of the RMW would have been sufficiently strong to power clay dislocation, and it is dislocated clay-sized particles that changed the HC in the Pitch Lake soils of this study.

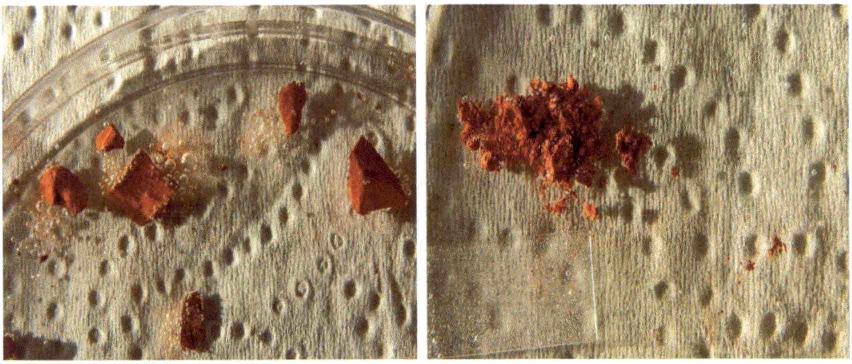

Fig. 8 *Picture #1* shows air bubbles from explosive reaction when dried untreated red mud makes contact with water, culminating in total dispersion and destruction after just five minutes in water (*picture #2*). The clay particles so released blocked pores in a hydrophobic Pitch Lake soil. Photo frame at right contains a microscope slide at lower left which supplies a scale

The difference between the dried gypsum-treated red mud waste at 0–15 cm depth and that of the 15–30 cm depth was still large even after 10 years. For example, it can be seen that an explosive reaction (air bubbles observed) occurred when the Na^+-rich red muds were placed in water (hydrated, Fig. 8), thus blocking pores in the PLS.

In contrast, the dried gypsum-treated red mud retained its integrity indefinitely (Fig. 9) and thus had no effect on the hydraulic conductivity of the Pitch Lake soil. These samples had been in water not for 5 min but for five weeks.

These results suggest that the RMW released components which blocked hydraulic channels in the PLS. As described in previous studies, the negative effects of Na^+ are most pronounced in soils which are high in clay, especially high charge-density clays (Frenkel et al. 1978; Beulow et al. 2015). In the current study, the water-repellent soils were more responsive to soil with higher Na^+ quantities, than the less reactive gypsum-treated soil. Though the original bauxite soils are mainly kaolinite, red mud wastes contain an increased number of fine clay particles compared to that of the original soil. These provide extra negatively charged sites for adsorbing hydrated Na^+ cations. Similarly, the tendency of Na^+ ions to facilitate the formation of multiple layer hydrates in smectite clays leads to greater swelling and reduction of HC. On the contrary, Frenkel et al. (1992) showed that smectites did not release clay particles when added to a sandy soil unless anions were added because the HC of smectite clay-sand mixtures decreased only following the addition of the various anions. Dispersed clay appeared in their effluent only upon addition of citrate or hexametaphosphate. Therefore, at 56 % intrinsic organic matter (Table 1), such hydrocarbon contamination would have greatly increased the negative charges of the Pitch Lake soil. This could have increased repelling forces amongst the thoroughly mixed-in clay particles, thereby blocking hydraulic channels.

Fig. 9 Dried gypsum-treated red mud waste (GRMW) had no observable reaction with water, retaining its stability even after many weeks. Scale is supplied by 1-mm-diameter indented circles in the background

Further, this result conforms with the very high sensitivity of illite to even small amounts of exchangeable Na (Oster et al. 1980). They found that, for a given ESP, the critical flocculation concentrations (CFC) of illite was much larger than for smectite, and hypothesized that the explanation lies in the irregular nature of illite particles, which prevent good contact between edges and planar surfaces, thereby decreasing the potential for inter-particle attraction. Similarly, the red mud particles, having undergone beneficiation (extreme comminution), would have been altered to more irregular shapes. Compared with Sintering red mud, the Bayer Process (used in the Caribbean alumina processing) produces a relatively small particle diameter. The particle diameter of Bayer red mud is between 0.8 and 50 μm with an average value of 14.8 μm (Wang and Dong-Yan 2012). In the case of the smectite cited by Frenkel et al. (1992), HC is decreased through partial blocking of pores by short distance migration of dispersed particles, provided the particle size of the sand (or host soil) is sufficiently fine to retain these particles. Being a clay-loam, the particle size of the PLS is smaller than that of fine sand and therefore would be susceptible to pore blockage by the fine clay particles released from the dried red mud.

5 Conclusion

By repelling water from pore channel walls, hydrophobicity can markedly reduce resistance to movements of aqueous liquids through soils. Finely crushing from <2 to <1 mm produced an at least 4-fold increase in the effective surface area of the RMW. Soil water repellence is a function of soil surface chemistry. More specifically, it is a function of the free energy of the solid/gas interface in soil (γSG). In contrast to the low Na^+ RMW, the high Na^+ RMW produced dislocated clay particles which blocked hydraulic channels.

With a high energy surface exhibited by rapidly decreasing the initial advancing contact angle (θ), the high Na^+ RMW either expanded, or released clay particles, either of which could have blocked hydraulic channels in the HCS.

6 Geotechnical Applications

6.1 *Field Application*

Whereas a change in soil HC due to swelling is essentially a reversible process, changes due to dispersion and particle movement are irreversible (Frenkel et al. 1978). Therefore, the effectiveness of HC reduction by clay-particle blockage is likely to be long-lasting. For field applications, it is postulated that the above-mentioned treatment can be applied in a vertical layer in a thin barrier trench. It is postulated that the trench could then periodically be re-packed as needed with a combination of the water repellant soil and red mud waste as applied above. On the other hand, clay addition by soil inversion can not only be effective, but cost-effective (Blackwell et al. 2014) by reducing haulage costs. In the latter process specific subsoils could be utilized, though less effectively, when red muds are unavailable.

6.2 *Edaphic Effects*

For water repellence in agricultural soils, residue retention with no-till, minor changes to seeding boots, wetting agents, rotary spading can remove hydrophobicity, according to Blackwell et al. (2014). The most profitable management options will depend on the scale of water repellence.

6.3 *Testing for Hydrophobicity*

Testing for hydrophobicity is done with the MED test. Also, the method of Lourenco et al. (2015) applied "the residual signature of hydrocarbon contamination" evident particularly in electrical conductivity. This indicates that natural soils were inherently wetting and that hydrocarbon-induced hydrophobicity could be transient. This non-persistence was attributed to high decomposition rates stimulated by tropical conditions and nutrients added to promote revegetation.

Water repellence can also be measured using the Molarity of Ethanol Droplet (MED) test (King 1981), where a value of 0 indicates a wetting soil, and greater than 3.0 indicates severely water repellent. The molarity of ethanol droplet (MED) test is a popular rapid method for assessing soil water repellence under field and laboratory conditions. Roy and McGill (2002) justifies this test as follows: "Soil water repellence is a function of soil surface chemistry. More specifically, it is a function of the free energy of the solid/gas interface in soil (γSG). Because γSG is not directly measurable in the laboratory, soil water repellence must be assessed using thermodynamically related parameters such as the initial advancing contact angle (θ) or the work of wetting (WW). The MED test can be used to determine θ, and in turn WW." Water repellence was measured at air dry because King (1981) found that at soil water contents between air dry and wilting point (pF 4.2), the infiltration rate of water was either unchanged or decreased in different soils. It then increased rapidly and reached a constant value near field capacity (pF 2.5). Moisture content of the soil had large and variable effects on the repellence tests. He thus recommended that repellence tests be made on oven- or air-dry soils.

6.4 Building Construction Activities

Soil dispersion adversely affects not only agriculture, but residential and commercial buildings by destabilization. Indicators of hydrophobicity can be used to predict dispersive behaviour in building foundations. One such indicator is soil dryness despite water having passed though the soil as shown in the above-mentioned laboratory experiment. Though the hydrophobic soil remained dry, water running through it completely destroyed observable cohesiveness of macro-aggregates such that only a crumbled, loose entity with the consistency of sand remained (Fig. 7).

6.5 Containment of Aqueous Contaminants

A groundwater cut-off wall comprised of impermeable material can prevent the further migration of aqueous contaminants. The impermeable fill material for the trench can be inert and relatively unreactive such as concrete. However such containment is inactive, in that the contaminant remains in the enclosed zone without being removed. If the wall is comprised of reactive, dislocating clays such as that of this study, aqueous contaminants can be held in the material to be removed on a progressive basis.

References

Arienzo M, Christen EW, Jayawardane NS, Quayle WC (2012) The relative effects of sodium and potassium on soil hydraulic conductivity and implications for winery wastewater management. Geoderma 173:303–310

Bardossy G (1982) Karst Bauxites. Elsevier Scientific Publishing Co

Blackwell PS (2000) Management of water repellence in Australia, and risks associated with preferential flow, pesticide concentration and leaching. J Hydrol 231:384–395

Blackwell PS, Hagan J, Davies SL, Bakker D, Hall DJM, Roper MM, Ward PR, Matthews A (2014) Smart no-till furrow sowing to optimise whole-farm profit on non-wetting soil. GRDC Perth Crop Updates. Perth: Department of Agriculture and Food, and Grains Research and Development Corporation

Buelowa MC, Steenwerthb K, Parikha SJ (2015) The effect of mineral-ion interactions on soil hydraulic conductivity. Agric Water Manag 152(2015):277–285
Churchman GJ, Skjemstad JO, Oades JM (1993) Influence of clay–minerals and organic-matter on effects of sodicity on soils. Aust J Soil Res 31:779–800
DeBano L (1981) Water repellent soils: a state of the art. USDA Forest Service. General Technical Report PSW-46
Dekker L, Ritsema CJ, Oostindie K, Moore D, Wesseling JG (2009) Methods for determining soil water repellence on field-moist samples. Water Resour Res, vol 45, w00d33. doi:10.1029/2008WR007070
Emerson WW (1967) A classification of soil aggregrates based on their coherence in water. Aust J Soil Res 5:47–57. doi:10.1071/SR9940173
EPA (2012) Radiation protection. https://www.epa.gov/radiation. Accessed 13 June 2016
Essington M (2004) Soil and water chemistry: an integrative approach. CRC Press, Boca Raton (0 8493 1258 2)
Frenkel H, Goertzen JO, Rhoades JD (1978) Effect of clay type and content, exchangeable sodium percentage and electrolyte concentration on clay dispersion and soil hydraulic conductivity. Soil Sci Soc Am J 42:32–39
Frenkel H, Levy GJ, Fey MV (1992) Clay dispersion and hydraulic conductivity of clay-sand mixtures as affected by the addition of various anions. Clays Clay Miner 40(5):515–521
Goldberg S, Glaubig RA (1987) Effect of saturating cation, pH, and aluminium and iron oxides on the flocculation of kaolinite and montmorillonite clay. Clay Miner 35:220–227
Hall D (2009) Water repellence. In: 'Managing south coast sandplain soils to yield potential' pp 49–63 (Bulletin 4773, Department of Agriculture and Food: Western Australia)
Harris MA (2008) Structural improvement of age-hardened gypsum-treated bauxite red mud waste using readily decomposable phyto-organics. Environmental Geology 56:1517–1522. doi:10.1007/s00254-008-1249-5
King P (1981) Comparison of methods for measuring severity of water repellence of sandy soils and assessment of some factors that affect its measurement. Aust J Soil Res 19(3). doi:10.1071/SR9810275. http://www.researchgate.net/publication/240505934_. Accessed 2 Nov 2015
Klute A (1965) Laboratory measurement of hydraulic conductivity of saturated soil. In: Black CA (ed.) Methods of Soil Analysis. Part 1. 210–221
Kopittke PM, So HB, Menzies NW (2006) Effect of ionic strength and clay mineralogy on Na–Ca exchange and the SAR–ESP relationship. Eur J Soil Sci 57:626–633
Lourenco SDN, Wakefield CF, Morley CP, Doerr SH, Bryant R (2015) Wettability decay in an oil-contaminated waste-mineral mixture with dry-wet cycles. Environ Earth Sci 74(3):2563–2569. doi:10.1007/s12665-015-4276z
McNeal BL, Coleman NT (1966) Effect of solution composition on soil hydraulic conductivity. Soil Sci Soc Am Pro 30:308
O'Callaghan WB, McDonald SC, Richards DM Reid RE (1998) Development of a topsoil-free vegetative cover on a former red mud disposal site. Alcan Jamaica Rehabilitaion Project Paper
Oster JD, Shainberg I, Wood JD (1980) Flocculation value and gel structure of Na/Ca montmorillonite and illite suspension. Soil Sci Soc Am J 44:955–959
Pietruszczak S, Oulapour M (1999) Assessment of dynamic stability of foundations on saturated sandy soils. J Geotech Geoenviron Eng 125(7):576–582
Pinnock WR (1991) Measurement of radioactivity in Jamaican building materials and gamma dose equivalents in a prototype red mud house. J Health Phys 61(5):647–651
Quirk JP (1986) Soil permeability in relation to sodicity and salinity. Philos Trans Roy Soc Lond Ser A 316:297–317
Quirk JP, Schofield RK (1955) The effect of electrolyte concentration on soil permeability. J Soil Sci 6:163–178
Roper MM, Ward PR, Keulen AF, Hill JR (2013) Under no-tillage and stubble retention, soil water content and crop growth are poorly related to soil water repellence. Soil Tillage Res 126:143–150

Roy JM & McGill WB (2002) Assessing soil water repellence using the molarity of ethanol droplet (Med) test. http://www.researchgate.net/publication/232200700_Assessing_Soil_Water_Repellence_Using_the_Molarity_of_Ethanol_Droplet_(Med)_Test. Accessed 2 Nov 2015

Takawira A, Gwenzi W, Nyamugafata P (2014) Does hydrocarbon contamination induce water repellence and changes in hydraulic properties in inherently wetting tropical sandy soils? Geoderma 235–236:279–289

Tuffour HO, Adjei-Gyapong T, Abubakari A, Melenya C, Aryee D, Khalid AA (2015) Assessment of changes in soil hydro-physical properties resulting from infiltration of muddy water. Appl Res J 1(3):137–140. ISSN: 2423-4796

Walden LL, Harper RJ, Mendham DS, Henry DJ, Fontaine JB (2015) Eucalyptus reforestation induces soil water repellence. Soil Res 02/2015; 53:168 doi:10.1071/SR13339

Wang P (2012) Dong-Yan L (2012) Physical and Chemical Properties of Sintering Red Mud and Bayer Red Mud and the Implications for Beneficial Utilization. Materials 5:1800–1810. doi:10.3390/ma5101800

Ward P, Roper MM, Jongepier R, Micin S, Davies S (2015) On-row seeding as a tool for management of water repellent sands. In: Building productive, diverse and sustainable landscapes proceedings of the 17th ASA conference, 20–24 Sept 2015, Hobart, Australia. Website www.agronomy2015.com.au

World Aluminium (2015) Bauxite Residue Management: Best Practice. www.world-aluminium.org.

Geotechnical Stability of Two Gypsum-Treated Bauxite Red Muds and "Marl" as a Road Base Under Submerged Conditions

Mark Anglin Harris

Abstract Under rain, rapid erosion of soft limestone rocks (L) used as road bedding from a quarry in Jamaica causes frequent destruction of extensive road sections. Therefore the stability for road bedding of widely available pre- and post-lithified bauxite wastes was studied and compared with soft limestone. Durability of (1) gypsum-treated atmospherically lithified carbonated red mud wastes (G) and (2) atmospherically lithified red mud waste subjacent to the gypsum-treated red mud waste (R) was examined using a slake-durability apparatus through wet/dry cycles. These cycles induce dissolution, recrystallization, and lead to the deterioration of the stone, and, finally, affect the strength of these rocks. The California Bearing Ratio of the rocks was also studied. The results showed clear differences between, and among all the rocks. Based on these properties, the studied rocks were classified into two relative types (weak and strong). The slake-durability test indicated weight losses ranging from 0.1 % for the gypsum-treated atmospherically lithified bauxite waste (G), to 99.5 % for the atmospherically lithified red mud (R). The soft limestone marl (L) lost 14.0 % of its mass. This test caused also a reduction of compressive strength for these rocks (G = 8 %, L = 50 %, RW = 100 %) when wet/dried with distilled water after 5 week-long cycles. Slaking reduced the strength most greatly for the soft limestone and the subjacent red atmospherically lithified red mud. The two strong rocks were found to be the microcrystalline limestone and the gypsum,-treated pre-lithified red mud. For the post lithified (non-gypsum-treated) bauxite waste (R), this low level of durability and strength after slaking was attributed to the large amount of residual Na^+ in the unflushed waste, which, despite post-lithification, still exhibited increased particle dispersion typical of the post-Bayer Process.

Keywords Bearing ratio • Red mud waste • Road-base • Slake-durability

M.A. Harris, *Geobiotechnological Solutions to Anthropogenic Disturbances*,
Environmental Earth Sciences, DOI 10.1007/978-3-319-30465-6_5

Highlights

- Less gypsum is required than is currently used for large increase in strength and durability of bauxite red muds.
- Gypsum-treated bauxite waste within a certain particle size range is at least geotechnically equivalent to soft limestone marl as a road base.

1 Introduction

1.1 Requirements of Road Fill

The existence of potholes requires two conditions; water and traffic (Youngtae and Ryu 2015). Water infiltration through the surface or from below can result in a progressive loss of strength of the pavement materials and consequent deterioration of the road.

Therefore, one of the principal aims of road materials design is to minimize the ingress of water (Woodbridge 1999). Road-building material should be hard but crush readily, yet not crumble too readily (USDA 1942). Tuffs, for example crumble too readily, whilst other volcanic rocks are usually too hard for crushing or are not indurated enough (USDA 1942). The more compact-grained and heavier a stone, the harder it is (USDA 1942). Due to alternate wetting and drying the resulting crushing strength can be reduced even up to 30–40 % (Khattab and Othman 2012). When dry, stones exhibit more crushing strength than when wet (Khattab and Othman 2012). The majority of limestones contain admixture of other substances, of which the commonest are clay and sand. Clayey or argillaceous limestones frequently occur in thin or thick beds alternating with shales, as in the Lias of England (the marlstone series). Friable argillaceous fresh-water limestones are called "marls." These impurities tend to lower the strength of the limestone (Khattab and Othman 2012).

1.2 Bauxite Waste as a Potential Road Base

Bauxite is a member of the family of lateritic rocks (Sutar et al. 2014), characterized by a particular enrichment of aluminium-hydroxide minerals, such as gibbsite, boehmite and/or diaspore. Produced by secondary mineralization, bauxite forms by weathering of aluminous silicate rock (to lateritic bauxite) and less commonly of carbonate rock (to karst bauxite) mainly in tropical and sub-tropical climates. Such conditions favour retention of alumina and leaching of other constituents of the parent rock. Bauxite rock has a specific gravity between 2.6 and 3.5 kg m^{-3}. It is usually, an amorphous or clay-like substance, but it is not plastic. The usual colour of bauxite is pink but if of lower iron content it may tend to become whitish in color and with increase in iron it is reddish brown in color (Geologydata 2016).

Aluminium mining leaves behind 120 million metric tons per year of salty, highly alkaline, heavy-metal-laden material, according to the International Aluminium Institute, a London-based trade organization (World Aluminium 2015). The aluminium industry has long tried to find ways to recycle the environmentally problematic red mud, but so far there have been few safe and economical large-scale applications (World Aluminium 2015). Red mud repurposed annually does not exceed 2 million metric tons (World Aluminium 2015).

Red mud carbonated with CO_2 from industrial gas streams has been used in road construction (World Aluminium 2015), the resulting "red sand" being also used to make cement (World Aluminium 2015). Qi (2005) suggested using red mud (from the sintering process) as road material. The sintering process of extracting alumina does not require the application of caustic soda (NaOH). This avoids the hydrated effects of the sodium ion. Based on his research, a 15 m wide and 4 km long highway using red mud as a base material was constructed in Zibo, Shandong Province, China. According to Yang et al. (2006), subsequent Government tests concluded that the sub-grade strength of the red mud base road meets the strength requirements of the highway. High-grade road base material using red mud from the sintering process is therefore promising, and may lead to large-scale consumption of red mud.

1.3 Soft Limestone "Marl"

Fill material for road surfaces must be free of organics, meet specific bearing ratio standards, and have a low plasticity index (Woodbridge 1999). In Belize and Mexico, soft limestone, erroneously called marl, was rejected as a road base in 1973 because of the low wet strength, poor grading, and relatively high plasticity; crushed stone was used as road base instead (Woodbridge 1999). The factors affecting rock strength and durability may include mineral compositions, microstructure (size, shape and geometry of grains), degrees of alteration (bonding, density and porosity) and texture (Sousa et al. 2005). Water plays a fundamental role in the stone deterioration, where it is well known that porous building materials absorb and desorb water as a function of weather conditions (temperature, relative humidity, and rainwater). Furthermore, fluxes of water within the stone affect the behavior of the material and can be responsible for its deterioration (Beck et al. 2003), through dissolution and decomposition of certain minerals.

1.4 Soft Limestone Road Bases in Belize, Northern Caribbean

Hard marl and limestone have been used in large quantities for roads and airstrips in Government areas during World War II (USDA 1942), and as mineral aggregate in concrete both for buildings and roads. Three marls tested in Belize (Woodbridge 1999) comprised high-purity carbonate materials containing mainly silt-sized particles and fell outside the grading, plasticity, and strength specifications normally required for road bases. These marls, each with slightly different characteristics, were substituted as road base for crushed stone.

1.5 Marl Road Bases in North-Eastern Jamaica

Therefore, as a road base with an adequately protective top dressing, untreated limestone marl performs satisfactorily in the long term (Woodbridge 1999) and is cheaper than crushed stone by a factor of more than 2:1 (Woodbridge 1999). Jamaica, though at similar latitudes to Belize, has two rainy seasons (McPherson 1990) compared to one in Belize (Woodbridge 1999). Further, in the wetter north-eastern areas of Jamaica, annual rainfall, at 3500 mm (McPherson 1990), doubles the average in Belize. To avoid rapid cracking from the pressure of heavy motorized traffic, asphalt top-dress should be of sufficient thickness. Yet, the top-dress asphalt under northeast Jamaica high-intensity rainfall locations frequently is at times <30 mm thick (Fig. 1). This is much thinner than that used in the less rainy Belize (Woodbridge 1999) thereby exposing the soft limestone bedding to rain within months of application (Fig. 1). Geotechnical failures occur when water erodes finer particles internally and washes them out of a less cohesive layer. Fine-grained unconsolidated material can also become saturated when inundated by floodwater. Here, deep gullying of roads rapidly follows sub-aerial exposure of the road sub-stratum. The addition of Portland cement, though substantially increasing the durability of the soft limestone under rain (Woodbridge 1999), is almost as costly as applying a road base of crushed stone (Woodbridge 1999). To avoid extensive, expensive road repairs at such locations in the absence of sufficiently thick asphalt top-dressing, a more durable road base than untreated marl, yet not more costly, is required.

1.6 Resiliency, and Puerto Rican Cretaceous Limestone

The California Bearing Ratio (CBR) test is a simple strength test that compares the bearing capacity of a material with that of a well-graded crushed stone (ASTM 1989). Soil plasticity is the soil water content at which the soil becomes 'plastic'; that is, capable of being deformed when external force is applied, and the introduction of water to an underlying marl stratum with low CBR values weakened this supporting material (USDA 1942). This characteristic normally causes marls (Fig. 1) to be unsatisfactory for road base (Woodbridge 1999).

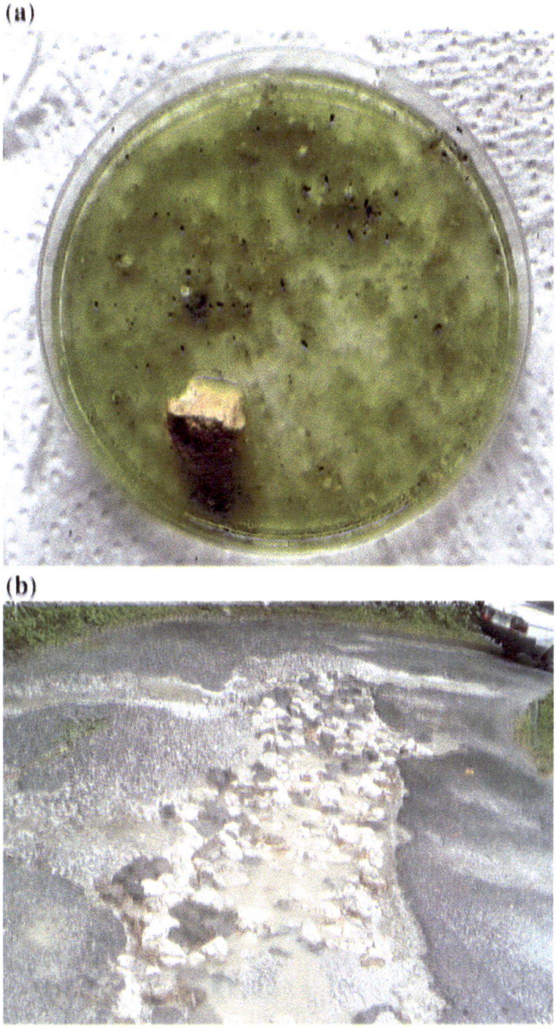

Fig. 1 **a** Decaying organic matter producing organic acids in stagnant conditions as that seen in picture can increase the dissolution rate for soft limestone "marl" (seen in watch glass) used as road beds. **b** Section of a poorly constructed road at Barry Hill Road, Portland, Jamaica, where rainfall averages >3000 mm p.a. The *whitish colour* of the water indicates the high solubility levels of the soft limestone base. Note the soft limestone visible at lower left corner under the thin (<30 mm), un-protective asphalt layer which on cracking admits water. It can be seen that the thickness of even the side moulding (30 mm wide) on the car at *top right* exceeds that of the asphalt top-dressing *Photo* Herve Anderson

Periods of prolonged wetting occur frequently in Caribbean locations adjacent to, and north of elevations exceeding 1500 m (McPherson 1990). In such areas, road fill material requires a high retention of dry CBR values after prolonged wetting. Materials other than marl are therefore needed where top-seals (such as thin asphalts) are inadequate. For example, the Cretaceous limestones of Puerto Rico have high value as road-building material as they are hard, yet break readily and have better binding qualities than most other rocks (USGS 1992). Therefore, if a particular bauxite waste has similar characteristics to the Cretaceous limestones of Puerto Rico it is a satisfactory alternative to marl as a road sub-stratum.

1.7 *Bauxite Waste: Health Concerns*

Nevertheless, the storage of red mud in lakes or ponds occupies huge areas of land, and the storage of dry red mud can lead to dust pollution which is a serious health problem for the people living near the red mud storage ponds. The cost of red mud disposal is expensive, accounting for about 2 % of the alumina price (World Aluminium 2015). For example, the alumina price is about US$439 per ton in China, so the disposal cost of red mud would be nearly US$9 per ton of alumina production (World Aluminium 2015).

The trace radioactive material remaining in bauxite residue gave rise to some concerns in Jamaica but an investigation showed that using 100 % bauxite residue gave a dose equivalent to just over 2 mSv yr^{-1} and was judged to be acceptable (Pinnock and Gordon 1992). However, other work on Hungarian bauxites has recommended a maximum addition of 15 % bauxite residue to avoid exceeding a level of 0.3 mSv yr^{-1} (World Aluminum 2015). Moreover, the material used in this study is comprised of gypsum-treated lithified muds, where the pH has been lowered to a value of 8.6 (Harris 2008).

1.8 *Hypothesis: Red Mud Waste*

Though marl treated with Portland cement is quite durable as a road base, the cost of such treatment is high, compared with that of lithified red mud waste. Further, red mud requires no fossil fuel for lithification (Harris 2008). However the durability of large masses (>2 mm) of such material after prolonged wetting and drying is undetermined. Based on its apparently high strength, and physical and chemical uniformity, it is thus hypothesized that G (gypsum-lithified bauxite waste) may exhibit greater durability after prolonged exposure to rainfall, and hence be a more resilient road bedding material than soft limestone marl (L). Moreover, as Woodbridge (1999) showed that with an effective surface seal, a soft limestone road bedding performed well for at least 19 years, testing of the materials seemed necessary to resolve the uncertainties regarding the potential road bases.

2 Methods

2.1 *Treatments*

A naturally de-watered red mud waste pond containing gypsum-treated red mud was selected for this study. The treatment had been done, and described in O'Callaghan et al. (1998). As the plough depth to which the gypsum had been laid was 15 cm, the effect of time (10 years) on the strength and durability of

(1) the 0–15 cm and (2) the subjacent 15–30 cm were to be studied. To preclude contamination, samples from both layers were removed with a sharp-pointed stainless steel implement. To pre-harden samples from the viscous, un-lithified subjacent layer, samples were stored until no further loss in water occurred (air-dry).

2.2 Impact/Shatter Test

One of two methods (named below) using irregularly shaped specimens (particles) were adopted in this study, as they present certain advantages over the alternative standard tests (compression and tensile tests) in which regularly shaped specimens (cylinders, cubes) are required. The alternative methods present added cost and much extra time for performing the tests.

In this study, the Aggregate Impact Value Test (AIV) is chosen because it is justified by Tsakalakis (2006). He defines impact testing as measuring the energy required to break a specimen by dynamically applying a load. Impact strength is one of the most commonly tested and reported properties of brittle materials in which the aggregates also belong. This is because the AIV method is simpler in its application than the Protodyakonov's (except that the latter requires no particle size preparation of the test sample). He ranked the AIV Test as more reliable due to the greater mass of the sample tested (at least 15 times that used in Protodyakonov's test).

For the AIV test, the following procedure was adopted as follows (after Tsakalakis 2006):

The material used consisted of irregularly shaped particles of the (a) 14, and (b) 10 mm size fraction. The weight of the specimen is *A*. The two (2) specimens tested were subjected to 15 blows of the hammer dropping 380 mm, at an interval not less than one second. The crushed material was removed from the mould and sieved over a 2.36 mm sieve. The mass of the material passing (*B*) and the retained fraction on the sieve (*C*) are weighed to the nearest 0.1 g. If $A-(B+C) > 1$ g, the result was discarded and the test is repeated with fresh material. The AIV value was calculated from the average of the two tests:

$$AIV = B/A \times 100(\%) \tag{1}$$

Figure 2 depicts the AIV apparatus. The mass of the material tested was approximately 500–800 g (depending on the SG) to fill the hardened steel mould. The number (N) of impacts was 15.

2.3 California Bearing Ratio Test

Strength tests were done (1) on dry samples and (2) after samples were soaked under water for 10 days (5 slaking cycles). This test was done, firstly, because soft

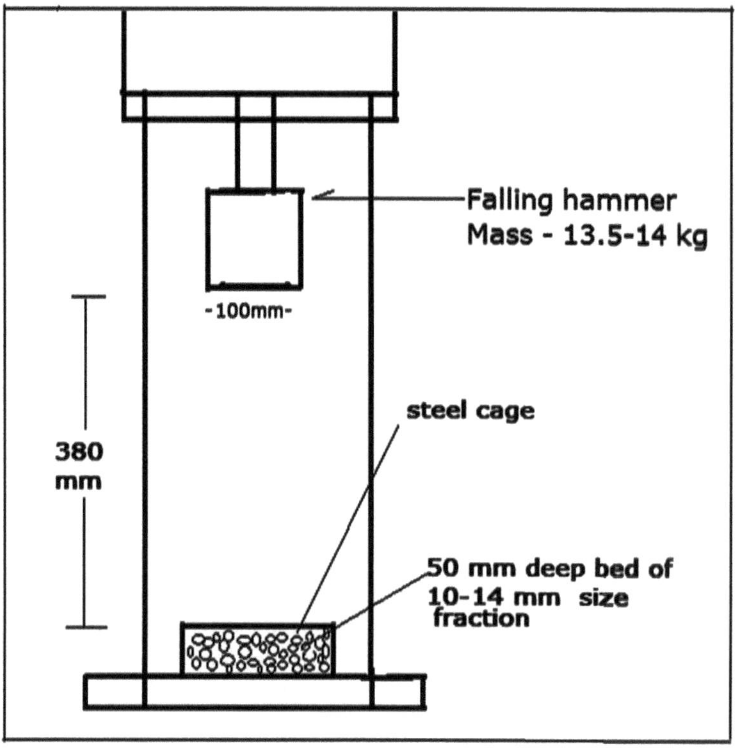

Fig. 2 Apparatus used for the Aggregate Impact Value (AIV) test (after Tsakalakis 2006)

limestone bedding rapidly dissolves and erodes under fluids with a sub-neutral pH such as rain (Fig. 1). Dissolution leads to cavitation, and cavitation in the road substratum causes collapse of the top-dressing (Fig. 3) and more erosion of soft limestone beddings. Hence, under an inadequate seal, the rate of dissolution of a road bedding determines the stability and longevity of the road.

The California Bearing Ratio test is a penetration test meant for the evaluation of subgrade strength of roads and pavements. As the most widely used method for the design of flexible pavement (Khattab and Othman 2012), the results obtained by these tests are used with the empirical curves to determine the thickness of pavement and its component layers. In this study, CBR was to be done with an Eijkelkamp pocket penetrometer, with a spring constant of 2.142 kg/cm (12 lb/in.). An 8 mm compression of the spring, as shown on the scale is equivalent to 1 kg cm^{-2} (ton ft^{-2}). Therefore, a compressive force of 1.71 kg (3.78 lb) on the foot is required to read 1 kg cm^{-2} (ton ft^{-2}). The equivalent of 1.71 kg (3.78 lb) on a 0.316 cm^2 (0.049 in.) (1/4" Diameter foot) is 5.41 kg cm^{-2} = 530 kPa = 53 N cm^{-2} or 5.53 tons ft^{-2} (Maart 2014). The penetrometer can be used to measure to a depth of 80 cm (to offset variation in readings of this device, ten results were to be taken).

Fig. 3 Cracking and collapse of asphalt top seal laid down on 15/12/15 at Seaman's valley, Portland, Jamaica. Note cavitating soft limestone bedding, dissolution of calcium carbonate in acidic rain water, and cracks formed around it in the asphalt surface seal. *Photo* Herve Anderson 30/12/15. *Source* usgs.gov

Under an ineffective surface seal, alternative road bedding requires greater durability than that of soft limestone. As the rate of comminution determines eventual particle size, and dissolution varies inversely with initial particle size, in addition to dissolution tests, the marl and alternative road bedding materials (the red mud wastes), were crushed and sieved to pass through a sieve of 2-cm diameter. This ensured a range of particle size conducive to compaction. The standard material for this test is crushed California limestone, which has a value of 100. Thus:

$$CBR = (p/ps)100, \tag{2}$$

where:

p = measured pressure for site soils (N/mm^2),
ps = pressure to achieve equal penetration.

2.4 Porosity

The importance of porosity determinations exists because pores are natural holes in the stones which allow fluids like rainwater to enter and leave the fabric. Associated with rock porosity is permeability. Very high porosities may allow excessive volumes of corrosive fluids such as acid rainwater to enter and cause severe damage to the rock. Being the ratio of pores (micro-voids) in the stone to its total solid volume, porosity can directly determine dissolution rates in rocks. Porosity was measured by a standard procedure of crushing the samples and determining the difference in water displaced beforehand by the uncrushed sample.

2.5 Density

Based on the findings of Harris (2010) which showed an inverse relationship between rock density of a shale and extent of internal corrosion, density determinations were made by comparing equal volumes of water with mass of the samples.

2.6 Slake Resistance

To test physical reactions to soaking, treatments consisted of exposing samples to 5 episodes of 5-day submergence under pure rain water followed by 5-day periods when the water was removed and the samples dried naturally to air-dry. Treatment#1 was carried out on bauxite waste collected from a ten-year-old gypsum-treated dried out red mud pond (O'Callaghan et al. 1998), and soft limestone marl. The test procedure and data reduction were similar to that of the standard practice (ASTM D4644 1989), except that the tests were performed up to 5 cycles, instead of two cycles as specified by the standard.

Two depths were sampled in the bauxite waste: surface (0–15 cm) and subsurface (15–30 cm). The gypsum application mentioned above occurred at 0–15 cm depth (G), but components could have leached subsequently downwards into the ensuing years to the 15–30 cm zone (R). Whereas the G is a hard rock, The R has the consistency of a stiff mud. After removal from the 15–30 cm depth in the red mud waste, the R samples were hardened by exposure to air for one year, whereupon they acquired the apparent hardness of the G samples. G, R, and L samples were separately crushed and passed through a sieve of diameter 2-cm.

3 Results

3.1 AIV Test

Results of the AIV Test are shown in Table 1. It can be seen that the limestones are more resilient than the carbonated lithified red muds, particularly that which was gypsum-treated. Under impacts, the toughness of the subjacent red mud was unexpected, considering the rapid slaking which it undergoes when exposed to water (deionized or non-deionized). In other words, slaking of some materials is not equated with soil (or rock,) strength. Harris and Rengasamy (2004) observed extremely rapid slaking of high-strength sodic subsoils at air-dry, compared to a reference soil. Though the toughness value of the subjacent red mud is almost as

Table 1 Aggregate impact value test results for limestones and lithified red muds

Rock* dimension (cm)	Test	Condition	L1 Hard limestone	L2 Soft limestone	G Gypsum-treated red mud	R Red mud subjacent to gypsum treatment
10–14	Aggregate impact Value (AIV)	Dry and saturated	99.6^{d} and 95^{d}	97^{d} and 45^{b}	60^{c} and 55^{c}	95^{d} and 0^{a}

*Values followed by different letters are significantly different according to Tukey's HSD ($P > 95\ \%$)

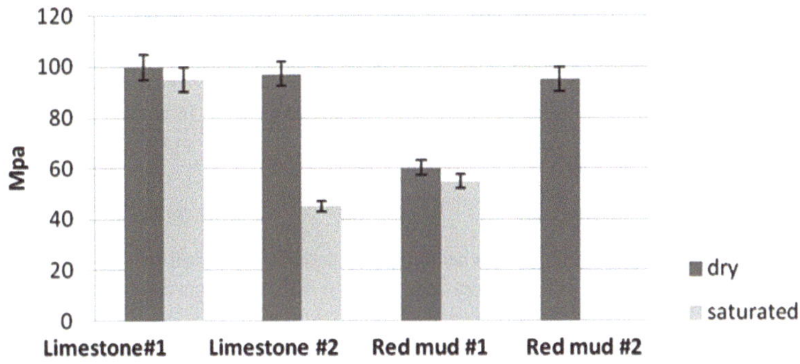

Fig. 4 Aggregate Impact Value test for limestones & lithified red muds

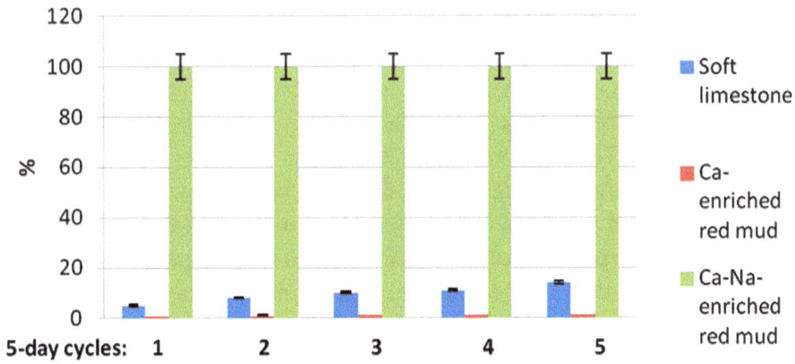

Fig. 5 Weight loss of rocks after slaking cycles in rain water

high as that of the limestones, it can be seen that strength is non-existent on wetting. It would be unsatisfactory therefore, in wet conditions as a road base. The gypsum-treated layer however is more resilient after wetting, having lost very little strength after prolonged soaking (Table 1 and Figs. 4 and 5).

3.2 Slake Resistance: Slaking Durability Test

Rate of dissolution was measured by the mass of sediments released in water from the soaked samples.

The main objectives of the slaking durability test was to determine the impact of water on weathering and degradation, and hence to assess long-term durability of the rocks under rain. The rate of weight loss after 5 cycles in the slake-durability test apparatus was 2.3, 14.8, and 100 % respectively for the G, L, and R (Table 2).

Table 2 Weight loss (%) after slaking cycles in rainwater (pH 5.6) of limestone and red mud bauxite wastes

Rock*	5 days (1 cycle)	10 days (2 cycles)	15 days (3 cycles)	20 days (4 cycles)	25 days (5 cycles)
L	5^{c}	8^{d}	10^{d}	11^{d}	14.8^{e}
G	0.8^{a}	1.1^{a}	1.3^{b}	1.3^{b}	1.3^{b}
R	100^{f}	100^{f}	100^{f}	100^{f}	100^{f}

**L* soft limestone, *G* pre-lithified bauxite waste, *R* post-lithified unflushed Na^{+}-rich bauxite waste. Numbers with different letters are significantly different according to Tukey's HSD at ($P<0.05$)

Therefore, the gypsum-treated carbonated lithified red mud waste proved far superior to all other rocks tested. The soft limestone marl, being a less pure rock than those from the bauxite waste, have often eroded or pitted surfaces, showing bands or patches rendered more resistant to the action of the weather by the presence of less soluble materials, which are probably sand or chert (Fig. 1).

According to Peng and Zhang (2007), soft limestone is highly porous compared with the known average range between (2–20 %) and (1–10 %) for limestone and dolomite rocks. The values of dry unit weight for every stone depend on the porosity, cracks and the existing metals. Conversely, as shown in Table 2, the values of dry unit weight decrease with increasing porosity. As found by Khattab and Othman (2012), absorption would increase with increasing porosity and connected porosity. In a study of highly weathered shales in north-eastern Jamaica, Harris (2010) similarly observed an inverse relationship between porosity and dry unit weight (shown in Table 3).

3.3 CBR Test

As shown by Pavement Interactive (2015) the load penetration CBR test is based on the equation:

$$\text{CBR}\ \% = 100(x/y)$$

Where x = material resistance or the unit load on the piston (pressure)
y = standard unit load (pressure) for well-graded crushed stone
= 2.54 mm (0.1 in. penetration = 6.9 MPa (1000 psi)

Table 3 Porosity and density of rocks as a potential road base

Rock*	Porosity (%)	Standard deviation (SD)	Density ($g\,cc^{-1}$)	SD	Absorption n (%)	SD
L	8	2.2	2.5	0.21	8	1.4
G	1.0	0.1	3.8	0.1	0.10	0
R	0.5	0.1	3.9	0.04	1.0	0

**L* soft limestone, *G* gypsum-treated bauxite waste, *R* unflushed subjacent Na^{+}-rich bauxite waste

Table 4 Mechanical properties of the rocks as a potential road base

Rock* dimension (cm)	Test	Condition	L1	L2	G	R
<2	California bearing ratio (CBR)	Dry and saturated	2 and 0.5	75 and 67	18 and 15	14 and 0
	SD for CBR	Dry and saturated	0.2 and 0.1	5.2 and 4.3	1.6 and 0.7	0.3 and 0

SD standard deviation
**L1* soft limestone, *L2* hard microcrystalline limestone, *G* gypsum-treated bauxite waste, *R* unflushed Na^+-rich bauxite waste

= for 5.08 mm (0.2 in. penetration) = 10.3 MPa = (1500 psi)

In this study, the CBR Test showed that the soft limestone (marl) had a normal strength of 4.2 % with water content at air-dry, and a "poor" value of 2.8 % after 24 h of submergence in de-ionized water. The less pure limestone rocks have often eroded or pitted surfaces (Fig. 3). Both limestones exhibited contrasting CBR values (Table 4).

4 Discussion

4.1 Slake Durability

The G outlasted all other samples by several-fold when subjected to 10 cycles of slaking under distilled water. This result apparently contradicts Harris (2009), who observed immediate slaking and breakdown of G aggregates during wet-sieving after he remoulded them following incubation at 60 % water content for 24 weeks. Agglomerates he produced in that study were composed of <2 mm particles of G. After incubation, these agglomerates were not resistant to physical pressure or electrolytic forces of high quality water and broke down immediately on contact with water. Moreover, this was not just a case of the <2 mm particles merely separating from the loose bonds which linked them loosely together during incubation: saturated hydraulic conductivity was measured at 0 cm/hr after incubation. This means that micro-pores immediately sealed up after saturation point was reached. Subjecting the <2 mm particles to a dispersive test in water, Harris (2008) noted that half of each of the R aggregates spontaneously disintegrated in water (level 3 dispersion) forming a colloidal cloud from the some of the individual <2 mm particles which had actually broken apart to release colloidal-size particles.

O'Callaghan et al. (1998) noted that coarse G agglomerates of 1500 mm diameter prevented the excessive release of Na^+ embedded in them. Using a tractor, they broke up the ground to a depth of 15 cm into 2–5 cm particles. Harris (2009) noted that such large agglomerates, as remnants of dried red mud which underwent gypsum-treatment, may not have been thoroughly penetrated by Ca^+ ions from the

O'Callaghan (1998) $CaSO_4$ treatment. Harris (2009) also observed that the G samples taken from the field site at Kirkvine Pond 6 were impermeable. Moreover, neither did O'Callaghan et al. (1998) nor Harris (2009) in his laboratory experiments, flush out with water at any stage the sodium sulphate resulting from the substitution reaction:

$$CaSO_4 + Na_2CO_3 \rightarrow CaCO_3 + Na_2SO_4 \quad (3)$$

Therefore, the G, though gypsum-treated in the year 1996 (O'Callaghan et al. 1998), had not undergone a systematic process to remove abundant Na^+ artefacts from the Bayer process. Yet, in the year 2007, the Na^+ level in the G (0–15 cm depth) was only 860 ppm compared to that of the RW (15–30 cm depth) at 22,000 ppm—a > 20-fold disparity (Harris 2009). And where sodicity and ESP is high, exposure to water causes clay dislocation, swelling, and a drastic decrease in infiltration. However, crushing of the G to <2 mm diameter by Harris (2009) greatly increased the surface area of exposure, causing more contact of hitherto hidden Na + with water. This increased the proportion of hydrated (Na^+) cations in the <2 mm entities. Thus by changing the location of the Na + ions (i.e., from internal to external surfaces), the same relatively low level of Na^+ in the G caused a reversion to a more slake-prone condition. This argument gains credence when it is seen that in this study, G entities of >20 mm in diameter did not slake, or exhibited a very low rate of slaking even after 10 cycles of slaking under distilled water. To test the veracity of this sub-hypothesis in this study, G agglomerates of >10 mm <20 mm were submerged in water and incubated for 24 weeks. This was a more extreme treatment than that inflicted by Harris (2009), who incubated <2 mm crushed G samples at just 60 % of field capacity for 24 weeks). Yet, at the end of 24 weeks the >10 mm <20 mm-diameter G samples remained largely intact with no observable signs of slaking. Stress tests revealed no change in the resilience of the >10 mm <20 mm G agglomerates after 24 weeks of submergence. These results can explain the high level of slake durability for 20 mm diameter G samples in this study.

Atmospheric carbonation of red muds is a slow process and it takes about 30 days to initiate the reaction and formation of stable carbonate phases (Muduli et al. 2012), being a cyclic path of sulfatization and carbonation which form polymorphs of carbonates (Muduli et al. 2012). In some cases up to 36 % of carbonate-C appears to be derived from atmospheric in-gassing of CO_2 (Muduli et al. 2012). In contrast to the discoloured water pools around soft limestone marls (Fig. 1), water collecting in pools in the G samples appeared relatively clear in colour even after 18 months (Fig. 5). This suggests a high resistance to the solubilizing of pre-carbonated red muds. Harris (2009) measured high strength of air-dry atmospherically carbonated gypsum-treated bauxite red mud tailings.

On the other hand, the low slake durability of the R at >10 mm <20 mm-diameter which contained >20-fold Na^+ than the G, seems logical due to (1) a >20-fold greater number of Na^+ exposed to water and (2) a much higher level of permeability. Nevertheless, despite gypsum application in the G in the 0–15 cm zone, it can be seen (Table 5) that the R (15–30 cm) subjacent to the G, exhibited an anomalously

Table 5 Strength of red mud 10 years after surface-applied treatment

Depth (cm)	No. of drop impacts before shattering
0–7	25
7–14	25
14–21	06

high level of Ca^+. The Bayer process cannot explain such a high level of Ca^+ at any location in red mud waste. Yet, despite surface application of gypsum, there is an almost 5:1 ratio of such Ca^+ in the 15–30 cm zone compared with the 0–15 cm zone (Harris 2008). The only explanation seems to be dissolution and vertical leaching of the gypsum during rainfall episodes over several years after the surface application. The similarly low Na^+ levels in the 0–15 cm zone when compared to the 15–30 cm zone supports a natural-leaching and illuviation explanation, especially as the original red mud waste had emerged from the spigot as a thoroughly homogenized slurry. Had the build-up of Na^+ in the 15–30 cm zone been avoided, and with almost 5-fold the number of Ca^+ ions, it is reasonable to suggest that the slake durability of the R would have been higher, and even have exceeded that of the G. The main determinants of slake durability for the Bayer-processed red mud waste therefore, are (1) gypsum treatment (2) particles of sufficient size, which should substantially exceed 2 mm, and (3) flushing out of Na_2SO_4 in a horizontal direction.

4.2 Gypsum Requirement for Bauxite-Waste Road Sub-Stratum

As stated above, the 5-fold level of gypsum in the post-lithified 15–30 cm zone compared to the pre-lithified 0–30 cm zone occurred after the gypsum treatment of the 0–15 cm zone. Yet, this relatively vast concentration of Ca^{2+} ions in the 15–30 cm zone could not have leached downwards after carbonation in the 0–15 cm zone. This is because the pH of the 0–15 cm zone after the gypsum treatment remained above pH8 (but below pH9) for all ensuing years (Harris 2008), and in such alkaline conditions $CaCO_3$ is insoluble. It is therefore reasonable to suggest that the downward leaching of the Ca^{2+} occurred after gypsum treatment of the overlying (G) layer but before pre-lithification of the G layer. As evidenced by the highly viscous consistency of the material in the 15–30 cm zone, this preponderance of Ca^{2+} ions did not carbonate. More than 80 % of the Ca^{2+} ions in the gypsum added to the 0–15 cm zone were therefore wastefully relocated dislocated to the unresponsive, continuously viscous 15–30 cm zone. These relocated Ca^{2+} ions were therefore totally unnecessary for the atmospheric carbonation which caused the induration of the 0–15 cm zone. Therefore the maximum quantity of gypsum required for carbonation would have been a fraction of the amount that had been applied. Moreover, the lithified 0–15 cm layer prevented movement of CO_2 from the atmosphere into the subjacent layer from the atmosphere. Furthermore, had the 0–15 cm layer been removed immediately after its atmospheric carbonation (induration), the 15–30 cm zone would have then been exposed for several years to the weather including many years of rain (as was the 0–15 cm zone). Since a substantial fraction of the applied gypsum reacted and disintegrated while it was intact, yet accomplished sharply defined lithification where it was applied, it stands to reason that an excess had been applied. Therefore, if the rate of gypsum application is appropriately adjusted, a lower rate might have produced similar levels of pH reduction and acidity. For example, a lower rate staggered over a longer time period could possibly have similar results while reducing the leaching of excess Ca^{2+}. This would have concurrently leached Na^{+} ions while carbonating other cations. Figure 5 therefore, shows that, providing Ca^{2+} leaching can be restricted, the process of indurating bauxite red mud waste does not require as high a rate of gypsum as was applied by O'Callaghan et al. (1998).

4.3 Effects of Slake/Durability Cycles on Rock Strength

When the rocks are subjected to the effect of water, through its transmission by capillary force, gravity force & absorption, this leads to dissolution of crystalline salts inside rock. The EC readings of the water containing slaked samples prove that salts were washed out and brought into solution. This increases the porosity and hence decreases the density and strength for these rocks (Ali 2011). Having a much higher concentration of soluble salts compared to the G samples, this could partly explain the rapid breakdown of the R samples in water. In a study by Tri (2008) it was found that point load strength index increases at the earlier cycles, followed by a reduction in strength. This could be attributed to the deposition of an amount of the salts inside the stone pores by crystallization after water evaporation. He found that such changes were followed by a reduction in the strength with the increase of cycles, due to increased dissolution/recrystallization of salts in the confined pore spaces. This process leads to a loss both in mass and strength (Trinh et al. 1997).

4.4 Dissolution of Rocks

The CBR Test showed that the soft limestone (marl) had a normal strength of 4.2 % with water content at air-dry, and a "poor" value of 2.8 % after 24 h of submergence in de-ionized water. The less pure limestone rocks have often eroded or pitted surfaces (Fig. 6), showing bands or patches rendered more resistant to the action of the weather by the presence of insoluble materials such as sand, clay or chert. When free from impurities, limestones are white. If very pure, they may have smooth rounded surfaces, These surfaces are often recognized from the crust of hydrous oxides of iron produced by the action of the atmosphere on any ferriferous ingredients of the rock; a thin layer of gritty sand grains may be left on the surface of limestones which are slightly arenaceous.

Fig. 6 **a** Slaking of rocks. The mass of material slaked and removed from the soft limestone (whitish, located in laboratory watch glass) by the water was 5 times that removed from lithified red mud (G). **b** Lithified red mud (G)

This reduction in strength was due to lowering of the free surface energy upon saturation and upon contact with water. Thus the rocks strength decreased in the "saturated" state for all the rock types except for the G, and G was found to have the lowest porosity.

5 Geotechnical Applications

5.1 Accelerating the Lithification of Red Mud Waste

As found in this study, to indurate bauxite red mud waste (Fig. 7), a substantially lower rate of gypsum is required than was applied by O'Callaghan et al. (1998). This would have concurrently leached Na^{+} ions while carbonating other cations. Even so, the rate of carbonation, and hence the required completion time for carbonation is yet to be ascertained. Moreover, Muduli et al. (2012) noted that atmospheric carbonation of red mud waste is a slow process. Thus, as found above, leaching of applied Ca^{2+} is faster than its carbonation. One way of ascertaining

Fig. 7 Relatively minute mass of slaked particles (compared to that of soft limestone—see Fig. 4) around pre-lithified gypsum-treated bauxite red mud tailings (G; 0–15 cm depth). Moreover, this photograph was taken after 78 weeks of continuous submergence, whereas the soft limestone had been submerged for less than 2 weeks

the rate of carbonation, could entail the use of periodic test drill cores. As the process of atmospheric carbonation would be expected to slow down with depth, as soon as induration solidified a thin layer (possibly of 2–3 cm thickness), that layer could be mechanically removed. The newly exposed surface layer would then be indurated at the "fast rate" compared with the "slow rate" at depth. Such a strategy could accelerate sub-surface induration of bauxite waste. As the transportation of large quantities of gypsum is a costly procedure, where road-building with bauxite waste is concerned, these are potentially vast reductions in transportation costs, as found in this study (Fig. 7).

5.2 Purifying of Variegated Limestone

As there are vast differences in the slake durability and strength of limestone from the same quarry, it is recommended that a separating apparatus is used to obtain the purity of the more durable limestone of this study. From such a device, the difference in density found in this study between the two limestones above could be economically exploited.

References

Ali HE (2011) Study the mechanism of deterioration on the rocks used in the historical building. College of Engineering, University of Mosul

Amadei B (1989) Strength properties of rocks and rock masses. CVEN 5768 Lecture Notes 8, 1 ceae.colorado.edu/~amadei/CVEN5768/PDF/NOTES8.pdf

ASTM (American Society for Testing and Materials) (1988) Standard definition of terms relating to natural building stones

ASTM (1989) American society for testing and materials. 4(8)

Beck K, AL-Mukhtar M, Rozenbaum O, Rautureau M (2003) Characterization water transfer properties and deterioration in tuffeau: building materials in the loire valley france. Downloaded from: http://www.google.com

Geologydata Info (2004) Info portal of geology with special reference to Rajasthan, India. Available: www.geologydata.info/bauxite_deposits.htm. Accessed 6 Jan 2016

Harris MA (2008) Structural improvement of age-hardened gypsum-treated bauxite red mud waste using readily decomposable phyto-organics. Environ Geol 56:1517–1522. doi:10.1007/s00254-008-1249-5

Harris MA (2010) Fragility of a dark grey shale in northeastern Jamaica: effects and implications of landslip exposure. Environ Earth Sci 61(2):369–377

Harris MA, Rengasamy P (2004) Sodium affected subsoils, gypsum, and green manure: Interactions and implications for amelioration of toxic red mud wastes. Environ Geol 45(8):1118–1130

http://www.gsa.gov/ Accessed 24 Nov 2015

Khattab, Othman (2012) Durability and strength of limestone used in building. Al Rafidain Eng 21(3):2013

Maart (2014) https://www.eijkelkamp.com/download.php?file=M10603e_Pocket Accessed 14 Feb 2016

McPherson J (1990) Caribbean lands. Longmans, London, New York, Sydney

Muduli SD, Nayak BD, Dhal NK, Mishra BK (2012) Atmospheric CO2 sequestration through mineral carbonation of industrial solid wastes and their utilization in plant growth. Institute of Minerals and Materials Technology, Bhubaneswar, India, 4(1):001–006, Jan 2014, ISSN:2276-7851

O'Callaghan WB, McDonald SC, Richards DM, Reid RE (1998) Development of a topsoil-free vegetative cover on a former red mud disposal site. Alcan Jamaica rehabilitaion project paper

Peng S, Zhang J (2007) Engineering geology for underground rocks. Springer, New York

Pinnock W, JN Gordon (1992) Assessment of Strength Development in Bayer Process Residues. Cement Concr. Compos. 18(6):71–379

Qi JZ (2005) Experimental research on road materials of red mud. University of Huazhong Science and Technology, Wuhan, China, p 2005

Sousa LOM, Suarez del Rio LM, Calleja L (2005) Influence of microfractures and porosity on the physico-mechanical properties and weathering of ornamental granites. Eng Geol 77:153–168

Sutar H, Subash Chandra M, Santosh Kumar S, Ananta Prasad C, Himanshu Sekhar M (2014) Progress of red mud utilization: an overview. Am Chem Sci J 4(3):255–279

Tri TV (2008) Vieillissement du Tuffeau en présence de sels: approche experimental et numérique. THÈSE, GÉNIE CIVIL, Discipline UNIVERSITÉ D'ORLÉANS PhD, Porte d'Orléans

Trinh C, Bucea L, Ferguson O (1997) sulfate resistance of cementitious materials mechanisms, deterioration processes, testing and influence of binder. Proc Concrete 97. Adelaide, Concrete Institute of Australia, p 263–268

Tsakalakis K (2006) Impact strength of brittle materials (rocks). Comparison of the Protodyakonov and AIV (aggregate impact value) methods. In: Proceedings at the meeting of the refractory experts more, T.U. Freiberg-Saxony, Germany, 4–5 May 2006

USDA (1942) Soil Survey of Puerto Rico. 1936(8):15. NRCS (Natural Resources Conservation Services). University of Florida Digital Collections

USGS (1992) Mineral Resource Assessment of Puerto Rico. pubs.usgs.gov/of/1992/0567/report.pdf. Accessed 25 July 2015

Woodbridge, ME (1999) Use of soft limestone for road-base construction in Belize. In: Seventh International Conference on Low-volume roads, Baton Rouge, Louisiana, USA. 23–26

May 1999. Department for International Development. Transport Research Laboratory, Crowthorne Berkshire RG45 6AU, United Kingdom. PA3450/99

World Aluminium (2015) Bauxite Residue Management. Best Pract. http://www.world-721aluminium.org

Yang JK, Chen F, Xiao B (2006) Engineering application of basic level materials of red mud high level pavement (In Chinese). China Munic Eng 5:7–9

Yasar E, Erdogan Y (2004) Estimation of rock physiomechanical properties using hardness methods. Eng Geol 71:281–288

Youngtae J, Ryu S (2015) Pothole Detection System Using a Black-box Camera. Sens 15(11):29316–29331. doi:10.3390/s151129316

Uniaxial Compressive Strength of Some Un-calcined Red Mud Mortars: Geotechnical Implications

Mark Anglin Harris

1 Introduction

From an ecological standpoint, the most challenging world problem in aluminium production is the creation of vast quantities of insoluble bauxite residue (RM). It is a suspension of very fine solid particles, with more than 90 % (v/v) being less than 75 μm (Schwarz and Lalik 2012). One ton of produced aluminium creates 2 tons of waste and it is estimated that 120 million tons of RM is produced worldwide (Red Mud Project 2011). Due to its noxious properties such as extremely high pH value, strong alkalinity, increased content of radioactive substances etc., it is a significant environmental burden now and in the foreseeable future (Paramguru et al. 2005; Snars and Gilkes 2009; Liu et al. 2011; Maddocks et al. 2004).

1.1 Utility of Red Muds

Due to suitable Fe oxides-Al oxides ratio, red mud is used in Portland cement production (Tsakiridis et al. 2004) or as an additive in special cements production (Singh et al. 1996). However the proportion used in cement is limited chiefly by its strong alkalinity (Schwarz and Lalik 2012). Therefore, proportions do not usually exceed 15 % (w/w). If therefore, red mud alkalinity could be reduced without adversely affecting its cementitious character, the potential for greater re-purposing of such waste could be realized.

1.2 Pozzolans

The group of pozzolanic materials, which, after mixing with $Ca(OH)_2$ (lime) in the presence of water, harden and form stable and durable compounds, include red muds (Yalçin and Sevinç 2000). Thus in producing ceramic glass, several

M.A. Harris, *Geobiotechnological Solutions to Anthropogenic Disturbances*,
Environmental Earth Sciences, DOI 10.1007/978-3-319-30465-6_6

workers discussed the addition of red mud to clay compounds used (Sglavo et al. 2000; Yalçin and Sevinç 2000; Pratt and Christoverson 1982). According to Muller (2005), clay minerals and cement hydrates are the major constituents of clays and hardened cements. Such building materials are usually composites of coarse well graded grains, held together by the fine-grained binding agents. The coarser grains are e.g. the sand in loam, or the aggregate in concrete. Portland cement always contains lime; it is thought to be the major reaction partner of pozzolans in cements. However, pozzolanic materials consume primarily the calcium supplied by the lime of the hydrated Portland cement and, only to an unknown small extent, the calcium from other cement hydrates. Moreover, for any substitution of cement by a pozzolan, the greater the substituted fraction of cement is, the lesser the lime content formed during cement hydration will be (Muller 2005).

A renaissance of supplementing cementing materials in Portland cements may replace part of the clinker in order to enhance the performance of the hydrated cement. According to Buraev and Kushnir (1986), replacing one-third of the content of silicate fractions in concrete compounds by red mud, hardens concrete such that it gains greater compression strength than concrete obtained by the exclusive use of quartz sand. Using RM amounts ranging from 1 to 15 %, Kohno et al. (1998) found that hardening of concrete compounds under high pressure helped to improve some final product properties including compression strength, water resistance, and fire resistance.

Some of the described uses of such concrete compounds include a filling for construction materials, materials used in dump construction, production of concrete blocks, filling dam structures or filling of building constructions (Klein 1998; Di San Filippo 1980; Browner 1995).

1.3 Stabilization of Limestone-Marl with Portland Cement

Under tropical rain, an alternative to a potential red mud-composite road base is limestone marl infused with artificial cements. Woodbridge (1999) stabilized a limestone-marl with ordinary Portland cement. After constructing the road pavement on an embankment to ensure good drainage, a good quality surface dressing seal was applied and maintained. Followed by 19 years of traffic measured at 1.3 million equivalent standard axles, the marl road bases performed at least as well as crushed stone. He concluded that the cement-stabilized marl road base performed extremely well, noting that the performance of the cement-stabilized marl road base in Belize has been excellent. A number of cracks developed in the early years but they had not increased. Low values of rut depth and deflection testified to the high strength of the road base. The samples he tested in 1992 gave UCS values averaging 9 MPa for the fresh stabilized marl beneath the surface dressing and 5 MPa for the partially carbonated marl from the road shoulders (Woodbridge 1999).

1.4 Previous Red Mud Tests

At a ratio of 60/40 mix proportion of bauxite residue and Portland cement, Kehagi (2008) conducted penetration tests to determine compaction characteristics and bearing capacity of the materials. California Bearing Ratio (CBR) test on dry specimens revealed a significant bearing capacity and increase in strength. Soaked specimens, on the other hand, revealed the high water-susceptibility of bauxite residue because soaked specimens rapidly disintegrated and exhibited no mechanical strength. The liquid limit and plastic limit of soils (along with the shrinkage limit) are often collectively referred to as the Atterberg limits. The Atterberg Limits (Liquid Limit LL and Plastic Limit PL), are a basic indication of suitability for road construction. Harris (2009) observed a very low value of plasticity index (PI), and, as found by Kehagi (2008) noted that the untreated red mud readily slakes, breaks down and disintegrates.

Nevertheless, in Jamaica, bricks made from Bayer-processed red muds by a silicate bonded system and red mud pozzolanic cement gave good characteristics to date (Pinnock and Gordon 1992) and the building was still in use by workers without health complaints (Pinnock and Gordon 1992). However, though the proportion of red mud in the mix was up to 30 % (w/w), the high lime fraction used meant that the energy requirement for calcining was high. The cost was not competitive, and the radioactive levels they report were too high. Though these are serious health concerns, such a material applied as a road base is far less harmful than it is when used indoors. Moreover, dilution, for example with anhydrite or gypsum can significantly reduce the concentration of radioactive elements in the red muds.

1.5 Justification, Aim and Hypothesis

Despite huge amounts of mentioned applications for RM re-use, the low profit rate for the goods manufactured from red muds results in this material remaining, in many cases, in the waste dumps (Schwarz and Lalik 2012). The drawbacks to the environment include high alkalinity and the high energy costs involved in calcining. Therefore, a non-calcining method of increasing the tensile strength and durability of red muds for industrial construction applications would be less costly and more sustainable.

Gypsum-treated red muds which had been hardened by natural carbonation exhibited substantially lowered pH values (Harris 2009). Moreover, increased levels of lime result when gypsum (anhydrite) combines with components in Bayer-treated red muds according to the following reaction:

$$CaSO_4 + 2NaOH \rightarrow Na_2SO_4 + CaOH_2 \quad (1)$$

Therefore, it is hypothesized that gypsum-treated red muds may be more pozzolanic than the untreated counterpart. Increased pozzolanic properties cause greater cementitious activity without increasing the level of calcining. The aim of this study is to increase the proportion of red muds above 15 % in a non-calcined mixture with Portland cement, without appreciably decreasing the tensile strength of the product.

1.6 Correlating Load-Bearing Tests

Although laboratory testing is the most reliable and direct method for estimating uniaxial compressive strength (UCS), direct determination of UCS in laboratory is time-consuming and expensive (Nazir et al. 2013). In addition, in the direct method of UCS determination, having sufficient number of high quality rock samples is a prerequisite. For the testing of red muds, restricted access to sites reduces the amount of material required for UCS tests, thereby limiting the number of samples.

Various correlations for UCS prediction have been compiled which relate other rock index parameters such as point load index, rebound number of Schmidt hammer, and indirect tensile strength of the rock to UCS (Cargill and Shakoor 1990; Sharma and Singh 2008; Nazir et al. 2013). According to a study by Sheorey (1997), the compressive strength of the rock is approximately 10 times its tensile strength. Further, as direct tensile test on a rock sample is relatively difficult to undertake, the Brazilian test (Brazilian Tensile Strength or BTS) offers an indirect method to measure tensile strength. They remind that the term 'indirect tensile' implies that load is applied under compression. Moreover, Nazir et al. (2013) noted that a tensile strength test is to verify tensile strength of rock samples or its resistance against fracturing.

Testing the effectiveness of the BTS in approximating 10 % of the corresponding UCS of a rock, Farah (2011) conducted a comprehensive study to obtain some correlations between UCS and other physical properties of weathered Ocala limestone. According to his study, indirect tensile strength (BTS) has a better correlation with UCS than that of point load strength. In order to discover correlations between UCS and other physical properties of the rock samples, Altindag and Guney (2010) researched a wide range of strength values encompassing UCS (5.7–464 MPa) and BTS (0.5–30.5 MPa). They found a strong correlation between UCS and BTS of the different rock types. Investigating the relationship between UCS and BTS of two different types of limestone in Pakistan, Din and Rafigh (1997) found that the UCS to BTS ratio is almost 7.5. Hence BTS is a reliable surrogate measurement for UCS.

2 Materials and Methods

The 0–15 cm layer (G) and the 15–30 cm subjacent zone (SG) of an oxide-rich gypsum-treated carbonated red mud from Kirkvine Pond 6, a red mud waste dump were sampled, crushed and treated after the method of Harris (2009). The gypsum had been added to a depth of 15 cm (O'Callaghan et al. 1998). To increase the reaction rates, <0.5 mm diameter particles were used rather than the <2 mm size used by Harris (2008). Further justification for crushing to <500 μm is that finer red mud particles tend to diminish the relative amount of capillary pores up to certain limits (Ribeiro et al. 2010) and, by increasing surface area, improves pozzolanic cementation. Table 1 depicts the properties of the 0–15 and 15–30 cm depth.

Table 1 Some properties of bauxite red mud waste in gypsum-treated layer and subjacent layer at Kirkvine Pond 6, Jamaica

Properties	Gypsum (0–15 cm)	Subjacent layer (15–30 cm)[a]
Red mud pH (saturated extract)	8.3	9.8
EC (saturated extract: mS cm^{-1})	2000	2500
Ca (ppm)	3700	15000
CEC (meq/100 g)	5.3	0.5
Na (ppm)	860	22,000
K (ppm)	288	–
Mg (ppm)	1	0

[a]Before gypsum treatment: After O'Callaghan et al. (1998)
– No data
Source O'Callaghan et al. (1998)

Table 2 depicts properties of the raw red mud waste (R). Like other red muds, the principal constituent of bauxite is iron oxide, and five major constituents, comprising Al_2O_3, SiO_2, TiO_2, Na_2O and CaO, plus small quantities of numerous trace elements.

At air-dry, crushed samples of each of the three lithified red muds were thoroughly mixed individually with Portland cement (P) and/or limestone aggregate (L) at 1 mm- or <1 mm diameter. Mixing continued until an even coloration was achieved. Mortars containing distinct additions of red mud (w/w) were prepared and tested. The mixing proportions were:

$$\text{G or SG or R} + (25- \mathit{or}\ 33\,\%\text{P}) + (\text{L} = \text{P})\text{or}(\text{L} = 0).$$

Two limestones—a microcrystalline and a limestone "marl" were similarly treated. Red mud controls included no L and/or no P inputs.

Table 2 Properties of untreated red mud waste at Kirkvine Pond 6, Jamaica

Properties	Values
Red mud pH (saturated extract)	12
EC (saturated extract: dS. m^{-1})	0.25
Organic carbon (%)	0.3
CEC (c molc kg^{-1})	40
Al_2O_3 (%)	16
CaO (% w/w)	7
Fe_2O_3 (%)	47
Na_2O (%)	3
P_2O_5 (%)	2
SiO_2 (%)	3
TiO_2 (%)	6
Bound H_2O (%)	14
Particle size	0.001-mm

Source O'Callaghan et al. (1998)

To attain the consistency required for moulding and shaping the material into discs, samples were mixed with water just up to the plastic limit of each combination of materials. This is because a lower quantity of free water in the mixture means lower porosity and higher mechanical strength of hardened paste and also better resistance to the penetration of aggressive agents such as chlorides and carbonation (Ribeiro et al. 2010).

Plastic limit was determined by rolling out a quantity of the fine portion of the bauxite waste on a flat, non-porous surface at a moisture content where its behavior was plastic, i.e., such that the thread retained its shape down to a very narrow diameter, after ASTM Standard D 4318 (2014). The plastic limit was determined at the moisture content where the thread broke apart at a diameter of 3.2 mm.

2.1 Brazilian Tensile Strength Test (BTS)

Sample is placed in a steel cradle fabricated with semicircular (Fig. 5), and then mounted in between the loading platens.

In the Brazilian test, a disc-shaped specimen of the rock is loaded by two opposing normal forces in the form of steel plates stationed at the disc periphery (Fig. 1). The sample rests in a semicircular opening having a larger diameter than the sample, such that a perpendicular force acts vertically on the top of the specimen, and similarly from the opposing point on the bottom of the sample. Using the theory of elasticity for isotropic continuous media, the equation below gives the

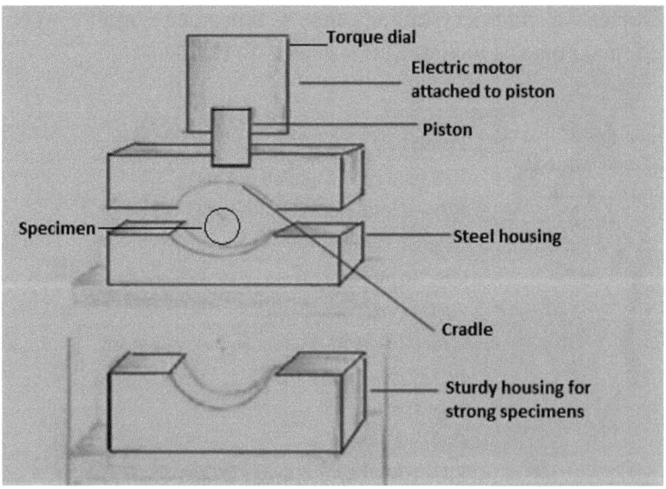

Fig. 1 Sketch of Brazilian Tensile Strength apparatus for indirect tensile testing. The piston forces the upper steel cradle downwards against the rock specimen with a constant force (F). The specimen disc is of a smaller diameter (less than 54 mm) than that of the semicircular housing. (After Gokhale 1960; ASTM D4318 2014)

tensile stress perpendicular to the loaded diameter at the center of the disc at the time of failure. All samples were isotropic (due to having been well-sorted by sieving, and small size of the grain particles). Therefore there were no planes of weakness from preferred orientation of minerals or stress history, as occurs in many rock types. Thus only one direction of compressive load was found necessary. The steel cradle ensured that the load acted tangentially on the disc-shaped sample and consequently induced tensile fracture along its vertical diameter. The top plate was movable at a constant rate in the vertical direction using a constant turning rate. The specifications of thickness/diameter ratio etc. (ASTM Standard Test Method D3967-81) were adhered to, and where the Brazilian test is performed on cylindrical "rock cores" having length-to-diameter ratio of 0.5 (Gokhale 1960). Special precautions recommended by Amadei (2015) were be taken at the contact between the rock and the loading platens to avoid crushing that could throw doubts as to the mechanism of failure in the test sample. Thus, strips of paper were used to replace the point load by a strip load over an angle not exceeding 10°.

The load was continuously applied at a constant rate until failure of the sample occurred. At disc failure, the tensile strength of the rock was calculated as follows: Having the sample dimension and maximum tensile load at failure (F), the Brazilian Tensile Strength (BTS) is obtained by using the following equation given by Gokhale (1960).

$$T_{BS} = 2F/(\pi DL) \tag{2}$$

where:

F Tensile load at failure
D Diameter of the specimen
L Thickness of the specimen

The uniaxial compression strength, which is the objective of the Brazilian test, is determined as follows:

$$C_{US} = F/A \tag{3}$$

where:

F Load at failure
A Cross-sectional area of the rock core

For standard results the length of the specimen is twice that of its diameter and loading shall be applied at a rate of 0.1–1 MPa/sec (Gokhale 1960).

2.2 Slake Durability Tests

Weight loss vs number of wet/dry cycles

The main objectives of slaking durability tests are to predict the long-term durability of the rock specimens, to establish weathering and degradation characteristics of each rock type, and to assess the impact of water on the rock degradation (Khattab and Othman 2013). The test procedure and data reduction were similar to that of the standard practice (ASTM D4644 1989), where two cycles were performed as specified by the standard. Justification for adopting the slake durability test is that the rate of weight loss increases with increasing the number of wetting/drying cycles (Khattab and Othman 2013). Such cycles induce dissolution and recrystallization, which lead to the deterioration (Khattab and Othman 2013).

2.3 CND Tests

Samples were dropped onto a concrete floor on the flat side from a pre-determined height. The number of drop impacts required to remove 50 % of mass chipped from the sample was recorded as an indication of the relative durability of the specimen.

3 Results

3.1 Brazilian Tensile Test: Red Muds

Specimen tension failure in the Brazilian test was determined by a fracture which originated near the center of the specimen and extended from the center to the loading contact points. Consequently, each specimen was divided into two approximately equal halves (Fig. 2).

The extent of resistance to fracture determined the category into which each sample was placed. Thus there were four categories: <6 MPa = weak, 6–7.1 MPa = medium, 7.2–10.5 = medium-strong, >10.5 = strong.

It can be seen that the most successful combined ingredients in these treatments are gypsum, components of red mud, and added limestone aggregate (Fig. 3). The tensile strength of all the red muds increased dramatically from 1.8–2.7 MPa up to 11.4 MPa (Fig. 3). For the most effective treatment, this is a >3-fold increase in strength.

It is those samples taken from the 0 to 15-cm horizon (previously gypsum-treated that exhibited the greatest increases in tensile strength (Fig. 3). The extra strength may have been caused by reactions common to a group of so-called pozzolanic materials (silicates, also found in red mud waste) which, after mixing with lime in the presence of water, harden and form stable and durable compounds (Sglavo et al. 2000). In this connection, G samples would have brought extra Ca^{2+} ions into the samples. Water reacts with dissolved cations such that the OH^{-} adheres to the cation, thereby leaving the hydrogen ions in the water and dropping the pH. The resulting increase in $Ca(OH)_2$ would have increased opportunities for

Fig. 2 Failure of sample in a direction of the applied load using the Brazilian Indirect Tensile Strength Test. All samples failed in this direction

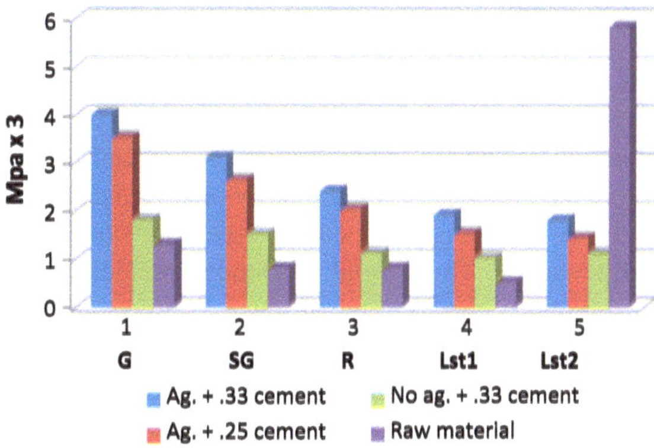

Fig. 3 Key Ag = 1–2 mm microcrystalline laimestone aggregate; *G* prior gypsum-treated red mud at 0–15 cm depth; *SG* 15–30 cm subjacent to G; *R* untreated carbonated air-dry red mud; *Lst1* soft "marl" limestone; *Lst2* microcrystalline limestone

combining lime with silicates in the red muds and the Portland cement. Several workers have discussed the additions of red mud into clay compounds when producing ceramic glass: (Sglavo et al. 2000; Yalçin and Sevinç 2000; Pratt and Christoverson 1982). When they added up to 20 % red mud, content of red mud did not affect the sample porosity. However, to acquire the increased density and flexural strength which they observed in final ceramic products, calcination at temperatures of 950 and 1050 °C was applied. Contrastingly, no heat was applied in the present study. Though red mud use as a component of building materials and its stabilization by Portland cement in ratio of 3:1 was proposed (Schwarz and Lalik 2012), they accomplished this by calcining the components.

According to Stillborg (1986), the BTS of limestone ranges from 5 to 20 MPa. Gordon et al. (1996) found similar values of strength (20 MPa) for red mud containing Portland cement composites. Majumdar et al. (1990) noted that hydrated lime reacts with the alumina left in red mud to produce calcium aluminates.

They affirm that these also hydrate to produce cementitious compounds, in reactions which are fairly well known from the fact that they are responsible for the strengths of high alumina cement mortars. In this study, gypsum would have supplied Ca^{2+} for the following reaction:

$$Na_2O + CaSO_4 \rightarrow CaO + Na_2SO_4 \quad (4)$$

This CaO, particularly in the presence of NaOH, would have increased the lime content of the Portland cement, thereby accelerating pozzolanic reactions.

Samples from the layer of red mud waste subjacent to the gypsum-treated layer (0–15 cm) also showed dramatic BTS increases, though less than those of the G samples. As for the G samples, residual Ca^{2+} cations from the dissolution and, in this case, illuviation, of surface applied gypsum could have caused pozzolanic activity within the subjacent (SG) layer samples when combined with Portland cement.

However, among the red muds, the weakest response to treatments was of those red muds which were neither treated with gypsum nor subjacent to the gypsum-treated locations (Fig. 3). In contrast to the subjacent samples these raw red-mud waste samples were crushed from the sub-aerial section of an air-dry red mud waste dump. The apparent lack of Ca^{2+} cations correlates with lowest increases in tensile strength (Fig. 3). Apart from the Portland cement, the 1–2 mm aggregates of crushed microcrystalline limestone were the common ingredient in all of the major strength increases of all samples. The method of the 1–2-mm limestone aggregates in increasing tensile strength is uncertain at this stage. However, though aggregates of good quality always increase strength of mortars, unless calcium carbonate is calcined, or otherwise converted to $Ca(OH)_2$, it is pozzolanically inactive. Throughout this study no calcining occurred. Yet, no vast increases in strength occurred in the absence of limestone aggregates (Fig. 3). Still, the substantial superiority of the G samples eliminates the limestone aggregates as the sole determinant in this study, of strength increases to red muds. Therefore, for the G samples, the only other source of increased strength seems to have been based on uncalcined pozzolanic activity resulting from excess Ca^{2+} cations (possibly as CaO) and residual lime in the Portland cement. However, on the assumption that Ca^{2+} cations made a significant difference to the strength increases, it seems surprising that mortar mixes with the SG layer which contained four times the amount of Ca^{2+} ions as the G layer (Table 1), exhibited smaller strength increases than those from the G layer. But the much higher ratio of Na^+/Ca^{2+} (1:0.6) in the SG samples (Table 1) compared to the G samples (1:4.5) may be the reason. After prolonged submergence for 8 days with the tops of the samples sub-aerially exposed, very little decreases in strength or slaking occurred for the majority of the treatments (Fig. 5). However, strength decreases substantially occurred for the raw hardened red mud samples (R), to a lesser extent, the SG (subjacent) samples, and all the limestone samples with the exception of the microcrystalline limestone (Fig. 4).

As water moves through porous material such as concrete or mortar, salts can be drawn with it. As evaporation of capillary flow occurs at the surface, the salt concentration there increases, thereby creating an osmotic gradient. Massive

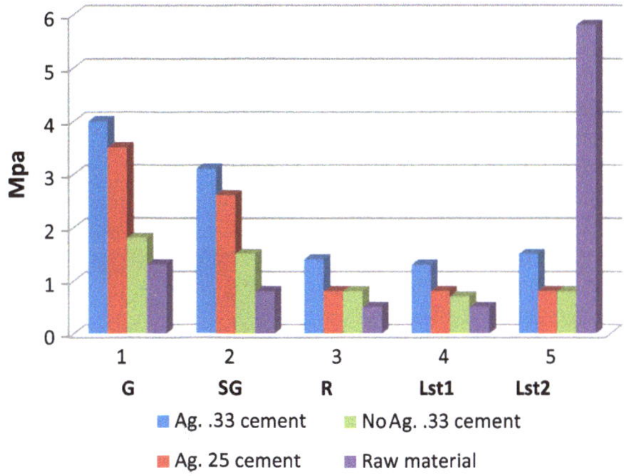

Fig. 4 Key Ag = 1–2 mm microcrystalline limestone aggregate; *G* prior gypsum-treated red mud at 0–15 cm depth; *SG* 15–30 cm subjacent to G; *R* untreated carbonated air-dry red mud; *Lst1* soft "marl" limestone; *Lst2* microcrystalline limestone

hydrostatic pressures can develop to levels exceeding the structural strength of concrete, which can be from 2000 to 3000 psi (Gromicko and Gromicko 2016). In this manner, the mortars containing high relative amounts of Na^+ can be weakened. The mortars mixed with the subjacent red mud all effloresced during partial submergence (Fig. 6), while no efflorescence occurred in G samples (Fig. 6). As stated earlier, the concentration of Na^+ in the subjacent layer at the time of treatment was several-fold that of the superjacent (0–15 cm) layer. This explanation is further strengthened by results of a shatter test, where repeated impact on the samples from a height depicted much greater brittleness amongst the SG samples as compared with the G counterparts at 0–15 cm (Table 3).

This suggests that in the layer subjacent to the plough-base depth, the gypsum treatment was not as effective as for the 0–15 cm treated zone, being severely weakened by the effects of excess sodicity.

3.1.1 Raw Carbonated Red Mud Samples

Of the three red mud wastes used in this study, the mortars made of naturally hardened red muds (R samples) are the lowest in indirect tensile strength (Figs. 3 and 4), and by extension, also in uniaxial compressive strength. These samples, never having been treated with gypsum, would have been expected to contain relatively low concentrations of Ca^{2+} plus high concentrations of Na^+. Even if this subjacent level contained high levels of Ca^{2+} illuviated from the superjacent gypsum-treated layer over several years, the high Na^+ concentration would have diluted the potential effect of the leached Ca^{2+}. Collectively, in the sense of pozzolanization, this is the worst condition of the three red mud wastes of this study.

Nevertheless, as shown previously, a high pH value is, in one way, beneficial when such materials are added to Portland cement matrices such as mortar and concrete, because strong alkaline environments favour hydration reactions (Gordon et al. 1996). For several days during curing of the mortars, the moisture

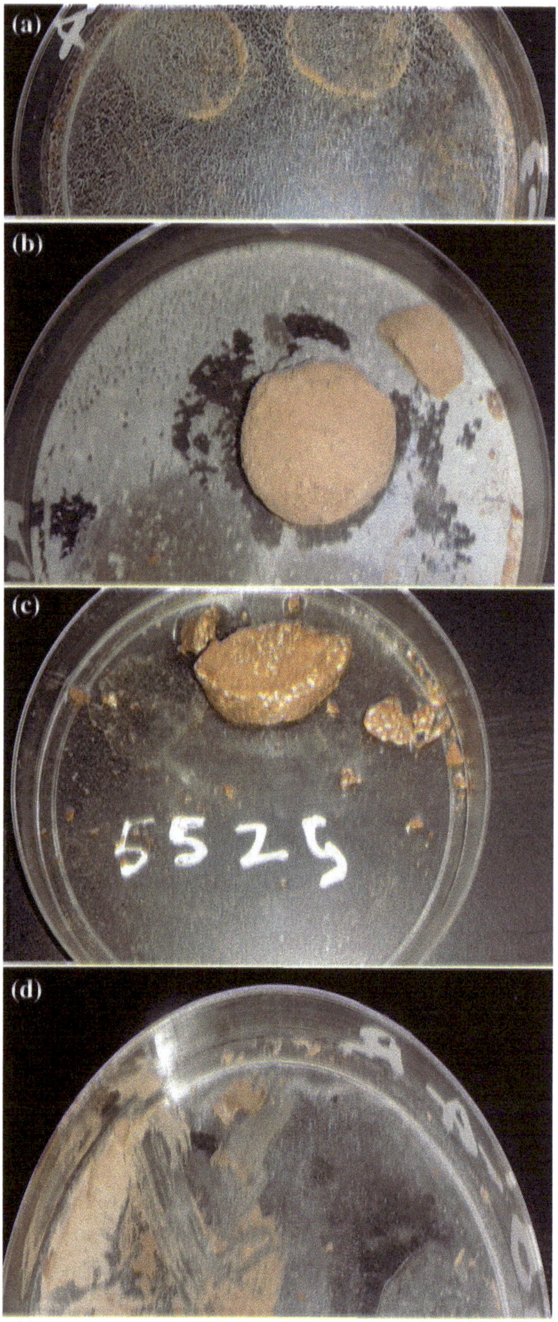

Fig. 5 **a** Efflorescence of salt crystals from subjacent (SG) mortar after partial submergence, **b** efflorescence from subjacent mortar samples after partial submergence in water, **c** no efflorescence from any of the G (0–15 cm gypsum-applied red mud) mortars after partial submergence, **d** efflorescence from subjacent mortar samples of the untreated red mud waste (R) after partial submergence in water

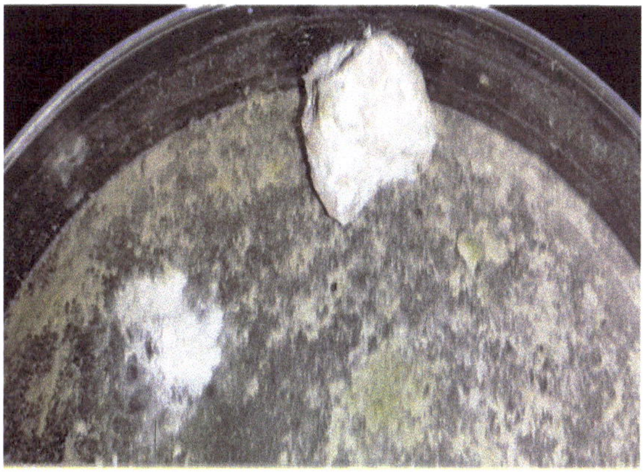

Fig. 6 High slaking of soft limestone during partial submergence

Table 3 Shatter test of three red mud wastes 7 years after surface-applied gypsum treatment, or with no gypsum

Source and depth (cm)	No. of drop impacts before shattering	S.D.[a]
0–15 superjacent (G)	25	6
15–30 subjacent (SG)	6	2
0–15 cm no treatment (R)	7	3

[a]Standard deviation

levels were continuously high. R samples, and samples taken from the >15 cm depth had a pH value of 12 (Table 1). These conditions could have increased hydration of the CaO (Table 2), thereby strengthening the samples. Yet, any such strengthening attributed to a high pH of the R samples was diminished, probably by the high sodium levels, but, for reasons stated above, more-so, for the naturally lithifies untreated red mud waste (R samples).

3.1.2 Limestones

As it was for all samples, increasing the rate of Portland cement increased tensile strength. Of the two limestones, this was more marked for the soft limestone marl. Nevertheless the final strength values did not approach those of the gypsum affected red muds (Fig. 3). Moreover it can be seen that, in contrast to the red muds, the original crystalline limestone is far stronger than any mortar made from it or from the other limestone.

3.2 *Effects of Slaking*

For unmixed samples, the slaking process occurred immediately for the subjacent layer (R), such that complete disintegration occurred after just one slaking cycle (Table 4). Soft limestone continuously lost mass by slaking throughout the period of 25 days (Table 4; Fig. 7). On the other hand, the gypsum-treated red mud resisted slaking to have lost <1 % after 25 days. This amount remained unchanged even after one year of submergence. Similarly, the mass of the hard limestone changed very little during the 25 days of 5 slaking cycles (Table 4). For the mortar samples, five slaking cycles were applied. Slaking durability of most mortars was high because few mortar samples indicated a loss in mass even after five slaking cycles (Table 5).

This was against expectations, for the samples with red mud proportions below 30 %, because saturation causes dissolution and decomposition of certain rock-forming minerals and water weakens the cementing material binding rock crystals and grains (Romana and Vasarhelyi 2006). Capillary water absorption (sorptivity) of mortars is very important. Excessive absorption can induce proliferation of fungi and bacteria and, in extreme conditions it might generate mortar detachments (Ribeiro et al. 2010). Apparently, the high density of de-watered red mud decreases

Table 4 Loss of mass (%) for unmixed Earth materials after slaking cycles

Rock	5-days (1 cycle)	10 days (2 cycles)	15 days (3 cycles)	20 days (4 cycles)	25 days (5 cycles)
M	0.5^{b}	0.8^{c}	1.0^{c}	1.1^{cd}	1.8^{e}
G	0^{a}	0^{a}	0^{a}	0^{a}	0^{a}
SG	100^{f}	100^{f}	100^{f}	100^{f}	100^{f}
L	0	0	0	0	0
R	0	0	0	0	0

M soft limestone, *L* microcrystalline limestone, *G* gypsum-treated red mud, *SG* subjacent layer to G, *R* naturally lithified red mud

Values followed by different letters are significantly different based on Tukey's HSD ($P < 0.05$)

(a) (b)

Fig. 7 **a** SEM image of fractured hardened cement paste, showing plates of calcium hydroxide and needles of ettringite (micron scale). *Source*: Wikipedia. **b** Sketch of crystals of calcium hydroxide and effringite

Table 5 Loss of mass (%) for mixed Earth mortars after slaking cycles

Earth mortars	5-days (1 cycle)	10 days (2 cycles)	15 days (3 cycles)	20 days (4 cycles)	25 days (5 cycles)
M	0^a	0^a	0^a	0^a	0.8^b
G	0^a	0^a	0^a	0^a	0^a
SG	0^a	0^a	0^a	0^a	0^a
L	0	0	0	0	0.5^b
R	0	0	0	0	0

M soft limestone, *L* microcrystalline limestone, *G* gypsum-treated red mud, *SG* subjacent layer to G, *R* naturally lithified red mud
Values followed by different letters are significantly different based on Tukey's HSD ($P < 0.05$)

capillary action. As stated above, replacing one-third of the content of silicate fractions in concrete compounds with red mud, Buraev and Kushnir (1986), found that this procedure hardened concrete such that it gained greater compressive strength than concrete obtained by the exclusive use of quartz sand.

3.3 Plastic Limit

The plastic limit of the R samples was low, and lower than that of the G samples. Only a few drops of water were required for plastic limit of R samples. Kehagi (2008) found that the composition and concentration of soluble salts in a soil affect the values of the liquid and plastic limits as well as the water content values of soils.

4 Conclusions

Strength of red muds increased at least 2-fold up to the level of medium-strength limestone and far exceeding that of soft limestone marl. In water, no slaking occurred among the stronger treatments even after 10+ days of submergence. This is far superior to marl limestone which slakes and readily loses substantial amounts of mass. The combination of gypsum-treated red mud at up to 40 %, Portland cement, and 1–2 mm diameter fine limestone aggregate without artificially supplied heat vastly improves the strength of all types of red mud but particularly the type previously hardened by gypsum treatments. This is attributed to the extra Ca^{2+} which leads to increased pozzolanic reaction with hydroxides contained in Portland cement.

The effects of prolonged submergence in water and of several slaking cycles on the strength, shatter resistance and durability of some red mud mortars was minimal. Hence, the use of such mortars as road bases under high-intensity rainfall is promising. The most effective mortar combinations always contained a limestone aggregate. Without the inclusion of aggregate, the BTS index of a mortar tended to be in the "weak rock" category.

5 Geobiotechnological Applications

The high resilience under indirect tensile strength of some mortars containing particular red mud wastes of the above study indicates a potential for uses requiring durability in wet conditions such as road beds in humid areas.

Red mud additions to mortars inhibited the capillary uptake of water. Limestone marl/cement/red mud combinations may similarly inhibit water absorption in limestone marls. However, though reports are that Jamaica red muds have lower levels of radioactivity further research is required to determine the radioactive levels of the varieties of red mud mortars used in this study. Further, Bayer red mud has a high content of hazardous elements like As, Pb and Hg.

Also, it has been shown that a highly alkaline environment as accompanies some red mud wastes helps to inhibit corrosion of steel bars in reinforced concrete (Ribeiro et al. 2010). This feature could be exploited considering the fact that buildings in hurricane-prone locations are increasingly reinforced with steel bars embedded in concrete.

However, based on its role in the production of effringite (Fig. 7), at 16 % (w/w), the heavily represented Al_2O_3 in Jamaica red muds presents concerns of durability. Nobst and Stark (2003) note that a high sulphate resistance is required if cements are to be used in sulphate bearing waters and soils especially under conditions favouring thaumasite formation. Merlini et al. (2008) noted that in hydrated Portland cement, the reaction of calcium sulphate with calcium aluminate produces ettringite (Fig. 4), which is a hexa-calcium aluminate tri-sulphate hydrate, of general formula:

$$(CaO)_6(Al_2O_3)(SO_3)_3 \cdot 32H_2O \tag{5}$$

Effringite is characterized as colorless to yellow, crystallizing in the trigonal system (McCarthy et al. 1992). Prismatic crystals are typically colorless, turning white on partial hydration (McCarthy et al. 1992). As ettringite contains large amounts of sulphate, and Nobst and Stark (2003) note that stoichiometric amounts of Ca, Al, SO_4 and H_2O (all present in gypsum-treated red muds) can produce ettringite, there are concerns regarding the long-term effects of Al_2O_3-rich cements. Their experimental results confirm that thaumasite (similar to ettringite) formation can be accelerated by Al_2O_3 bearing components in cements. However, they also observed that thaumasite formation is also possible without active participation of Al^{3+}. Thus assessment of sulphate resistance of cements only from the chemical point of view apparently gives results which are contrary to the field experience. Therefore the conditions for the formation of ettringite in red mud mortars require specific studies before further inferences.

References

Altindag R, Guney A (2010) Predicting the relationships between brittleness and mechanical properties (UCS, TS and SH) of rocks. Sci Res Essays 5(16):2107–2118

Amadei B (2015) Principles and procedure of Brazilian Test. CVEN 5768 -Lecture Notes 8. www.ejge.com/2013/Ppr2013.168clr.pdf. Accessed 5 March 2016

Amritphale SS, Patel M (1987) Utilization of red mud, fly ash for manufacturing bricks with pyrophyllite. Silic Industriels 52(3–4):31–35

ASTM (1989) American Society for Testing and Materials, vol 04–08

ASTM D4318 (2014) ebookdig.biz/ebook/q/pdf/**astm-d-4318**.html. Accessed 28 Feb 2016

Brown ET (1981) Rock characterization, testing and monitoring. In: Brown ET (ed) International society for rock mechanics suggested methods. Pergamon Press, Oxford, pp 119–121

Browner RE (1995) The use of bauxite waste mud in the treatment of gold ores. Hydrometallurgy, 37(3):339–348, ISSN: 0304-386X

Buraev MI, Kushnir LI (1986) Facing tiles obtained from hydromica clays and red mud. Kompleksn Ispol'z Miner Syr'ya 7(1986):66–69

Cargill JS, Shakoor A (1990) Evaluation of empirical methods for measuring the uniaxial compressive strength of rock. Int J Rock Mech Min Sci Geomech Abstr 27:495–503

Dass A, Malhotra SK (1990) Lime-stabilized red mud bricks. Mater Struct 23(4):252–255

Di San Filippo A (1980) Riutilizzo del fango rosso. Rendiconti del seminario della Facolt`a di Scienze dell'Universit`a di Cagliari, vol L, No. 3–4, Cagliari, Italy

Dimas DD, Giannopoulou IP, Panias D (2009) Utilization of alumina red mud for synthesis of inorganic polymeric materials. Miner Process Extr Metall Rev 30(3):211–239

Din F, Rafiq M (1997) Correlation between compressive strength and tensile strength/ index strength of some rocks of north- west frontier province (limestone and granite). Geol Bull Univ Peshawar 30:183

Farah R (2011) Correlations between index properties and unconfined compressive strength of weathered Ocala limestone. UNF Theses and Dissertations. Paper 142

Franklin JA, Dusseault MB (1989) Rock engineering. McGraw Hill, USA

Geotechdata (2008) Geotechdata.info—updated 20.12.2014 ASTM Standard Test Method D3967-81 www.geotechdata.info/geotest/brazilian-test.html. Accessed 2 Feb 2016

Gokhale KVJK (1960) Experiments in engineering geology. Tata McGraw-Hill, New Delhi, pp 23–32, 47–49

Gordon JN, Pinnock WR, Moore MM (1996) A preliminary investigation of strength development in Jamaican red mud composites. Cement Concr Compos 18(1996):371–379

Gromicko B, Gromicko N (2016) Efflorescence for Inspectors. InterNACHI. www.nachi.org/efflorescence. Accessed 24 March 2016

Harris MA (2008) Structural improvement of age-hardened gypsum-treated bauxite red mud waste using readily decomposable phyto-organics. Environ Geol 56:1517–1522. doi:10.1007/s00254-008-1249-5

Kehagi F (2008) An innovative geotechnical application of bauxite residue. Electron J Geotechnical Eng 01/2008; 13 www.researchgate.net/publication/242296762_An_Innovative. Accessed 28 Feb 2016

Khattab S, Othman H (2013) Durability and strength of limestone used in building. Al Rafidain Engineering 21(3)

Klein M (1998) German patent, 3633413, chemical abstracts 108, 191683

Kohno K, Sugimoto A, Kashiwai T (1998) High-strength concrete containing finely ground silica and red mud. Semento Gijutsu Nenpo 42:136 (1998); Chemical Abstracts 111: 083124 (1998)

Liu W, Yang J, Xiao B (2009) Application of Bayer red mud for iron recovery and building material production from aluminosilicate residues. J Hazard Mater 161:474–478. ISSN 03043894

Liu Y, Naidu R, Ming M (2011) Red mud as an amendment for pollutants in solid and liquid phases. Geoderma 163(1–2):1–12, ISSN 0016-7061

Lombi E, Zhao FJ, Zhang G, Sun B et al (2002) In situ fixation of metals in soils using bauxite residue: chemical assessment. Environ Pollut 118(3):435–443, ISSN 0269-7491

Lombi E, Hamon RE, McGrath SP, McLaughlin MJ (2003) Lability of Cd, Cu, and Zn in polluted soils treated with lime, beringite, and red mud and identification of a non-labile colloidal fraction of metals using istopic techniques. Environ Sci Technol 37(5):979–984, ISSN 0013-936X

Maddocks G, Lin C, McConchie D (2004). Effects of Bauxsol™ and biosolids on soil conditions of acid-generating mine spoil for plant growth. Environ Pollut 127(2):157–167, ISSN 0269-7491

Maddocks G, Reichelt-Brushett A, McConchie D, Vangronsveld J (2005) Bioaccumulation of metals in *Eisenia fetida* after exposure to a metal-loaded bauxsol™ reagent. Environ Toxicol Chem 24(3):554–563, ISSN 1552-8618

Majumdar AJ, Singh B, Edmonds RN (1990) Hydration of mixtures of cement aluminous cement and granulated blast furnace slag. Cem Concr Res 20(2):197–208

McCarthy GJ, Hassett DJ, Bender JA (1992) Synthesis, crystal chemistry and stability of ettringite, a material with potential applications in hazardous waste immobilization. Mater Res Soc Symp Proc 245:129–140

Merlini M, Artioli G, Cerulli T, Cella F, Bravo A (2008) Tricalcium aluminate hydration in additivated systems. A crystallographic study by SR-XRPD. Cem Concr Res (Elsevier) 38(4):477–486. doi:10.1016/j.cemconres.2007.11.011

Muller CJ (2005) Pozzolanic activity of natural clay minerals with respect to environmental geotechnics. PhD dissertation. ETH No. 16299, dissertation submitted to the Swiss Federal Institute of Technology Zurich

Nazir R, Momeni E, Armaghani DJ, Amin MFM (2013) Correlation between unconfined compressive strength and indirect tensile strength of limestone rock samples. Edge 18(2013): 1737–1746 (Bund. I)

Nobst P, Stark J (2003) Investigations on the influence of cement type on thaumasite formation. Cement Concr. Compos 25(8):899–906. doi:10.1016/S0958-9465(03)00118-5

O'Callaghan WB, McDonald SC, Richards DM, Reid RE (1998) Development of a topsoil-free vegetative cover on a former red mud disposal site. Alcan Jamaica Rehabilitaion Project Paper

Paramguru RK, Rath PC, Misra VN (2005) Trends in red mud utilization—a review. Mineral Process Extr Metall Rev 26(1), ISSN 0371-9553

Pinnock W, Gordon JN (1992) Assessment of strength development in bayer process residues. Cem Concr Compos 18(6):71–379

Pratt KC, Christoverson V (1982). Hydrogenation of a model hydrogen-donor system using activated red mud catalyst. Fuel 61(5):460–462, ISSN 0016-2361

Red mud project (2011) http://www.redmud.org/home.html. Accessed 28 Feb 2016

Ribeiro DV, Labrincha JA, Morelli MR (2010) Use of red mud as addition for Portland cement mortars. J Mater Sci Eng 4(8) (Serial No.33), ISSN 1934-8959, USA

Romana M, Vasarhelyi B (2006) A discussion on the decrease of unconfined compressive strength between saturated and dry rock samples. http://www.google.com

Schwarz M, Lalík V (2012) Possibilities of exploitation of bauxite tailing from alumina production, recent researches in metallurgical engineering—from extraction to forming. Dr Nusheh M (Ed.), ISBN: 978-953-51-0356-1, InTech, Available from: http://www.intechopen.com/books/recentresearches

Sglavo VM, Campostrini R, Maurina S, Carturan G et al (2000) J Eur Ceram Soc 20(3):235–244, ISSN 0955-2219

Sharma PK, Singh TN (2008) A correlation between P-wave velocity, impact strength index, slake durability index and uniaxial compressive strength. Bull Eng Geol Environ 67:17–22

Sheorey PR (1997) Empirical Rock Failure Criteria. Rotterdam: A.A. Balkema, 176p.

Singh IB, Singh DR (2002) Cr(VI) removal in acidic aqueous solution using iron bearing industrial solid. Environ Technol 23(1):85–95, ISSN 0959-3330

Singh AP, Singh PC Singh VN (1993) Cyclohexanethiol separation from kerosene oil by red mud. J Chem Technol Biotechnol 56(2):167–174, ISSN 0268-2575

Singh M, Upadhayay SN, Prasad PM (1996) Preparation of special cements from red mud. Waste Manage 16(8):665–670, ISSN 0956-053X

Snars K, Gilkes RJ (2009). Evaluation of bauxite residues (red muds) of different origins for environmental applications. Appl Clay Sci 46(1):13–20, ISSN 0169-1317

Stillborg B (1986) Professional users handbook for rock bolting. Transtech publication, Germany 144 pp

Tsakiridis PE, Agatzini-Leonardou S, Oustadakis P (2004) Red mud addition in the raw meal for the production of Portland cement clinker. J Hazard Mater 116(1–2):103–110, ISSN 0304-3894

Woodbridge ME (1999) Use of soft limestone for road-base construction in Belize. In: Seventh International Conference on Low-Volume Roads, Baton Rouge, Louisiana, USA. 23–26 May 1999. Department for International Development. Transport Research Laboratory, Crowthorne Berkshire RG45 6AU, United Kingdom. PA3450/99

Yalçin N, Sevinç V (2000) Utilization of bauxite waste in ceramic glass. Ceramics Int 26(5):485–493

Part III
Acidic Mine Water: Bioremediation

Bioremediation of a Stagnant Polluted Acid Mine Drainage Using a Clay-Lime Spiked Sludge and Bacterial Degradation

Mark Anglin Harris and Santo Ragusa

Abstract Effects of a clay-lime spiked sewage sludge and fresh decomposable ryegrass on the mitigation of an acid drainage were studied in the laboratory. Treatments (dry ameliorant weight/leachate ratio) were: (1) sludge (air-dried) at rates of 0, 8, 16 and 24 %, (2) ryegrass at 0, 1, 1.5 and 2 % (dry weight), (3) sludge (at the above-mentioned rates) and 1.5 % ryegrass mixture. Measurements of mitigation (according to the criteria of changes in pH, Fe, S, Al and heavy metals) made every 10th day for 100 days showed ryegrass/sludge combination the most effective while sustaining mitigation longest, with or without the influence of sulphate reducing bacteria (SRB). sulphate and Fe in the acid drainage decreased in the order: sludge-ryegrass > sludge ryegrass by 180, 40, 19; and 96, 83 and 54 % respectively, compared with controls. An 11-fold decrease in soluble Al was caused by the highest rate of the combined sludge-ryegrass treatment but Al was doubled by the sludge-only treatment and only minimally affected (2 % reduction) by the ryegrass-only treatment. For the sludge plus ryegrass treatments at the highest rate of application, pH levels increased significantly, from 2.3 to 17 units and within 20 days of SRB activation, the concentration of Co, Cu, Mn, Ni and Zn decreased respectively: 3-, 15-, 90-, 3- and 50-fold.

Keywords Acid-drainage · Bioremediation · Sewage-sludge · Sulphate-reducing-bacteria

Highlights

- Clay-lime spiked sewage sludge and fresh decomposable ryegrass greatly decrease heavy metals in acid mine drainage
- SRB activation occurs after acclimatization to 2.5 pH AMD.

Original article: **Bioremediation of acid drainage using decomposable plant material and sludge**. *Environmental Geology*, Vol. 40, Issue 1/2, pp. 195–215.

M.A. Harris, *Geobiotechnological Solutions to Anthropogenic Disturbances*, Environmental Earth Sciences, DOI 10.1007/978-3-319-30465-6_7

1 Introduction

1.1 Water Acidification: Effects

Mosquitoes thrive in acidic water. When sulfide minerals undergo chemical and biological oxidation processes (Fig. 1), the environment can become contaminated with acid drainage characterized by low pH and high levels of sulphate and metal ions. Un-weathered (or un-oxidized) mine-soils which contain a significant amount of pyritic-S in excess of their neutralizers (carbonates) will rapidly drop the pH to a range of 2.2–3.5 after exposure to water and oxygen (QGRM 2002).

Acidic water may subsequently leach other metals in the rock, resulting in the contamination of surface and groundwater. Waste rock piles, other exposed waste, mine openings, and pit walls are often the source of acidic effluents from a mine site. The process may occur rapidly and will continue until there are no remaining sulfides. This can take centuries, given the large quantities of exposed rock at some mine sites (Bell et al. 1989). Among the least stable minerals when exposed to weathering are halite, gypsum, calcite, dolomite and pyrite. The most stable are clay minerals, quartz, aluminium oxides (bauxites) and iron oxides. Pyritic minerals (FeS_2), when oxidized to sulfuric acid drastically lower the pH, while carbonate ($Ca/MgCO_3$) bearing minerals and rocks tend to increase the pH as they weather and dissolve (Strosnider and Nain 2010). Jarosite is an iron-and sulfur-containing mineral that forms in oxidizing acid sulphate soil, with the formula

Fig. 1 The step-like pattern of open-pit sulfidic mines exposes pyrite to oxidation by a factor of at least 2:1. Source: QGRM (2002)

$KFe_3(OH)_6(SO_4)_2$. It can only form in environments where the pH is less than 3.7, so it is considered very strong evidence of acid sulphate soil oxidation.

1.2 Low-Cost Bioremediation

Developing countries require low-cost technologies to effectively remediate mine wastes. Sinha et al. (2009) evaluated the increasing role of bioremediation. He observes that since the late 1980s, after the chemical and mechanical treatments of lands and water bodies and thermal treatment (incineration) of hazardous wastes proved economically and environmentally unsustainable, focus shifted towards the biological methods which are cost-effective as well as environmentally sustainable and also socially acceptable. Bioremediation is a soft bioengineering technique to clean up contaminated lands/sites using microbes (bacteria or fungus), plants (terrestrial and aquatic) and earthworms. It is also a technique to stabilize the eroded lands and prevent soil erosion. Bioremediation works carried out by the microorganisms are called 'micro-remediation' while those performed by plants are called 'phyto-remediation' (Sinha et al. 2009).

Microbial technology treats the soil contaminants in the site itself, and avoids the high costs of excavating soil. Acid-ingesting bacteria when present, can accelerate such a process. Decreasing solubility in the "reduced state" often decreases the toxicity of heavy metals compared to that of the "oxidized state." Bacteria can reduce metals, where microbes utilize them (metals) as terminal electron acceptor for anaerobic respiration (Sinha et al. 2009).

Being an intrinsic process done in situ, it relies on the naturally occurring biological process carried out by native and indigenous microorganisms. It can be applied to sites with high water table and does not destroy the site that is to be treated (detoxified). Microbial remediation can handle all types of soil pollutants starting from rare metals to radionuclides (Sinha et al. 2009). Native microbes in any contaminated site are 'acclimatized' and are capable of transforming the toxic metals to their oxides or hydroxides (Sinha et al. 2009).

1.3 Role of Hydroxides

The chemical precipitation of hydroxides is termed hydrolysis. Water reacts with dissolved cations such that the OH^- adheres to the cation, thereby leaving the hydrogen ions in the water and dropping the pH. The precipitation of dissolved Fe^{3+} provides substantial acidity by the release of extra hydrogen ions in the water (Lottermoster 2003). Under acid conditions of <pH3, Fe^+ remains in solution, whilst at progressively higher pH values, precipitation of hydroxides occurs (Lottermoster 2003; Theng 1974).

1.4 Sulphate Reducing Bacteria

These oxidation processes are accelerated by activity of iron-oxidizing chemolithotrophic microorganisms. Acidophilic sulphate-reducing bacteria (aSRB) exist in anaerobic sediments in acid mine drainage environments, and can reduce sulphate to sulfide. Sen and Johnson (1999) employed neutrophilic SRB to remove metals and sulphate from acidic wastewaters; however, a pre-treatment step is necessary in order to produce the circum-neutral pH required by the bacteria. aSRB therefore have a potentially important role in bioremediation of AMD, in view of their ability to withstand low pH.

Direct reduction by biologically produced hydrogen sulfide (H_2S) by sulphate reducing bacteria can reduce and precipitate the metals. Microbes are adapted to thrive in 'adverse conditions' of high acidity/alkalinity/toxicity and high temperature. They can develop 'biological resistance' against any toxic substance in the environment due to special 'jumping genes' (Sinha et al. 2009). Hence, while a number of them may be killed due to high toxicity, some resistant microbes survive and are cultured for further use. Under favourable conditions of growth e.g. pH, temperature and moisture and adequate supply of nutrients like vitamins, magnesium, manganese, copper, sulfur, potassium, phosphorus and nitrogen, microbes can biodegrade/bio-transform the complex hazardous organic chemicals into simpler and harmless ones.

Research has shown that, with an added energy source, SRB can cause a reduction of deleterious effects of AMD by raising its pH and removing metal ions with the toxic metals being locked into metal sulfides (Bechard et al. 1994). Biosorption is sequestration of the positively charged heavy metal ions (cations) to the negatively charged microbial cell membranes and polysachharides secreted in most of the bacteria on the outer surfaces through slime and capsule formation. From the surface the metals are transported into the cell cytoplasm through the cell membrane with the aid of transporter proteins and get bio-accumulated (Sinha et al. 2009).

The sulfur reducing bacterium is strictly anaerobic and gram-negative. It acquires its energy from sulfur respiration and completely oxidizes acetic acid (organic acid) with sulfur to CO_2 (Sinha et al. 2009). Reduction of sulfur (S) produces hydrogen sulfide (H_2S) which reacts with heavy metal ions to form less toxic insoluble metal-sulfides. Furthermore, these bacteria are also able to enzymatically reduce and precipitate these heavy metals (Bruschi and Goulhen 2006). They observe that pilot plants developed by Shell Research Ltd. and Budelco BV, France, using an undefined consortium of SRB (sulphate reducing bacteria), have been used successfully to remove zinc (Zn) and sulphate. The metals were precipitated as sulfides. Acetic acid, produced by the SRB was removed by the 'methanogenic bacteria' present in the consortium. This technology has been scaled up and now capable of treating 7000 cubic meter (cum) of contaminated soil per day. Bruschi and Goulhen (2006) successfully carried out in situ bioremediation of uranium-contaminated sites with *Desulfosphorosinus* spp. and *Closteridium* spp.

Even "phosphor-gypsum", a hazardous waste from fertilizer industries has been removed by SRB, and an anaerobic bioreactor with SRB community culture removed other metals and reduced "lead wastes" to less hazardous PbS (Galena) (Bruschi and Goulhen 2006). Additionally, sulphate reducing bacteria can metabolically precipitate heavy metals and insoluble sulfides indirectly (Ramasamy et al. 2006).

Traditionally, removal of heavy metals from polluted water have included mainly their precipitation by NaOH, $CaCO_3$, or Na_2CO_3 (Kalin et al. 1993) and $Ca(OH)_2$ (Elliott et al. 1998; Strosnider and Nain 2010). However, with respect to a time scale measured in decades such as that for persistent AMD, such above-mentioned materials could incur high cumulative costs. Further, Larsen and Schierup (1981) highlighted the inadequacy of such methods because they do not remove the metals quantitatively. In other words, such methods do not guarantee against re-hydroxidation and re-solubilizing when wetter, more acidic conditions prevail (Elliott et al. 1998; McCullough and Lund 2011), thereby again leading to the pollution of groundwater supplies.

Where sources of lime are not readily available, improvements in the quality of AMD can occur after treatment with organic wastes (Hard et al. 1997). Charles (1998), for instance, investigated the effects of dumping raw sewage for several years into an acid-sulphate lake. His reported changes included an increase in pH values towards neutrality, lower levels of suspended colloids, and the return of several major aquatic animal species. He attributed these ameliorative results to the phosphates in the eutrophic water containing raw sewage, which bind to the iron oxide in the acid-sulphate water. These then form compounds that fall to the bottom of the lake as sediment, chemically locking away these reactants (it is iron oxide that causes the characteristic reddish-brown coloration in acid drainage). This confirms an observation of Gillman (1985) that the variable charge surfaces of oxides are capable of adsorbing a range of poly-anions such as phosphates in competition with other organic poly-anions. The fact that Gillman (1985) also found that, as organic matter content decreased with depth in soils, the phosphate sorption capacity increased, suggests that phosphates may be highly competitive against other organic poly-anions for spaces on oxide surfaces. Microbial alteration has also been studied by several workers (Tuttle et al. 1969; Schindler et al. 1980; Mills 1985; Gibert et al. 2002), who showed that sulphate reduction (SR) to sulfide in anoxic zones of aquatic environments may increase because of the action of sulphate-reducing bacteria (SRB). Other research found that SR in water overlying freshwater lake sediments would increase alkalinity and precipitate metallic sulfides (Berner et al. 1970; Ben-yaakow 1973; Figueroa et al. 2004). However, according to Havas (1990); Bechard et al. (1994), AMD's that have a $pH < 3$ are not uncommon; yet SRB cannot normally survive in acidic environments of $pH < 5.5$ (Alexander 1977a, b). Because SRB operate within a much higher pH range of 5–9 than commonly exists in AMDs (often with $pH < 4$), and commonly have their optimum pH around neutral to slightly alkaline (Alexander 1977a, b), there have been few successful attempts at AMD remediation using SRB. For example, when Lyew et al. (1994) introduced new AMD of pH 3.5 into a continuous flow reactor (with a prior AMD pH of 4.8 containing actively operating SRB colonies) SRB activity ceased. Gyure et al. (1990) found that organic acids can be very potent against SRB, and SRB activity also stopped when they introduced organic acids of pH 3.8 into anoxic sediments. Hard et al. (1997) found that of all their isolates, only one strain of *Desulfovibrio salexigens* grew in a Postgate medium at a pH as low as 4.0. Further, Connell and Patrick (1968) observe that conditions of low oxidation–reduction potential (ORP), which are necessary for enhancing the activities of SRB (Alexander 1977a, b; Lyew et al. 1994), are usually limited to a range of pH values of 6.0 or above.

It is also known that, to supply energy for the metabolic processes of SRB microorganisms, a decomposable substrate is needed (Bechard et al. 1993; McCullough and Lund 2011). Sewage sludge is the by-product of wastewater treatment and its nutrients include nitrates, phosphates and organic matter. However, because of public resistance in many municipalities, untreated sewage is legally unavailable for such purposes. Aerobically treated sewage sludge is more resistant to decomposition than is raw sewage because in the activated sludge process, air is mechanically mixed in with liquid sludge to increase bacterial oxidation (Boyle 1990), eventually leaving a higher proportion of resistant residues. Therefore, in areas affected by AMD, use of the oxidized form may not be as effective a mitigating alternative as the raw form, where raw sewage is unacceptable. Yet even though oxidised sludge may contain substantial proportions of resistant fractions, such as lignins and polyphenols, which are un-degradable by anaerobes, anaerobic decomposition of such materials seems to occur in submerged conditions. Various studies have shown that even under anoxic conditions, anaerobes can degrade plant-derived materials such as straw (Larsen and Schierup 1981) wood debris (Tuttle et al. 1969) and hay (Stark et al. 1988; Koschorreck et al. 2002). Nevertheless the exclusive application of such hard-to-degrade substrates under submerged conditions does not lead to complete or satisfactory mitigation of AMD. At 3.5-, 7- and 35-day retention times, Bechard et al. (1994) found that none of several such added decomposable substrates (alfalfa hay, straw, timothy hay, applied singly or in combination) significantly decreased the Fe concentration in a simulated AMD. Further, even for some nitrogenous treatment combinations (urea flowing through straw, urea flowing through timothy hay) they reported that no change in pH, Fe, Al, and SO_4^{2-} was observed. One implication of that study on the biological mitigation of Fe-rich AMD waters is the need for further work with different forms of those and other organic additives.

Sulphate-reducing bacteria are unable to use complex organic substrates such as starch, cellulose, proteins, and fats. Yet Chang et al. (2000) tested different organic waste materials as electron donors for sulphate reducers and found that cellulose polysaccharides were the main components of the waste materials consumed in the reactors. SRB are therefore dependent on other microbes that degrade these complex substrates and ferment them to products that can serve as substrates for SRB (Figueroa et al. 2004; Muyzer and Stams 2008).

Kumar et al. (2013) note that a rapid increase in pH seen in all their treatments at Day 1 was too rapid for it to be derived from SRB activity; instead it is believed to be due to alkaline substances within the sewage that directly neutralized the water. The increases in pH was proportionate to the amount of sewage added. However, with the increased quantities are embedded a larger amount of the "active" ingredient. Thus a review by Gibert et al. (2002) found that the specific composition of the organic matter was a primary determinant of the efficacy of the passive SRB-based treatment system; in particular, the lability and biological availability of the material. However, the ability of the organic matter to be decomposed over a prolonged period is likely more important to bioremediation of AMD than initial lability and bioavailability.

Waybrant et al. (1998) observes that sewage sludge causes the highest levels of sulphate reduction compared with seven other organic carbon sources. However, multiple studies have noted that mixtures of sewage sludge with multiple organic carbon sources generally promote higher sulphate reduction rates than single sources (Waybrant et al. 1998; Harris and Ragusa 2000; McCullough et al. 2006).

Given time for adaptation to acidity, and a suitable energy source, SRB have been re-activated in AMD at pH < 4. With added lactate (readily assimilated) as the energy source, Elliott et al. (1998) reactivated SRB in an influent having a pH of 3.2, thereby removing 38.0 % of influent sulphate from solution, and increasing effluent pH to 5.8 units. They found, however, that longer periods of adaptation for the SRB were required for increasingly lower influent pH values. Thus, at pH 3.5, 5 days elapsed before significant SRB activity resumed, compared with 2 days for pH 4.8. Since the AMD from the decommissioned Nairne Pyrites Mine in South Australia is even more acidic (<pH 2.3) than the above-mentioned examples, a substantially longer adaptation time for SRB activation would be anticipated. Moreover, little research has been done as to whether or not SRB can be activated from a dormant state in such highly acidic liquids (that is <pH 2.5). However, should microenvironments of lower acidity levels be created within the physical interstices of an incorporated sediment medium that has a suitable energy source, it may be possible that changes in SRB activity within these pore spaces may then modify the chemical characteristics of the overlying AMD water layer. Though raw untreated sludge contains more nitrates and phosphates than contained in old sludge (Boyle 1990), raw sludge lacks soil derived additives of substantial proportions used in its treatment, including lime (Horvath and Koshut 1981; Maiti 2003) or clay (Grindley 1998).

1.5 Role of Inorganic and Organic Colloids

Metals are more bioavailable at acidic pH values. Solubility of metals is dependent on soil characteristics and is strongly influenced by pH of the soil and the degree of complexation with soluble ligands (Sinha et al. 2009). Generally, only a fraction of soil metal is readily available (bio-available) for the plant uptake. In the soil, the organic matter and the clay mineral content are important factors that can reduce metal bioavailability. Metal bioavailability is often low in soil systems. Clays, with high cation exchange capacities, such as montmorillonite, appear to reduce metal bioavailability and toxicity. In a study used to investigate the effect of cadmium (Cd) on microbial biodegradation of toxic organic compound ‘phenanthrene’, a total of 394 mg cadmium (Cd) per kg of soil was added, but only 3 mg cadmium (Cd L^{-1}) were actually bioavailable. Similarly, only 1 % of the total zinc (Zn) used in a study was present in the aqueous phase (Sinha et al. 2009). The bulk of the metal in soil is commonly found as ‘insoluble’ compounds unavailable for transport into the plant roots from the aqueous phase.

It is reasonable to postulate that these additives can cause changes in the physical and chemical behaviour of the sludge itself (hence their application in the first instance). For example in a study involving kaolinites, bentonites and illites submerged in acid solutions (0.1 and 0.5 M), Yeoh and Oades (1981) found that phosphoric acid reacted with clay minerals to release aluminium, which precipitated with phosphate to form insoluble aluminium phosphate, and that the quantities of released Al were similar regardless of the kind of clay. Therefore, such treated sludges may also modify the chemical characteristics of the AMD into which they may be placed. In contrast to reactions involving organic acids, mineral acids (such as H_2SO_4 in AMD) reacting with clays may not result in the complexing of released Al and its removal from solution.

Thus Yeoh and Oades (1981) found that after an initial rapid, acid dissolution (by not only phosphoric, but also nitric acid) of aluminium (within the first 3 to 5 days) from both 2:1 and 1:1 clays had occurred, the Al releases persisted for up to 40 days. However, because Gillman (1985) found that poly-carboxylic materials are specifically adsorbed on a range of amphoterically charged oxide surfaces, it is also possible that clays such as kaolinite, which have relatively weak negative charges, may likewise act as adsorption sites for organic material concurrently added to the AMD. Because Ahmad and Tan (1985) found that under highly acidic, waterlogged conditions (similar to those proposed in the present study) solubilized aluminium combined with functional carboxylic groups, they hypothesized that small unidentified Al species can interact electrostatically with the functional groups, thereby complexing Al ions out of solution. Probably, Al released from clays by AMD would be similarly complexed by organic additives, thereby mitigating the effects of metals inherent in the AMD, along with those metals released from any added inorganic ameliorants. Further, because acid neutralization reactions can be regarded as buffering reactions (Norton 1983), and the incorporation of material containing a substantial amount of clay particles into an acidic medium may delay equilibration of that acidity (Norton et al. 1990), it is conceivable that when added to AMD, sludges that contain particular additives from soil environments, such as secondary clay minerals and/or lime, may resist acidification. As reported by Herhily and Mills (1985), such conditions can initiate SRB activation, by raising pH to a critical level. This might have been the case when Lyew et al. (1994) established a population of SRB in a 15.24-cm-deep limestone-gravel bed impregnated with a mixed carbon-nitrogen source at the bottom of a bioreactor. Finally, some sediments have been shown to be an important medium for activating SRB, not only as a physical support system but as an aid for reducing acidity, possibly by physical means. The AMD water, being denser, tends to remain on top of the floor sediments and mixing with the overlying water is restricted, but because one of the effects of floor sediments in such environments is to restrict water circulation within the sediment body, the acidifying effects of mixing within the sediment interstices may be restricted. For instance, Herhily and Mills (1985) reported that SRB activity in an AMD lake was caused by higher pH environments in micro-niches that were physically protected by sediment particles. Because a physical support and attachment system is required for SRB activity, Elliott et al. (1998) utilized beds of acid-washed sand as a physical support for SRB colonies that mitigated AMD. It is, therefore, postulated here that SRB can survive and become active in minute pores within pH buffering sediments, if supplied with an energy source. Several workers have found that biodegradation of resistant carbon in soils low in soluble carbon have been initiated by priming with small amounts of soluble carbon such as glucose (Hayes 1980; Swift et al. 1979; Baldock et al. 1994). Aerobically digested sludges high in resistant carbon may probably be similarly activated because, compared with those that have low C:N ratios, readily available substrates with high C:N ratios have been shown to be more effective in improving the capability of resistant organics to mitigate AMD. For instance, in a reactor containing urea-amended straw as a substrate, Bechard et al. (1994) found that no change in pH, Fe, Al, and SO_4 concentrations occurred, yet with sucrose as the only additive, the pH increased significantly and large amounts of Al were removed. The addition of soluble C in the form of sodium lactate as an energy source, achieved substantial increases for SRB amelioration of an AMD (Elliott et al. 1998; Koschorreck et al. 2011). Therefore, in combination with a suitable, readily decomposable organic substrate to supply some nutrients lost in prior sludge treatment, it is probable that SRB mitigation of AMD may be accelerated by a clay-lime impregnated aerobically digested sewage sludge. Nevertheless, for most studies involving AMD of $pH < 4$, complete mitigation of AMD (that is, where pH and removal of all toxic metals occur concurrently) had not occurred (Tuttle et al. 1969; Gyure et al. 1990; Bechard et al. 1994; Lyew et al. 1994; Elliot et al. 1998; Muyzer and Stams 2008; La et al. 2003).

1.6 Sludge Rates

Sewage sludge contributed alkaline materials that directly neutralized AMD in proportion to the quantity added and therefore played a primary role in stimulating SRB bioremediation (Kumar et al. 2013). The lowest concentration of sewage sludge ($30\ g\ L^{-1}$) tested proved to be inadequate for effective SRB bioremediation. However, there were no measurable beneficial effects on SRB bioremediation efficiency when sewage sludge was added at concentrations of $60\ g\ L^{-1}$ (Kumar et al. 2013). They compared their results with existing literature data to develop a conceptual model for remediation of AMD in pit lakes through organic material amendments. The model indicated that labile organic carbon availability was more important to the bioremediation rate than AMD strength, so long as iron and sulphate concentrations were not limiting. Poor allochthonous carbon inputs into the pit lake from outside sources such as runoff from the vegetated catchment (McCullough et al. 2009); and reduced autochthonous carbon—i.e.—inputs from sources within the pit lake, such as microbial decomposition of particulate organic carbon or algal production (Peine and Peiffer 1998) in pit lakes typically limit sulphate reduction.

1.7 Cuba: Santa Lucia Mine Wastes

After conducting a detailed geochemical study at the inactive Zn–Pb mine of Santa Lucia, in western Cuba, Romero et al. (2010) found that the mine-wastes there are characterized by high total concentrations of potentially toxic elements (PTE), with average values of 17.4 % Fe, 5.47 % Ba, 2.27 % Pb, 0.83 % Zn, 1724 mg kg^{-1} As and 811 mg kg^{-1} Cu. Oxidation of sulfide minerals in mine-waste dumps and in the open pit produces acid mine effluents (pH = 2.5–2.6) were enriched in dissolved SO_4^{2-} (up to 6754 mg L^{-1}), Fe (up to 4620 mg L^{-1}) and Zn (up to 2090 mg L^{-1}). They recorded low pH values (2.5–2.8) and high dissolved concentrations of the same PTE in surface waters, up to 1500 m downstream from the mine.

These are substantially above the safe levels stated by both the USEPA and ANZECC (Australia & New Zealand Environment Conservation Council) safe water quality guidelines. Nevertheless, concentrations of As, Ba and Pb in acid mine effluents and impacted surface waters are relatively low: 0.01–0.3 mg L^{-1} As, 0.002–0.03 mg L^{-1} Ba and 0.3–4.3 mg L^{-1} Pb (Romero et al. 2010). Analysis by X-ray diffraction and electron microscopy revealed the occurrence of lead–bearing barite and beudantite and the more common solid phases, reported elsewhere in similar environments including Fe-oxyhydroxides, jarosite, anglesite and plumbojarosite. Because the reported solubilities for barite and beudantite are very low under acidic conditions, these minerals may serve as the most important control in the mobility of As, Ba and Pb. In contrast, Fe-oxy-hydroxides are relatively soluble under acidic conditions and, therefore, they may have a less significant role in PTE on-site immobilization.

1.8 Aim

The objective of this study is to examine the influence of fresh decomposable plant material on the effectiveness of aerobically treated sewage sludge to mitigate an Fe-rich leachate from an acid sulphate mine tailings heap. It is, therefore, hypothesized that given (1) a readily decomposable cellulose substrate in combination with a highly inorganically buffered sewage sludge, and (2) enough adaptation time, dormant SRB will be activated to mitigate an AMD that has a pH of <2.5. It is also hypothesized that after the establishment of active SRB colonies in such a medium, more efficient decreases in the levels of toxic metals and more substantial reductions of acidity would occur. The results of this experiment were used to help define the ideal proportions of additives, to reduce the hydraulic retention time (HRT) in an SRB driven constant flow AMD bioreactor.

2 Materials and Methods

2.1 Acid Drainage

Acid drainage from the decommissioned Nairne Pyrites Mine, South Australia was collected from a stagnant pool at the site. Some water quality criteria for Australia are shown in Table 1. The chemical characteristics of the Nairne Pyrites Mine drainage water, as determined by inductively coupled plasma atomic emission spectroscopy (ICPAES), are shown in Table 2. The pH was measured as 2.3, and EC as 7.5 dS m^{-1}. The amount of acidity was measured (Narwal et al. 1983) by adding powdered limestone (crushed to pass through a 1-mm sieve), as the amount needed to bring the AD to a neutral pH of 7.0.

2.2 Sludge

Dehydrated aerobically digested sludge was obtained from the Murray Bridge Treatment Plant in South Australia. Some relevant chemical and physical properties are shown in Table 3. Sludge containing passive SRB colonies was utilized.

Table 1 Some water quality criteria for the Australia and New Zealand Environment and Conservation Council (ANZECC)

Aluminium	<0.005 (if pH 6.5)
Arsenic	<0.050
Boron	No data
Cadmium	0.001
Cobalt	0.1
Chromium	0.002
Copper	0.002–0.005 (depending on hardness)
Iron	1.000
Magnesium	0.05
Manganese	0.01
Nickel	0.015–0.150 (depending on hardness)
Sulfur	0.2
selenium	0.005
Zinc	0.005–0.050 (depending on hardness)

Table 2 Chemical characteristics of acid drainage from Nairne Pyrite Mine AMD, South Australia

Aluminium	46
Arsenic	0.2
Boron	<0.1
Cadmium	<0.1
Cobalt	0.1
Chromium	0.01
Copper	0.1
Iron	2100
Magnesium	290
Manganese	110
Nickel	0.03
Sulfur	2700
Selenium	<0.01
Zinc	5.7
Molybdenum	<0.1
Sodium	220
Phosphorus	<1
Acidity as $CaCO_3$ (kg)	2006–2602

This is because the quality of organic matter in sewage affects the speed of bioremediation (Kumar et al. 2011). The sludge had previously been aerobically digested and air dried. This sludge had been selected against three other sludges, on the basis of the following attributes: (1) the substantial amount of kaolinite clay (40 %) and lime (2.5 %) that, on analysis, was found to have been added during its treatment, (2) it was derived from a residential area and, therefore, its anticipated low levels of toxic heavy metals, and (3) it contained a significant proportion of quartz sand (20 %). Several workers (Herhily and Mills 1985; Herhily and Mills 1985; Elliott et al. 1998) have found sand to be an effective physical means of attachment for SRB populations. In adidition, because at pH < 4.2, aluminium is a major buffering soil material, and Robertson et al. (1982) found that the amount of extractable heavy metals from sludges increased with the acidity of the liquid medium in which it was placed, the added advantages of applying a sludge buffered with clay prior to its use as an ameliorant for a low pH AD is apparent. A further reason kaolinite sludge was chosen rather than a 2:1 clay-impregnated sludge was because of the need for positive sites on the clay mineral surfaces to adsorb SO_4^{2-}, and negatively charged humic molecules. This is because several workers found that the alumino-silicate kaolinite readily adsorbed humic acid molecules because of its positive edge charges (Kobo and Fujisawa 1963; Schnitzer and Skinner 1964; Greenland 1965) and, therefore, may also adsorb SO_4^{2-} from the AMD. On the other hand, not only are there fewer exposed positive sites on 2:1 clay particles such as montmorillonite, but, according to Greenland (1965), X-ray evidence indicates that humic acid does not penetrate the inter-lamellar regions of 2:1 silicates either.

Table 3 Some chemical and physical characteristics of Murray Bridge sludge

Soil property (t ha^{-1})	Value
Organic carbon % by weight	1.0
pH (1:5 sludge:water ratio)	6.5
EC (1:5) sludge:water extract (dS m^{-1})	0.27
Total C (%)	1.24
Total N (%)	1.8
$CaCO_3$ (%)	2.5
Exchangeable cations (cmol (+)/kg)	
Ca	11.6
Mg	5.4
Na	0.43
K	0.95
Total	18.4
Clay (%)	40
Quartz sand (%)	39.2
Silt (%)	12.8
Gravel (%)	8

Table 4 Some metals extracted by 0.1 M HCl and 4 M HNO_3 from Murray Bridge sludge

Cations	0.01 M HCl extracted (mg L^{-1})	4 M HNO3 extracted (mg L^{-1})
Al	4.4	1224
Cd	<0.1	No data
Co	0.4	No data
Cu	0.3	4.2
Fe	<0.2	1164
Mn	19.3	12
Ni	0.3	No data
Pb	0.1	No data
Zn	0.6	4.9

Quantities of each heavy metal in the sewage sludge were determined (Table 4) by HNO_3 extraction (after Pierce et al. 1982). Two grams of sludge were thoroughly mixed with 20 ml of 4 M HNO_3 and heated at 70 °C for 24 h. After dilution to 35 ml with additional 4 M HNO_3, the mixture was shaken for 30 min and filtered with Whatman no. 43 paper to obtain the extract. This extraction process was repeated four times in the same manner. Toxic metals in extracts (Al, B, Cu, Fe, Mn, Mg, Zn) were determined by ICPAES (inductively coupled plasma atomic emission spectroscopy analysis). The above-mentioned metals (plus Cd, Co, Ni, and Pb) were also separately extracted less aggressively (Table 4) with 0.01 M HCl (Hinesly et al. 1977). The pH of samples was measured by a Watson–Victor glass pH electrode model 5003.

2.3 Cation Exchange Capacity

(CEC) was determined after extraction of the bases with neutral ammonium acetate (CH_3COONH_4). To clear the sludge of free salts, the samples were first leached with an ethanol and deionized water solution until free of salt. One aim of this experiment was to establish an initiation and adaptation time needed for SRB activation from dormancy in an AMD of pH < 2.5. Therefore, in this study, it was decided to use inactive rather than already actively operating SRB populations as an inoculant. Since SRB is widespread in wide-ranging soil environments (according to Lyew et al. 1994), a 'natural' inoculation of the sludge was adopted. Because SRB thrive under anaerobic conditions, the wet season (winter) was chosen for this proposed natural inoculation of the sludge. Thus, for 6 months before its addition to the AD for this study, the sludge was placed on a flat soil surface at the Brukunga iron pyrites mine-spoil site. To maintain a high soil–water content at the sludge/soil interface the sludge was not spread, but left in a 60-cm-high pile. This process allowed the natural inoculation of SRB upwards into the sludge from the relatively wet, anoxic soil surface. The dried sludge was then ground to pass a 1-mm sieve and thoroughly hand-mixed.

2.4 Plant Matter

Fresh ryegrass was harvested and chopped to <1 mm in a portable food processing machine operated at 1000 rpm because of the increased decomposition rate of finely divided material reported by several authors (Kroth and Page 1946; Alexander 1977a, b; Singh et al. 1992). For example, Amato (1983), after measuring the decomposition of different parts of the Medic plant found that the rate of decomposition in the field occurred in the following order: leaves > stems > roots > pods, but that ground pods were decomposed as easily as underground stems. Because of the difficulty in differentiating grass stems from the more easily decomposable leaves, all parts of the ryegrass except roots were applied. To compensate for the proportion of less readily decomposed stem fraction (approximately 50 % w/w) the rate of application of the ryegrass in this experiment was substantially higher than normally applied under a leaves-only green manure regime. The chopped material was then immediately placed into sealed polythene bags and refrigerated at 5–10 °C before incorporation into the AD 24 h later.

2.5 Incubation Procedure

The incubation was conducted under stagnant conditions because Lyew et al. (1994) found that the AD treatment process is governed by a relationship between the influent flow rate and the acidity of the water and that, after the introduction of lower pH water, SRB culture was able to recover quickest upon cessation of flow. Further, under stagnant conditions, they found that SRB culture was also always able to recover from perturbances. Transparent cylindrical polyacrylate cylinders of 8 cm inner diameter by 20 cm height were used as AD treatment vessels. Aerobically digested and air-dried sewage sludge was ground, passed through a 1-mm sieve and added to the polyacrylate cylinders. Treatment (1) consisted of aerobically digested sludge at the rates of 25, 50 and 75 g (referred to hereafter as CS1, CS2 and CS3 respectively), placed into cylinders containing 300 ml each of AD. Treatment (2) consisted of 20-g aliquots of ground ryegrass added to the above-mentioned sludge portions (referred to here as CS1/RY, CS2/RY and CS3/RY respectively).

Treatment (3) consisted of just ryegrass added to the AD at rates of 15, 20 and 30 g per container (referred to here as RY1, RY2 and RY3). Treatment (4) consisted of 300 ml of AD (control). All the organic additives (CS and RY) were placed at between 8 and 12 cm below the water surface in the cylinders. Each jar was sealed tightly with 0.1 mm thick polythene sheeting. Air spaces were left above the water surface. The jars were shaken every 10th day to ensure thorough mixing of AD with substrate. However, it was difficult to keep the plant material submerged and some of the ryegrass floated. The samples were incubated at a temperature range of 15–20 °C.

2.6 Sampling and Analysis

On every 10th day for 40 days, 20-ml water samples of the treated AD water were taken from the top 2 cm. These samples were then filtered through a Whatman no. 43 filter paper before further analysis. The pH of samples was measured by the previously stated method. The quantities of all elements in solution for the samples taken were determined by ICPAES analysis. Though measurements of pollutants occurred every 10th day for 100 days, the last measurement was made at 120 days, and monitoring of pH continued until the 140th day of incubation.

2.7 Electrophoretic Mobility

Because the measurement of particle charge and their mobility reveals the effect of the treatments on cations in solution, the zeta potential (electronegativity) was measured (using a Malvern Zetamaster particle electrophoresis analyzer). According to Pashley (1985), the zeta potential in mV, represents a surface charge effect that is directly related to electrophoretic mobility, such that the zeta potential is 12.85 mV per mobility unit. Organo-mineral complexing neutralizes negative charges on soil colloids, thereby lowering the electrophoretic mobility of particles. All potential measurements were done at a constant temperature (20 °C) and with a constant field strength of 80 V cm^{-1}. Measurements were carried out over a 30-s run time and the final values used were the averages of 10 runs.

3 Results

For every AD parameter examined, it can be seen that (1) fresh finely chopped ryegrass substantially increased the capacity of aerobic sludge to mitigate AD, and (2) the effects varied directly with the treatment rates (Figs. 1, 2, 3, 4, 5, 6, 7, 8, 9, 10, 11).

3.1 *Initial Colour Changes*

A reddish-brown transparent coloration of the AD controls, indicating the presence of Fe oxides, was observed throughout the incubation. Although all samples treated with RY also retained this coloration, after 5 days they lost a significant amount of transparency, and thereafter remained unchanged throughout incubation. For combined sludge-ryegrass treatments there was a dramatic colour change, with complete clarity and lack of colour observed in all of these treatments after just 12 days, indicating a lowered concentration of Fe in the treated AD. The liquid attained a clear, colourless appearance, on day 6 in the CS3/RY, on day 10 in the CS2/RY, and on day 12 in the CS1/RY samples, with all changes persisting for the remaining period of the study (i.e. 100 days). For the CS treatment regime, the sequence of colour loss was similar and varied directly with the rates, with the CS3 being the first to change the AD water to a light yellow after 10 days, gradually clearing until becoming colourless on day 20. This was followed in like manner by the CS2 and then CS1.

3.2 *Changes in Acidity*

The changes in pH of the samples are shown in Fig. 2. The pH increased substantially only for two treatments, both of which involved a combination of ryegrass and sludge (CS3/RY and CS2/RY). After 100 days of incubation pH increases were in the order: CS3/RY1CS2/RY1CS31CS1/RY1CS2pCS11RY31 (Table 5). Although at the highest rate (CS3) within the sludge-only treatment regime pH increased significantly from 2.3 to 3.0 (Fig. 2), there were no other substantial increases within the CS treatment regime. Further, for the RY treatment regime, AD pH actually decreased over time below that of the controls as incubation proceeded after day 10. On the other hand, at no time during this study did any of the pH decreases involving sludge-only (or in any combination with ryegrass), show a decrease to levels below that of the controls.

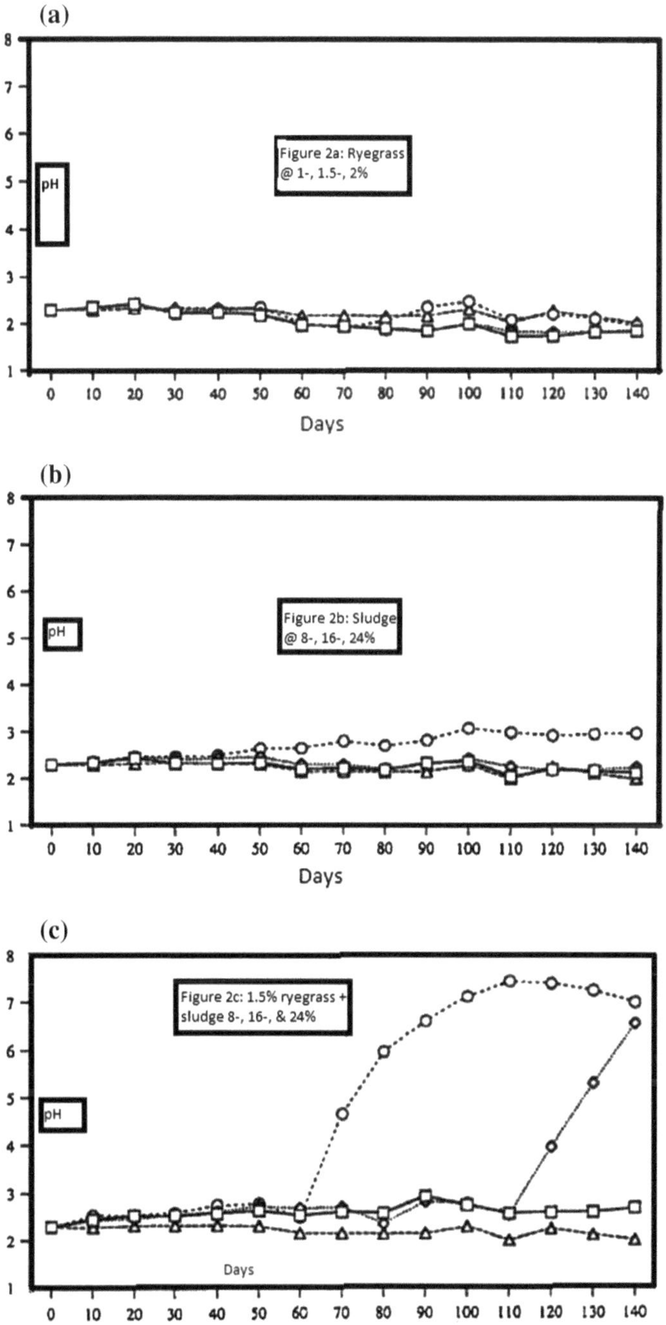

Fig. 2 The pH of Nairne Pyrite Mine AMD during incubation with phyto-organic materials. *RY1* Ryegrass (1 %); *RY2* (1.5 %); *RY3* (2 %); *CS1* sludge (8 %); *CS2* (16 %); *CS3* (24 %); *CS1/RY* sludge 8 % + ryegrass (1.5 %); *CS2/RY* (16 %)/(1.5 %); *CS3/RY* (24 %)/(1.5 %)

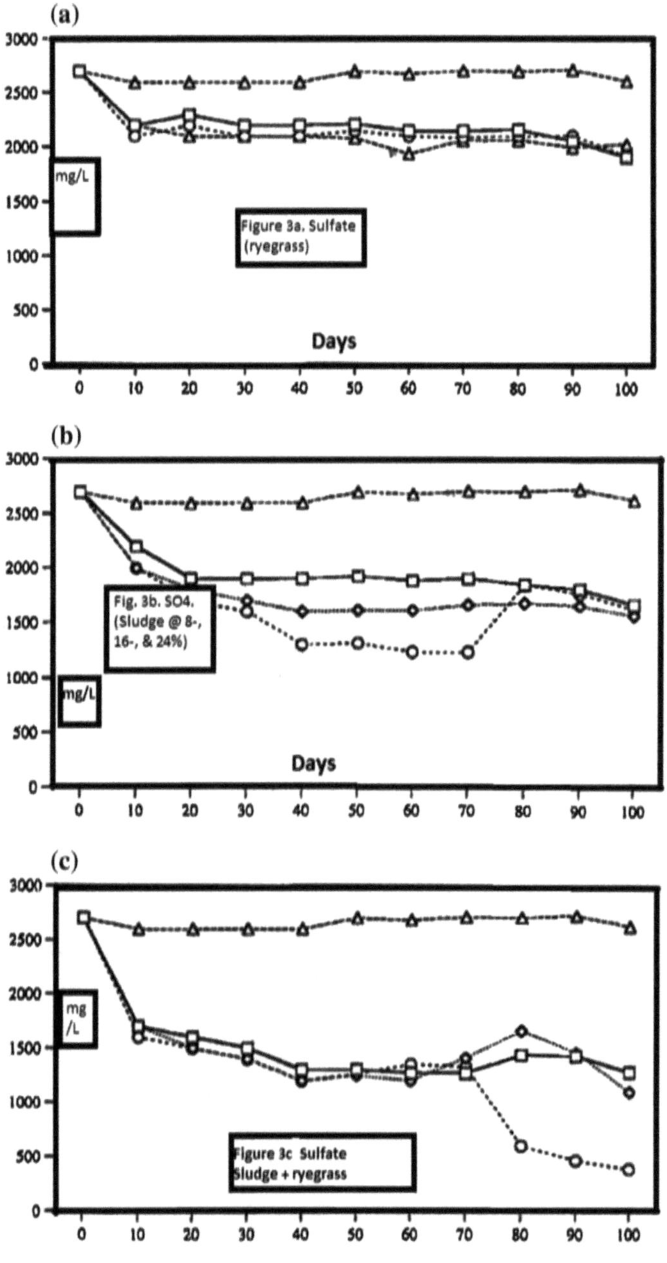

Fig. 3 Concentration of sulphate in the Nairne Pyrite Mine AMD during incubation with phyto-organic materials. *RY1* Ryegrass (1 %); *RY2* (1.5 %); *RY3* (2 %); *CS1* sludge (8 %); *CS2* (16 %); *CS3* (24 %); *CS1/RY* sludge 8 % + ryegrass (1.5 %); *CS2/RY* (16 %)/(1.5 %); *CS3/RY* (24 %)/(1.5 %)

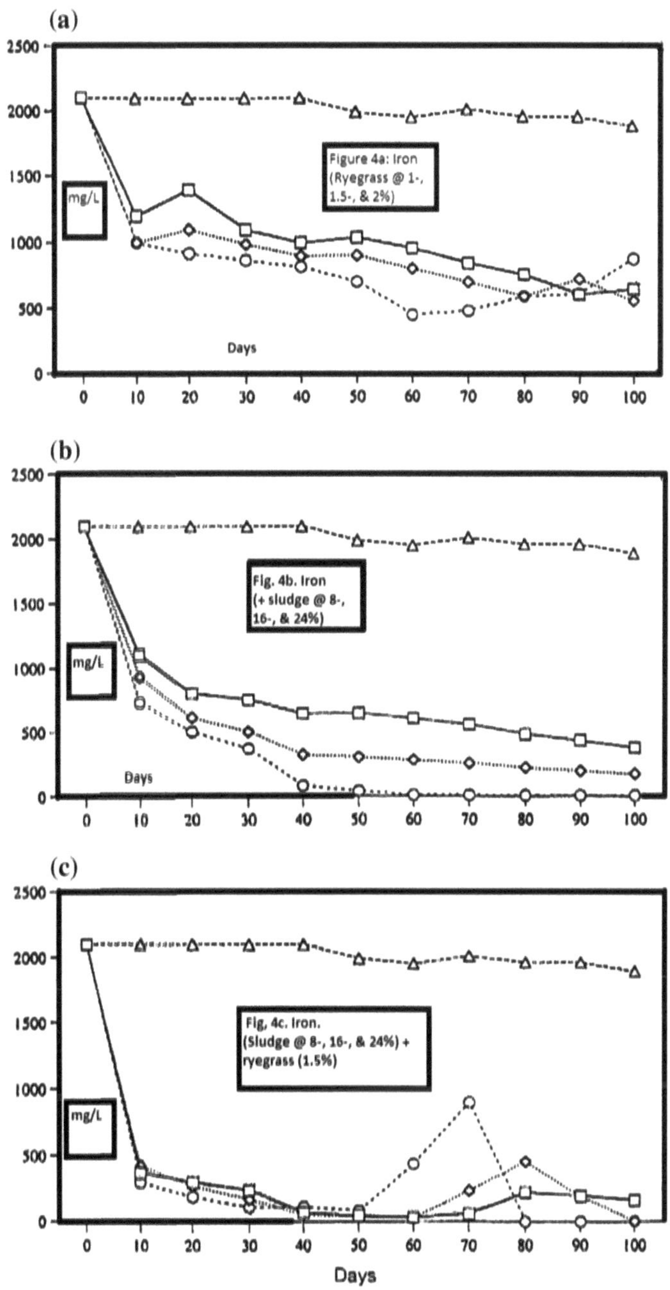

Fig. 4 Concentration of iron in the Nairne Pyrite Mine AMD during incubation with organic materials. *RY1* Ryegrass (1 %); *RY2* (1.5 %); *RY3* (2 %); *CS1* sludge (8 %); *CS2* (16 %); *CS3* (24 %); *CS1/RY* sludge (8 %) ryegrass (1.5 %); *CS2/RY* (16 %)/(1.5 %); *CS3/RY* (24 %)/(1.5 %)

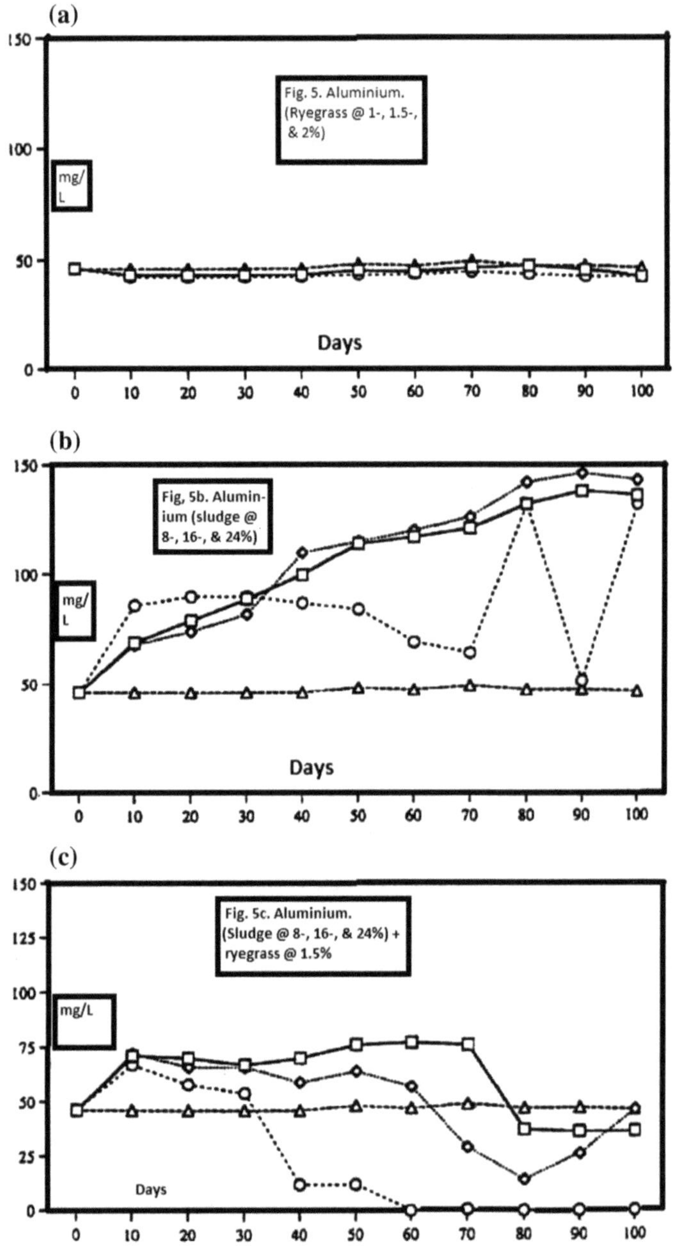

Fig. 5 Concentration of aluminium in the Nairne Pyrite Mine AMD during incubation with organic materials. *RY1* Ryegrass (1 %); *RY2* (1.5 %); *RY3* (2 %); *CS1* sludge (8 %); *CS2* (16 %); *CS3* (24 %); *CS1/RY* sludge (8 %) ryegrass (1.5 %); *CS2/RY* (16 %)/(1.5 %); *CS3/RY* (24 %)/(1.5 %)

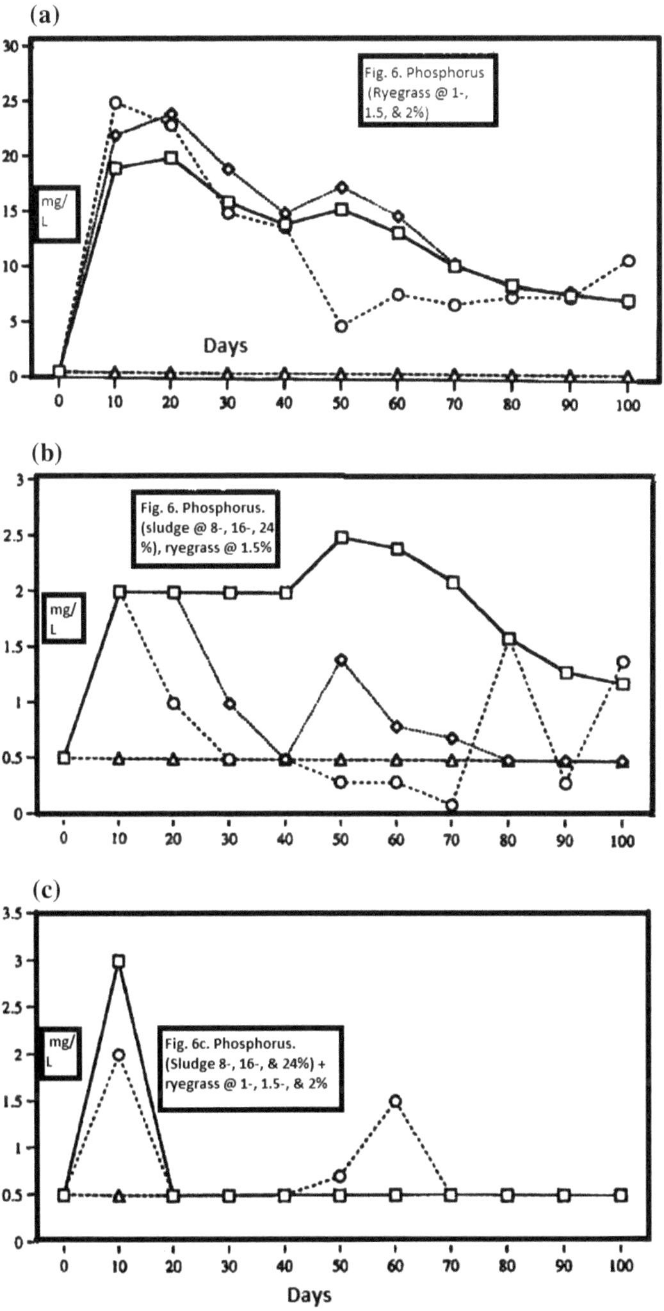

Fig. 6 Concentration of phosphate in the Nairne Pyrite Mine AMD during incubation with organic materials. *RY1* Ryegrass (1 %); *RY2* (1.5 %); *RY3* (2 %); *CS1* sludge (8 %); *CS2* (16 %); *CS3* (24 %); *CS1/RY* sludge (8 %) ryegrass (1.5 %); *CS2/RY* (16 %)/(1.5 %); *CS3/RY* (24 %)/(1.5 %)

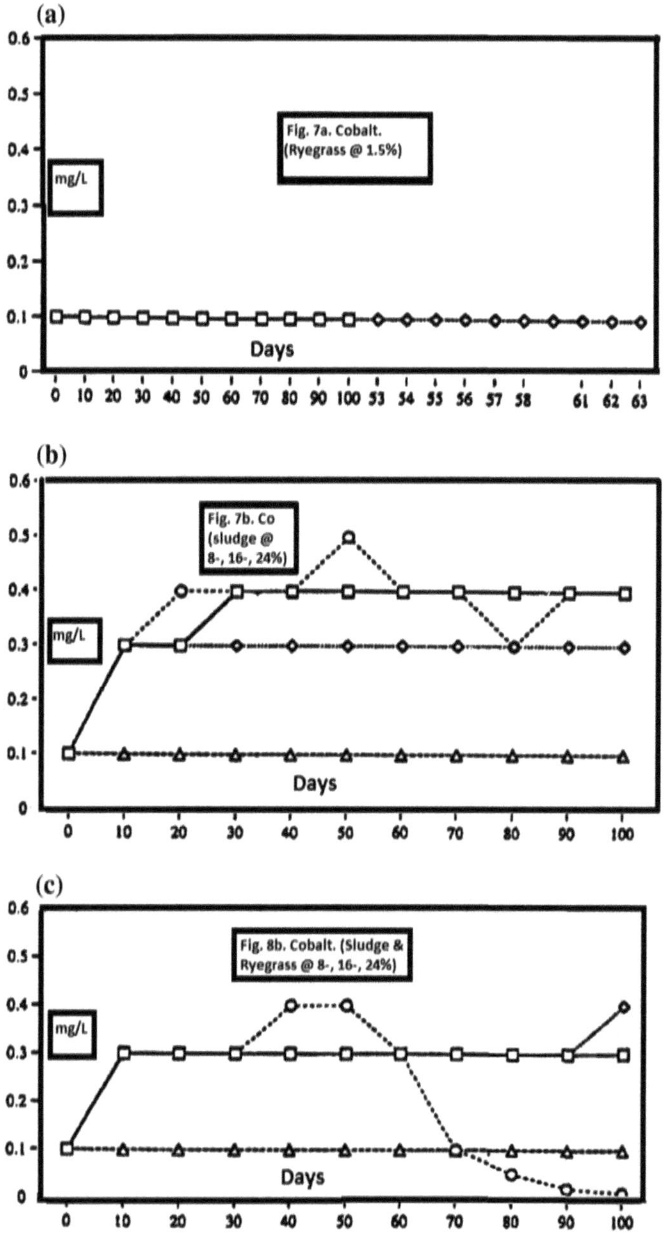

Fig. 7 Concentration of cobalt in the Nairne Pyrite Mine AMD during incubation with organic materials. *RY1* Ryegrass (1 %); *RY2* (1.5 %); *RY3* (2 %); *CS1* sludge (8 %); *CS2* (16 %); *CS3* (24 %); *CS1/RY* sludge (8 %) ryegrass (1.5 %); *CS2/RY* (16 %)/(1.5 %); *CS3/RY* (24 %)/(1.5 %)

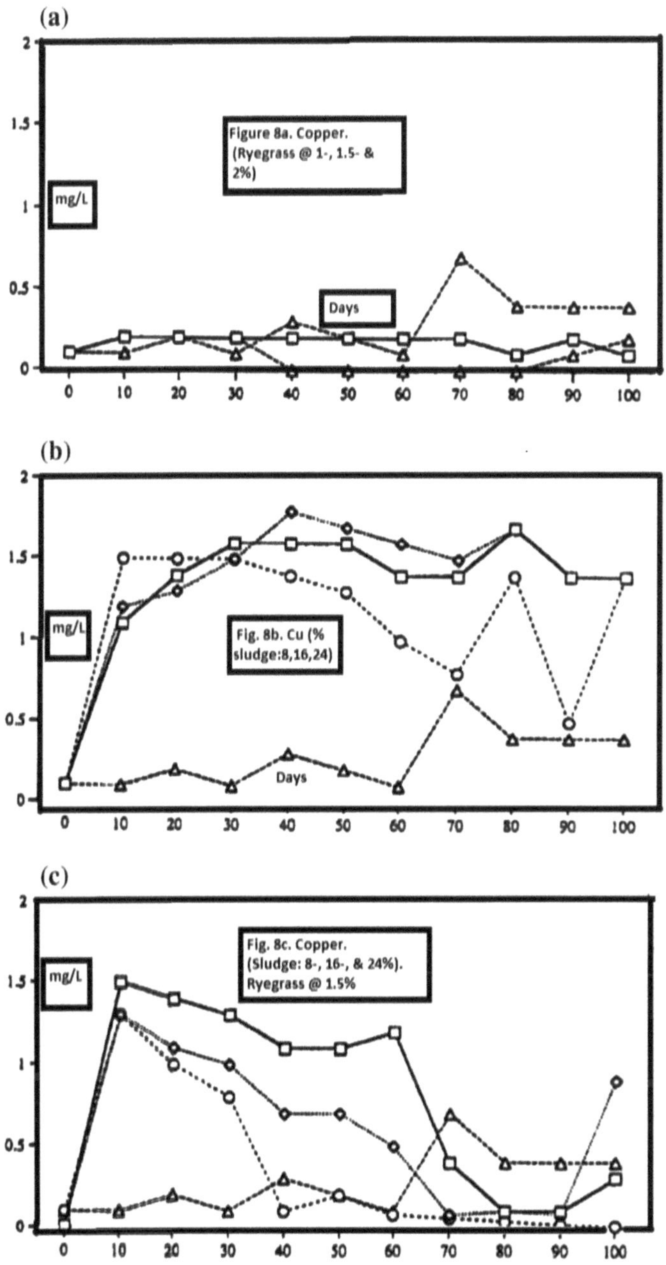

Fig. 8 Concentration of copper in the Nairne Pyrite Mine AMD during incubation with organic materials. *RY1* Ryegrass (1 %); *RY2* (1.5 %); *RY3* (2 %); *CS1* sludge (8 %); *CS2* (16 %); *CS3* (24 %); *CS1/RY* sludge (8 %) ryegrass (1.5 %); *CS2/RY* (16 %)/(1.5 %); *CS3/RY* (24 %)/(1.5 %)

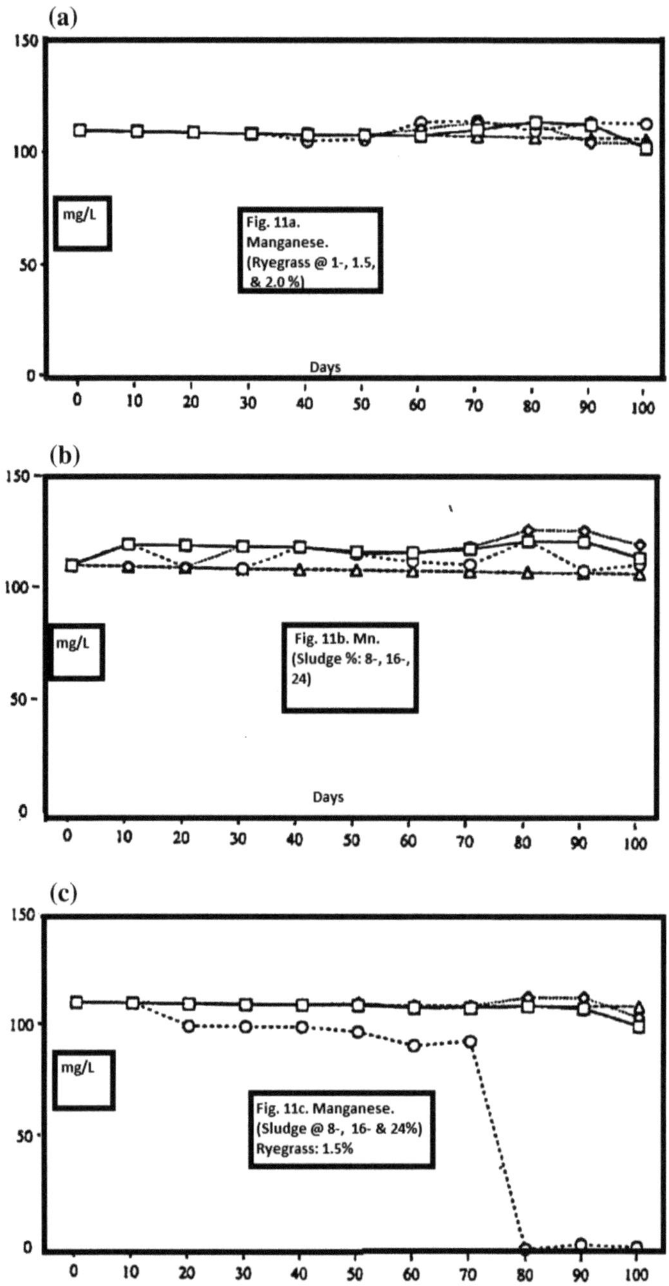

Fig. 9 Concentration of manganese in the Nairne Pyrite Mine AMD during incubation with organic materials. *RY1* Ryegrass (1 %); *RY2* (1.5 %); *RY3* (2 %); *CS1* sludge (8 %); *CS2* (16 %); *CS3* (24 %); *CS1/RY* sludge (8 %) ryegrass (1.5 %); *CS2/RY* (16 %)/(1.5 %); *CS3/RY* (24 %)/(1.5 %)

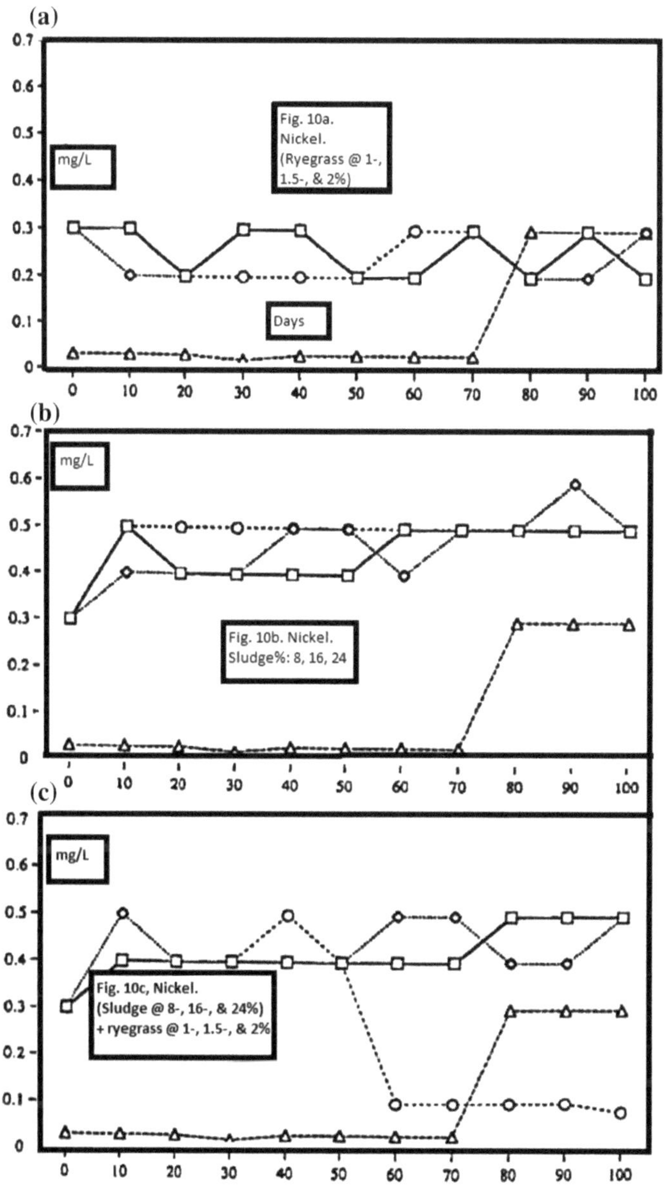

Fig. 10 Concentration of nickel in the Nairne Pyrite Mine AMD during incubation with organic materials. *RY1* Ryegrass (1 %); *RY2* (1.5 %); *RY3* (2 %); *CS1* sludge (8 %); *CS2* (16 %); *CS3* (24 %); *CS1/RY* sludge (8 %) ryegrass (1.5 %); *CS2/RY* (16 %)/(1.5 %); *CS3/RY* (24 %)/(1.5 %)

Thus, sludge-only as a component had a greater positive effect on pH than ryegrass-only did. In the early stages of incubation (days 0–10), pH increases were detected for all treatments. This increasing trend, though not large, was most

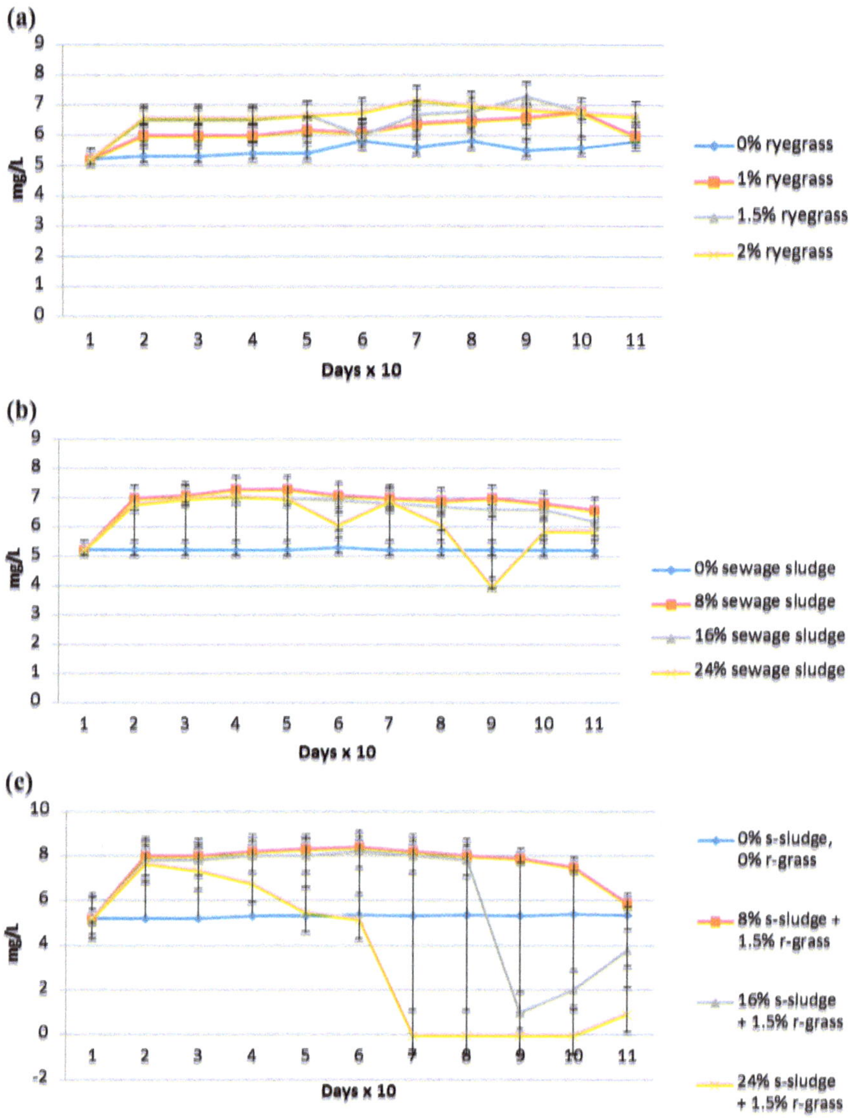

Fig. 11 Concentration of Zn in the Brukunga pyrites AMD during incubation with ryegrass (**a**), sewage sludge (**b**), sewage sludge + ryegrass (**c**)

substantial amongst the ryegrass/sludge treatments, particularly for the CS3/RY treatment on day 10, at 0.25 unit above that of the controls. Apart from a small transient decrease between days 50 and 70 for the CS/RY treatment regime, pH continued to increase until at least day 100, when CS1/RY and CS2/RY showed pH drops of 0.36 and 0.28 respectively. For the CS3/RY, which, as already stated, had the largest pH increases, such increases occurred in two phases: (1) days 0–50,

Table 5 Changes in pH of Nairne Pyrite Mine AMD after 140 days incubation

Treatment	Average pH change (units)
CS3/RY	+4.95
CSR2/RY	+2.98
CS1/RY	0.3
CS3	+0.66
CS2	−0.1
CS1	−0.13
RY3	−0.22
RY2	−0.5
AD control	−0.1

AD Acid drainage; *CS* sludge; *RY* ryegrass

(2) days 60–100, the latter increase being sudden and more substantial (Fig. 2). More will be said on this later.

The pH changes caused by the sludge-only treatment regime (CS) are qualitatively comparable with those for the sludge–ryegrass treatment regime because, like the CS/RY, its highest rate caused the largest pH increases. At various periods between days 50 and 80, a trend of lowered pH values was recorded for all treatments. For RY treatments, this trend, which began earlier at day 20, was maintained for the remainder of the incubation, culminating in pH values of 1.71, 1.81 and 2.06 at day 110 for RY1, RY2 and RY3 respectively. The RY3 pH value, though dropping several times below that of the AD control (after day 20), was at all times higher than the RY1 and RY2 pH values (Fig. 2). Transient pH decreases that occurred between days 60 and 80 for the CS/RY treatment regime were recorded only for the CS2/RY (day 80: decrease <0.35 units) and the CS3/RY (day 60: decrease <0.29), each lasting briefly (for less than 10 days), in contrast to those decreases for the other treatment regimes (which lasted for at least 40 days).

Therefore, whereas a sharp recovery of pH values was recorded for the sludge-ryegrass treatments, the recoveries for the other treatment regimes were either slower or non-existent (Fig. 2). CS3/RY showed a constant increase in pH values until day 120, after which values began to decrease (to 7.20 at day 130, and 6.99 at day 140). The above-mentioned recoveries of pH values were contemporaneous with the commencement of the signs of SRB activity in the CS3/RY samples. At the 55th day, a deep brown coloration appeared in the CS3/RY, and was the precursor for sharp pH increases. On day 58, the brown zonation appeared darker from the bottom, moving upwards to 2 cm below the water surface of CS3/RY and permeating all of the treatment material in the container by day 62. The pH value for this treatment had by this time climbed from 2.49 on day 60 to 3.8 on day 65. This darkened zone gradually took on the appearance of a black precipitate on day 65, sharply defined from the distinctly colourless but dwindling liquid phase (because of the continual prior extraction of 20-ml aliquots) above it. At day 70 the solid phase in the CS3/RY samples appeared totally blackened and had the characteristic rotten egg odour of H_2S gas. In these samples, pH values increased from 3.8 at day 65 to 4.15 at day 70, 6.6 at day 90, and 17 at day 100.

For the CS2/RY, a similar coloration pattern (culminating also in a black precipitate) began at day 70. However, whereas for the CS3/RY samples, only 15 days had been required before the black coloration advanced up to the top surface of the solid material, 27 days (until day 97) elapsed for CS2/RY. Further, for CS2/RY, it was found (day 70–97) that pH values did not concurrently increase with coloration changes as they did for the CS3/RY coloration phase. Thus for CS2/RY, its biggest pH increase did not occur before day 110 (from pH 2.54 to 5.6), that is, 13 days after the blackening of the substrate solids had been completed compared with a concurrent increase for the CS3/RY during its coloration phase. The other treatment regimes of the experiment, excepting CS3, CS2RY and CS3/RY, pH throughout the incubation was held to a value of <3. For CS3, pH was held at its highest value of 3.08 from day 90 until the end of incubation; this value was maintained for the remaining 30 days of the incubation. For the CS3, a 3-ml-narrow blackened zone was observed at the bottom of the solid phase at day 85, thereby suggesting the beginnings of an active SRB population. However, in contrast to that of the CS/RY samples, the blackened zone did not grow or extend upwards for the ensuing 30 days of incubation, but remained static.

3.3 Sulfur (as sulphate)

Figure 3 shows changes in sulphate concentration for all trials. sulphate was decreased in all samples except those of the controls. The smallest sulphate decreases occurred for the RY treatments. At day 10 the most substantial decreases occurred in the combined sludge–ryegrass (CS/RY) samples (Fig. 3c), followed by those of the CS treatments. After a large average initial decrease in concentration from 2700 to at least 1700 mg L^{-1} in the first 10 days (average 100 mg L^{-1} per day), sulphate concentration continued to fall at an approximate rate of 17 mg L^{-1} day^{-1} for the ensuing days of the incubation for the CS/RY treatments. A similar pattern of decrease prevailed for CS treatment regime, where the above-mentioned decrease for the first 10-day period of 650 mg L^{-1}, was followed by 13 mg L^{-1} for the ensuing 30 days. However, differences within the treatment suites for CS were greater than those of CS/RY. The substrate masses at the beginning of the treatments, approximately in the above order and corresponding to the drop in sulphate concentration, are shown in Tables 6, 7, 8. It can be seen that for sulphate removed within the CS/RY treatment suite up to day 10 (Table 6), there is little difference amongst the CS1/RY, CS2/RY and CS3/RY treatments.

Within CS/RY regime for the ensuing 30 days, CS1/RY and CS2/RY maintained the same rate of decrease (F100 mg L^{-1}) as that of CS3/RY (Table 7). On the other hand, during the same period of time (30 days), although the CS3 had removed just as much sulphate as the CS3/RY, this amount (110 mg L^{-1} day^{-1}) was significantly more sulphate than the CS1 (60 mg L^{-1} day^{-1}) or the CS2 (80 mg L^{-1} day^{-1}) had removed (Table 7). Up to this point in time, it can be seen that within the CSS/RY regime, each treatment removed a total of

Table 6 Sulphate removed from Nairne Pyrite Mine AMD by treatments in the first 10 days of incubation

Mass of substrate (g)	Amount of sulphate removed ($mg\ L^{-1}$)
95CS3/RY	1100
75CSR2/RY	1000
50CS1/RY	1000
70CS3	700
50CS2	700
35CS1	50

AD Acid drainage; *CS* sludge; *RY* ryegrass

Table 7 Sulphate removed by treatments in the Nairne Pyrite Mine AD after 40 days

Mass of substrate (g)	Amount of sulphate removed ($mg\ L^{-1}$)
95CS3/RY	1100
75CSR2/RY	1500
50CS1/RY	1400
70CS3	1400
50CS2	1100
35CS1	800

AD Acid drainage; *CS* sludge; *RY* ryegrass

Table 8 Sulphate removed from Nairne Pyrite Mine AMD by sludge-based treatments in the first 100 days

Mass of substrate (g)	Amount of sulphate removed ($mg\ L^{-1}$)
95CS3/RY	2700
75CSR2/RY	1600
50CS1/RY	1430
70CS3	1530
50CS2	1130
35CS1	1030

AD Acid drainage; *CS* sludge; *RY* ryegrass

1400–1500 $mg\ L^{-1}$, whereas, within the CS regime, the respective range was 800–1400 $mg\ L^{-1}$ (Table 7). Yet, the only difference between both treatment regimes (CS and CS/RY) was added ryegrass (the same mass of ryegrass was used for each CS/RY treatment): the same sludge rates (35, 50 and 75 g) were applied. For CS, the mass of sulphate removed from solution generally varied directly with the mass of ameliorant, but not for the CS/RY regime. Thus, the order of effectiveness at day 40 for the treatments were: CS3/RY = CS2/RY > CS1/RY = CS 3 > CS2 > CS1 > RY3 = RY2 = RY1 (Fig. 3). Further, this suggests that the role of the ryegrass could be crudely comparable to that of a catalyst because, at the lowest CS rate, no further decreases in sulphate occurred after day 10, whereas for the lowest rate of the combined treatment, decreases in sulphate concentration continued after day 10. During the middle phase of incubation (days 40–60), very little change in sulphate concentration occurred as a result of the treatments. However, by the end of incubation (day 100), sulphate concentration had been decreased much more greatly by the CS/RY regime compared with the CS regime (Fig. 3; Table 8). This stage was immediately preceded by detectable signs of SRB activity. After day 60, sulphate concentration for the CS treatments fluctuated and increased substantially, but fluctuations were not as extreme for the CS/RY (Fig. 3). The most dramatic decreases in sulphate concentration of this study was caused by the CS3/RY after day 70 (Fig. 3; Table 8). At the end of incubation, the amount of sulphate removed by CS3/RY was double that of the best sludge-only treatment (CS3) and treble that of the best RY (RY1, RY2) (Table 9). Table 10 shows a comparison between the periods 0–40 days and 60–100 days, of the proportion of sulphate removed from the AD. To show mainly the effect of SRB on SO_4^{2-} removal, the period 60–100 days was chosen (as opposed to 41–100 days) because activation of SRB occurred between day 60 and day 100 of

Table 9 Proportion of sulphate removed from Nairne Pyrite Mine AMD during 100-day incubation with sludge, sludge plus ryegrass, and ryegrass treatments

Treatment	Sulphate removed (mg L^{-1})	AD sulphate (%)
CS3/RY	2247	85
CS3/RY	1530	58
CS1/RY	1350	51
CS2	1060	40
CS3	1000	38
CS1	960	37
RY1	710	27
RY2	700	27
RY3	590	22

AD Acid drainage; *CS* sludge; *RY* ryegrass

Table 10 Sulphate removed from Nairne Pyrite Mine AMD by sludge-based treatments as a proportion of concentration at (a) day 40, (b) day 100 for two 40-day periods during incubation

Treatment	% Sulphate removed (0–60 days)	% Sulphate removed (60–100 days)
CS3/RY	55	72
CS3/RY	55	8
CS1/RY	52	0
CS2	52	0
CS3	41	11
CS1	30	2

AD Acid drainage; *CS* sludge; *RY* ryegrass

this incubation. It can be seen that by far the greatest reduction in SO_4^{2-} concentration was caused by the CS3/RY (72 %) during days 60–100. Although the RY treatments caused an average decrease of 500 mg L^{-1} for the first 10-day period, no further decreases occurred during the incubation. After day 10, sulfur levels did not fall below 2100 mg L^{-1} in any of the RY treatments (Fig. 3).

3.4 Iron

Figure 4 shows the changes in Fe concentration. Additionally, the graph for Fe (and sulphate) removal shows two distinct periods of decrease—the first for days 0–50, the second (more precipitous) for days 0–100 (Fig. 4). Thus, the lowering of Fe concentration in AD solution occurred in the (1) pre-SRB (0–50 days) phase and (2) the SRB phase (60–100 days). The descending order of treatment effectiveness was CS3/RY > CS2/RY > CS1/RY > CS3 > CS2 > CS1 > RY3 > RY2 > RY1.

In all treatments, Fe concentration decreased with time. In the combined treatments (CS/RY samples) this consistent decrease was most significant at the highest rate (CS3/RY) where Fe concentrations dropped spectacularly from 2100 mg L^{-1} at day 1 to 300 mg L^{-1} at day 10, with an average decrease in concentration of 180 mg l-1 per day for the first 10 days (Fig. 4). The average rate of decrease for the pre SRB stage of the incubation was 6 mg L^{-1} per day, at an average percentage per day decrease of 9.5 % (first 10 days) and 2 % (ensuing 40 days). At the highest CS rate, the average decrease was 137 mg l per day for the first 10 days, and 22 mg L^{-1} for the ensuing 40 days (average), representing a daily percentage decrease of 6.5 and 3.4 % respectively. For the CS/RY and CS treatment regimes, the decreases in Fe^{2+} concentration were, in general, greater the higher the rate of treatment. The least effective treatment regime was the RY because, although its average Fe^{2+} decrease was 90 mg day^{-1} for the first 10 days, this was followed by 6 mg day^{-1} for the ensuing 40 days. Before and after the SRB stage, the average daily percentage decreases were 4.2 and 0.6 % respectively (as already stated, the RY did not activate SRB). For identical masses of ameliorant (for example RY2 = 25 g = CS4), the RY treatment was always less efficient in this experiment compared with other treatment regimes, especially the combined sludge-ryegrass treatments. Thus, as the incubation proceeded, the RY decreased Fe^{2+} concentration more slowly than either the CS or CS/RY treatments did. It can be also seen that at any time during incubation, much more Fe than sulphate was removed. For example, for the most effective treatment (CS3/RY), the percentage of Fe removed at days 20 and 100 were 95 and 99 %, compared with 48 and 36 % of sulphate removed by the CS3/RY. Thus, whereas at day 100, CS3/RY removed all detectable Fe, its sulphate concentration was 383 mg L^{-1}. Therefore, it can be seen that sulphate removal lagged behind that of Fe during incubation. However, the second large decrease in Fe and sulphate is seen to be greater than the first (Figs. 3 and 4). This fall also coincides with the onset of SRB activities (days 60–80) and the most substantial rise in pH (up to 7 units, in the case of CS3/RY). Also, the largest decreases of both Fe, and sulphate occurred concurrently with the activation of the SRB in the CS3/RY samples. SRB seems further implicated because Zehnder and Stumm (1988) found that when both sulphate and Fe are available to SRB under reduction by microbial action, facilitated by a nutrient source, Fe will be used up first. The incubation was terminated at day 120 because fluctuations in heavy metal concentration, which began at day 90, continued (however, the last pH readings were taken at day 140). Some resolubilizing of metals as a result of the depleting SRB carbon nutrient source may have caused the above-mentioned fluctuations.

3.5 Aluminium

Figure 5 shows the following changes in Al concentration caused by the treatments. Aluminium inherent in the AD control remained at 44 mg L^{-1} throughout 100 days of incubation (Fig. 5). However, the CS and CS/RY treatments each increased aluminium concentrations by 150 % within the first 10 days, but, as the incubation proceeded, differences between the CS and CS/RY in Al concentration increased greatly. Thus, CS increased Al concentration from 44 mg L^{-1} at control levels to an average of 1130 mg L^{-1} on day 100, (a 275 % increase over the controls) at the end of incubation. Also, Al in the AD rose at average rates of 117.4 % for the CS but only 1 % for the CS/RY treatment suite. Further, at the lower CS rates (CS1 and CS2), the level of Al concentration in the samples remained at least 68 mg L^{-1} throughout the incubation, but mostly much higher (Fig. 5b), and climbed to 156 mg L^{-1} on day 120 (Table 11). This represented more than a threefold average increase (339 %) over the controls. On the other hand, at the highest CS rate (CS3), the Al concentration was not as high, and at day 100 this had dropped to just 46 % above the controls (that is 67 mg L^{-1}). Thus, increased quantities of sludge led to decreased Al concentration in the AD. For the RY treatment

Table 11 Concentration (mg L^{-1}) of ions in Nairne Pyrite Mine AMD after 120 days incubation

Treatment	Al	Co	Cu	Fe	Mn	Ni	S	Zn
CS3/RY	0.1	0	<0.1	0.6	<0.1	0.1	649	<0.1
CS2/RY	<0.1	0.1	0.1	95	84	0.3	1380	<0.1
CS1/RY	32	0.4	0.1	86	110	0.6	1423	1.2
CS3	67	0.6	0.5	3.5	123	1	1446	3.9
CS2	156	0.4	1.2	126	130	0.6	1650	6.0
CS1	155	0.4	1.0	321	128	0.6	1760	5.5
RY3	48	0.1	0.2	678	117	0.3	2164	6.4
RY2	47	0.1	0.2	567	116	0.3	2109	6.9
RY1	45	0.1	<0.1	730	112	0.3	2143	7.0
AD control	58	0.2	0.4	2160	127	0.4	3055	6.1

AD Acid drainage; *CS* sludge; *RY* ryegrass

suite, a decreasing trend that averaged 8 % for the first 10-day period of incubation was recorded, with no further increases or decreases after day 10. Because Al is generally toxic to most plants and, therefore, not normally taken up by most root systems in neutral to mildly acidic soils, the lack of any Al coming directly from decomposing ryegrass previously grown in a neutral soil was not unexpected in this study. Of all the treatments, only the CS3/RY caused a large decrease in Al following initial increases. Thus, at day 40, the total Al concentration in the CS3/RY treated AD was only 12 mg L^{-1}, representing an 11-fold decrease for the concentration of Al compared with the control. However, for the other two treatments of the CS/RY suite (CS1/RY and CS2/RY), the increases recorded at day 10 were followed by just a slight decreasing trend over the ensuing 30 days (Fig. 5). Thus, for three consecutive 10-day periods after the 10th day, the average decreases in Al concentration were 0.1, 4 and 4 %; and 8, 0 and 7.5 %, for CS1/RY and CS2/RY respectively, yet this still represented a net increase in Al over control concentrations. The effectiveness of the treatments up to that point were, therefore, in the order CS3/RY > RY1 > RY2 > RY3 > CS2/RY > CS1/RY > CS3 > CS2 > CS1. As in the case of SO_4^{2-} and Fe^{2+} for both the CS and CS/RY treatment regimes, it was the highest treatment rates that were most effective. Just before the first appearance of the signs of SRB activity (black coloration, H_2S odour) in the CS3/RY samples, dissolved Al had dropped suddenly from 56 mg L^{-1} at day 30 to 14 mg L^{-1} at day 40. By day 80, the CS1/RY, and CS2/RY had also decreased Al concentration from 70 L^{-1} and 57 mg L^{-1} at day 60, to 37 and 14 mg L^{-1} respectively. The CS/RY was, therefore, the only treatment regime in which net decreases in Al concentration were eventually detected. At day 100, concentration of Al for the CS3/RY was almost negligible (<0.1 mg L^{-1}), being more than a hundred times less than that of the controls and within the Australia and NZ Environment and Conservation Council (ANZECC) water quality guideline concentration (Table 1). The RY treatment regime exerted no significant effect on the Al concentration of the AD in this study indicating its ineffectiveness, when un-combined with sludge, to remove Al from AD solution. Therefore, decomposable ryegrass was effective in reducing Al concentration only when it was combined with sludge.

3.6 Phosphate

Phosphate levels are important in this study because, as mentioned earlier, phosphates in sludge can complex and form insoluble precipitates with Fe and Al. Figure 6 illustrates the following treatment effects on phosphate concentration. Throughout the incubation the P concentration in the controls were <1 mg L^{-1}. The results of treatments indicate that the combined treatment suite (CS/RY) was the most efficient at returning P back to levels seen in the controls after an initial increase. Most of the P in this study came from decomposing ryegrass and, after 10 days, RY3, RY2 and RY1 increased P concentrations up to 19, 22 and

25 mg L^{-1} respectively. For all other treatments, P concentration was only 2–3 mg L^{-1} after 10 days of incubation. Yet the CS/RY treatment suite contained the same mass of ryegrass as each RY treatment. At day 20, the amount of P in the three CS/RY treatments was reduced to <0.1 mg L^{-1}, i.e. to the concentrations of the controls, and this level was maintained for the remainder of the incubation. For the CS3 and CS2 samples (the two higher rates), P also decreased (but not as quickly as that of the CS/RY treatments) and varied inversely with the amount of sewage sludge contained in the samples (Fig. 6b). At the highest CS rates (CS3), P returned returned to the control levels at day 30, whereas for the second highest rate (CS2), that return time was not until day 40. For the lowest rate (CS1), P levels did not return to control levels, but throughout the remainder of the incubation remained at the day-10 levels (2 mg L^{-1}), that is, more than twice that of the control, and 100 times that of the maximum prescribed by the ANZECC (Table 1). It is of interest that, for most of the incubation, for the CS1 treatment, while P levels remained static, S levels also remained unmoved. Even after 40 days incubation, P levels attained by RY1, RY2 and RY3 treatments were 75, 79 and 65 % of their day 10 concentration respectively. This was still 15-fold higher than that of any other treatment at this stage and, at 100 days, the average RY-P concentration was at least 7 times higher than that for other treatments (Fig. 6).

3.7 *Heavy Metals*

CS at the rates used in this study introduced significant amounts of each of the elements Zn, Ni, Cu and Mn to the AD (Figs. 7, 8, 9, 10), above the defined environmentally safe limits (Table 1). Toxic metals that may have been in excessive amounts (Cu, Mn, Ni, Pb, Zn) and sulphate were removed to some degree from the AD, in decreasing order of treatment regimes as follows: CS/RY > CS1RY. In contrast to the CS and CS/RY regimes, RY had little effect one way or another on toxic heavy metals in the AD. For example, dissolved Cu native to the AD had been removed by CS3/RY, whereas it was increased in the CS. Even though Cu concentration was increased throughout the incubation by the CS treatment regime by up to 15 times that of the amounts in the AD controls, the CS3/RY treatment had, by day 60, cut Cu concentration to lower concentrations than even those of the control samples (that is, below pre-treatment levels), whereas RY had no effect on Cu concentration (Fig. 7). A similar pattern prevailed for Co, Mn, Ni and Zn. For Co, in particular, the RY regime had no net effect on Co concentration. All other treatments initially caused a threefold increase for Co concentration in the AD, but by day 70, only the CS3/RY had reduced the 3-fold increase to 0.1 mg L^{-1}, (that is, back to the control levels) (Fig. 7).

Further decreases by CS3/RY (to <0.1 mg L^{-1}) were achieved and maintained from day 80 and throughout incubation, while a higher concentration of 0.1 mg L^{-1} still existed in the controls. Although the CS2/RY was the only other treatment that eventually decreased initial increases in Co concentration in the AD (a four-fold decrease to 0.1 mg L^{-1}), this did not occur until day 120 (Fig. 7b; Table 11). Although Mn was not substantially increased by any of the treatments (Fig. 9), the only decrease that occurred between days 0 and 50 was achieved by CS3/RY (a 10 % decrease at day 20). For the CS3/RY, a trend of lowering values continued between days 50 and 70 but this was not significant. However, 10 days later at day 80, the CS3/RY caused the only major Mn decrease of the experiment. This was a sharp decrease in concentration from 94 mg L^{-1} at day 70, to <1 mg L^{-1} at day 80 (a 194-fold decrease).

The CS3/RY maintained this decrease in Mn concentration for the remaining period of incubation. For nickel, all treatment effects differed little from each other between days 0 and 60 (Fig. 9): they all caused just slight increases in Ni concentration in the AD (average of 25–30 %). The only large decrease occurred at day 60, when the CS3/RY caused a 3-fold decrease in Ni concentration. This decrease was not only maintained throughout the rest of incubation, but further decreased to <1 mg L^{-1} at day 100. Zn concentration increased up to day 30 for all treatments. Thus, average increases of 36, 17.5 and 8 % for the CS/RY, CS and RY regimes respectively, were observed in the first 30-day period (Fig. 10). At day 40, the Zn concentration in the CS3/RY samples had been reduced to control levels (5.6 mg L^{-1}), and at day 50 it had dropped a further 7 % to 5.1 mg L^{-1}. Further, at day 60 Zn had dropped to 0.1 mg L^{-1} for CS3/RY, a 50-fold decrease within 10 days. From day 70 onwards, the concentration of Zn for CS3/RY remained below 0.1 mg L^{-1}, until day 100 when it again increased to 0.8 mg L^{-1}. Only one other treatment (CS2/RY) brought Zn down below the control levels at any time during this study, this being a 7-fold decrease for CS2/RY that occurred on day 80, from 7.5 L^{-1} to 1.0 mg L^{-1}. Further, similar to CS3/RY on day 70, the CS2/RY decreased Zn to <0.1 mg L^{-1}on day 120, showing again the dominance of CS/RY to control toxic substances in the AD (Table 11). Throughout the incubation, the low levels of Cd, Cr, Mo and Se in the original AD controls (<0.1 mg L^{-1}) were unaffected by any of the treatments, that is, they remained below the detectable amounts required in solution by the ICPAES measuring system. Safe levels of soluble Pb in water should not exceed 0.005 mg L^{-1} (ANZECC, Table 1). Some treatments (CS1, CS1/RY and CS2/RY) increased high levels of Pb in the AD from 0.05 to 10.15 mg L^{-1} (Fig. 12). It is therefore clear that not only were the Pb levels in the sludge used too high, but that some treatments were ineffective.

It has been shown that decreases in toxic metals were most effectively accomplished by the CS3/RY and CS2/RY, in that order. However, the most significant decreases occurred in each case after day 50, that is, coincident with the first appearance in the samples of the signs of SRB activity. The CS treatment regime failed to remove the majority of toxic metals from the AD. The organic matter component inherent in the sludge-only thus failed to complex the Cu and other toxic metals inherent in the AD, and those introduced by the sewage sludge. Although RY caused no increases in the toxic heavy metals (sulfur and iron not included), neither did it cause decreases. For toxic heavy metals in the AD, the RY treatment regime was, thus, comparatively inert in its effect.

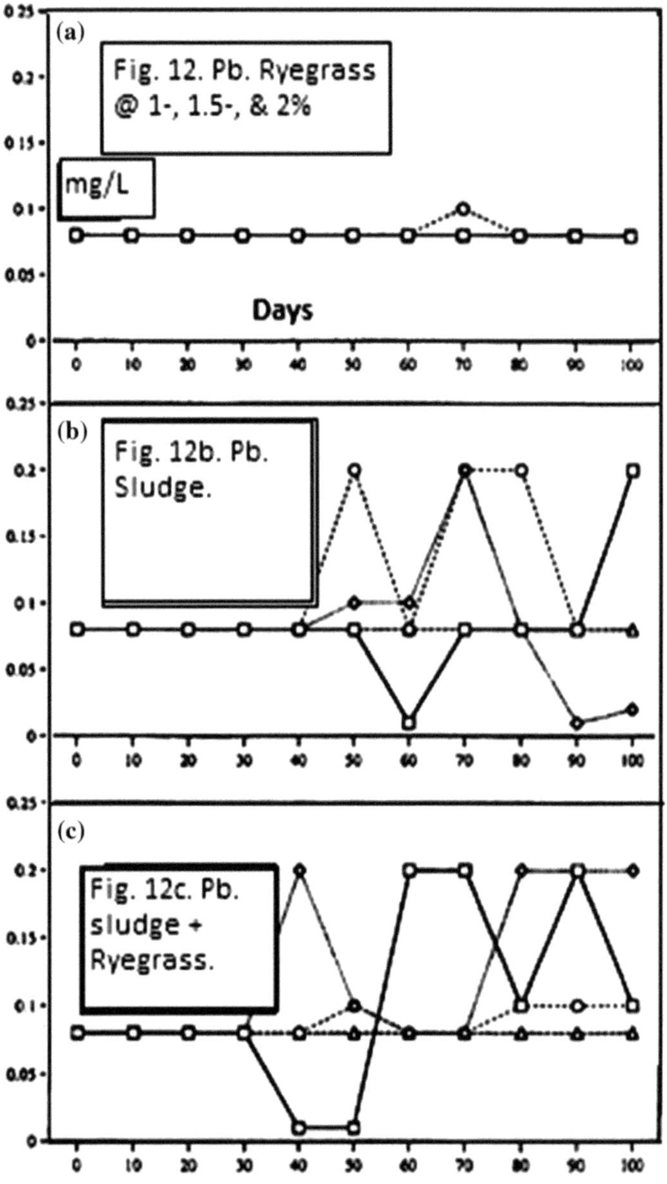

Fig. 12 Concentration of lead in the Nairne Pyrite Mine AMD during incubation with organic materials. *RY1* Ryegrass (1 %); *RY2* (1.5 %); *RY3* (2 %); *CS1* sludge (8 %); *CS2* (16 %); *CS3* (24 %); *CS1/RY* sludge (8 %) ryegrass (1.5 %); *CS2/RY* (16 %)/(1.5 %); *CS3/RY* (24 %)/(1.5 %)

4 Discussion

4.1 Aluminium, Iron, and Pre-SRB Sulphate

The physical conditions that resulted in successful partial removal of Al and Fe from the AD were similar to those prevailing in some low pH submerged soils. This is akin to an in situ method, noted by Sinha et al. (2009) as an environmentally sound and inexpensive alternative to pump and treat (for contaminated groundwater) or excavate and treat (for contaminated soils). In the field, microbial growth is induced in sub-surface zones by injecting substrates. The migrating metals are intercepted and immobilized by precipitation with biologically produced H_2S. Thus, the removal of cations from the AD would have been expected to depend on their interaction with the anaerobically biodegraded products of the ryegrass. The most likely cause of lowered Fe, Al and SO_4^{2-} concentrations in the AD seems to be related to large amounts of kaolinite added to the sludge. The positive sites on clay particle surfaces, such as those of kaolinite, which increase at low pH, could have attracted and complexed the negatively charged organic anions from the decomposing ryegrass. Gillman (1985) found that where oxide surfaces dominate, a net positive charge develops, and Gillman (1985) reported that sulphate is adsorbed by the variable charge surfaces (that is, the + parts) of oxides. As indicated by X-ray diffraction, the clay mineralogy of the sludge used in this study was dominated by kaolinite. Gillman (1974) also found that the amount of adsorbed sulphate increased with an increase of positive sites present on kaolinite, and Barrow et al. (1969) showed that sulphate sorption was positively correlated with soils derived from a range of parent materials that had undergone intense weathering (which leaves behind an increasing proportion of kaolinite). Such weathered residual soils are therefore usually rich in 1:1 alumino-silicate minerals. Bear (1968) states that because there is little isomorphous substitution of ions in kaolinite, the layers are electrically neutral, and Schofield and Samson (1954) found that, owing to its acceptance of hydrogen ions by its edge hydroxyls, kaolinite has a positive charge when the pH of the surrounding solution is low. Further, Quirk (1960) found that this positive charge may persist up to high pH's. On the assumption that this mechanism was a major cause of decrease in sulphate concentration, at least in the early stages of this study, there should be little difference expected in sulphate concentration between the sludge-only and sludge–ryegrass treatments prior to the activation of the SRB. Thus, it has been shown that, for the highest rate of CS/RY and CS, the decreases in sulphate concentration in the AD were very similar throughout the first 70 days of incubation, but were dramatically different after day 70 (that is, after the SRB became active in CS/RY) (Fig. 3b, c).

The rise in Al concentration in the AD water for the sludge-based treatments could have been caused by breakdown of the clay minerals by acid. This could have been the reason for the largest increases in aluminium concentration that occurred in the early stages of incubation for all treatments containing sludge. With a pH of 2.3, the AD solubilized substantial amounts of Al from the added sludge with and without the aid of organic acids from the decomposing ryegrass. However, because of all treatments, the CS3/RY's Al concentration was lowest not only after, but before SRB activation, the interaction and complexation of Al with decomposing organic materials seems a major cause of Al removal from the AMD in this study. Yeoh and Oades (1981), and Hue et al. (1986) found that acid attack released quantities of aluminium from kaolinite by dissolution of the clay. Hue et al. (1986) also found that many organic ions from organic acids have the ability to complex monomeric forms of Al released in acidic waters from soil layers, thereby reducing Al toxicity. Thus, in the present study, the AD attacked the clay fraction in the sludge, releasing components (including Al) that not only increased the pH of the AMD, but then probably reacted (and complexed) with those organic materials released from the decomposing ryegrass. Further, the premise that negatively charged organic matter (in this case anions from the ryegrass) is held at positive sites on mineral surfaces could largely explain the fact that the CS3/RY caused the greatest sorption of Fe^{2+}, Al^{3+} and other cations from the AMD, by having provided the adsorption sites for those cations. Ahlrichs (1962) showed that the adsorption of anionic polymers is greater in the presence of salts or cations, since these decrease the zeta potential of both the organic polymer and the clay, thereby indicating the lowering of repelling forces between sub-micron particles. Thus, in this experiment, low zeta potentials were similarly expected for the CS/RY treatments (at least for the CS/3RY). Therefore, the high zeta potentials found for the sub-micron particles in all CS/RY (especially the CS/2RY and CS3/RY samples) compared with those of the RY and CS regimes (Table 12) initially seemed contradictory.

However, this significant increase in negative charge for sub-micron particles, which occurred only for these combined ryegrass-sludge treatments, suggests an increase in the size, and hence the high charge, on organic polymers. Thus, if the older, larger, more resistant organic molecules inherent in the sludge had

Table 12 Zeta potential and mobility of sub-micron particles in Nairne Pyrite Mine AMD after 120 days incubation

Treatment	Zeta potential (mV)	Particle mobility (mm s^{-1} V cm^{-1})
CS3/RY	−12.3	−0.924
CS3/RY	−14.1	−1.062
CS1/RY	−1.0	−0.077
CS3	−0.8	−0.062
CS2	No data	No data
CS1	−3.9	−00.287
RY3	+3.8	0.279
RY2	+2.2	+0.166
RY1	+2.0	+0.151
AMD	+1.2	+1.2

CS Sludge; *RY* ryegrass

combined with the smaller, more recently formed organic molecules from the decaying ryegrass, an increase in negative charges would be expected. The mechanism for such a process would have to have forced these negatively charged particles into close proximity. Therefore, it is not surprising that, as the incubation progressed, the free, positively charged Al inherent in the AD (plus that released by the clay) was significantly removed from solution only by the sludge–ryegrass treatment regime, particularly the CS/2RY and CS3/RY (Fig. 5). The Al cations in solution were, therefore, attracted to all negative particles. The consequent weakening of the electrostatic repulsion between the negatively charged polymers and the clay particles would have also brought them closer together, thereby increasing physical interaction and complexing of the cations.

This probably explains the fact that, despite Al having been initially increased in the AMD in similar quantities from CS3/RY as it had been for the other treatments containing sludge, it was the CS3/RY that achieved the largest and quickest lowering of the cation concentration in the AMD. On the other hand, the low, positive zeta potentials for all RY treatments (Table 12) indicate a greater concentration of still un-complexed cations, and a larger amount of smaller molecules of negatively charged organic acids, which would be expected to have come from the readily decomposable plant material. Such charges from organic acids would not be as strong as those from larger anions, and thus not strong enough to completely neutralize un-adsorbed soluble cations in the AMD. The resultant negative charges would have been relatively weak whether or not complexation occurred there. In the present study, the amount of phosphate in the AMD control started and ended at <1 mg L^{-1}, and that the highest phosphate levels (a 20-fold increase) came from the fresh ryegrass-only treatment only. Whereas this 20-fold phosphate increase was largely maintained for at least 40 days for the AMD samples treated with the ryegrass-only, the phosphate levels in the samples treated with sludge when combined with fresh ryegrass showed only a small fraction of these increased amounts of phosphate throughout incubation. Phosphates had also been introduced into the AMD by the sewage sludge itself, yet only a small increase of P in solution was seen in those CS samples or the CS/RY treatment regime (Fig. 6).

The relatively low phosphate concentrations recorded when ryegrass was combined with sludge (Fig. 6) indicates either rapid complexation of ryegrass produced P by the sludge component in that treatment regime or sludge induced inhibition of P release from the ryegrass. However, even if the phosphates had been complexed in any of the treatments, phosphate anions introduced by ryegrass components are 1–2 orders of magnitude less than decreases in other cations such as Fe^{2+}. Further, even though the RY treatment regime removed Fe very effectively in the first 10 days of incubation, thereby attaining a 1000 mg L^{-1} (50 %) decrease in concentration at day 10, there were no further decreases between days 10 and 20. There remained a high phosphate concentration in the RY samples between days 10 and 20; the RY phosphate levels at day 20 were not significantly different to what they were on day 10 (Fig. 6). Thus, in the case of the RY treatments, although the phosphate was available in the AD, the Fe was not being complexed. Therefore, it would seem that the impact of phosphate on Fe concentration in this study was not a major one. Because, for example, in the present study the CS3/RY treatment contained just as much ryegrass as the RY2, it would have been expected that a comparable increase in phosphate concentration would have resulted from the CS3/RY treatment, but this did not occur (Fig. 6). Compared with the relatively large increase in the phosphate concentration caused by the RY after just 10 days of incubation, the concentration of phosphate ions detected in each treatment of the CS/RY regime after the same period of time was minuscule (Fig. 6). On the assumption that, after 10 days, approximately equal amounts of phosphate should have been released into the AMD by the RY2 and the CS3/RY, the only explanation for its immediate disappearance from the CS3/RY3 samples seems to be possibly its rapid complexation with Al before it was measured at day 10. This seems likely because there is reason to conclude that some 'missing' Al was immediately released from the CS3/RY and thereafter complexed before it could have been measured at day 10 (Figs. 5 and 6). Thus, it is seen that, at 10 days, CS3 (which had an identical sludge rate to that of CS3/RY, but no ryegrass) had by then caused a 20 % higher Al concentration in the AD than had the CS3/RY. Because no extra Al was contributed by the ryegrass component of any treatment (Fig. 5a), the amount of Al contributed by the CS3/RY should have been the same as that of the CS3 within the first 10 days of incubation; however, this was not the case. Therefore, the 'missing' 20 % of Al from the CS3/RY may be explained by Al combining with phosphate from the CS3/RY itself before the 10th day (and hence going undetected in the ICPAES measurement). Yeoh and Oades (1981), after precipitating the products of clay acidification by phosphoric acid, showed that an aluminium phosphate compound resulted that contained 2 mol of aluminium to each mole of phosphorus. Therefore, although the amount of phosphate supplied by the ryegrass component was relatively small, it may have combined in a 1:2 mol ratio with Al. The unaccounted for 20 % Al loss from the CS3/RY in the first 10 days can be explained by a much smaller amount of phosphate than Al being needed to remove the 20 % Al as aluminium phosphate.

In addition to the electrostatic attraction of SO_4^{2-} to positive edge charges on kaolinite, aerobic conditions prevailed in the decomposing ryegrass floating at the top of the cylinders, thereby depleting oxygen levels and causing oxygen starvation at the lower levels. Charles (1998) found that similar anoxic conditions caused the removal of S at greater water depths in a narrow lake because sulfur was bound directly to Fe under the anoxic conditions. With similar conditions prevailing in the present study, it is likely that such conditions removed sulphate from the AMD.

4.2 Acidity Levels Before SRB Activation

As the sewage sludge concentration increased, the bioremediation efficiency also increased, reflecting the higher organic carbon concentrations. Similarly,

McCullough and Lund (2011) found that the combination of high TC and TOC indicated that the sludge would potentially be able to combat the acidic environment and initiate and sustain SRB activity until remediation was achieved. In this study, sewage sludge contributed alkaline materials that directly neutralized the AMD in proportion to the quantity added and therefore played a primary role in stimulating SRB bioremediation. Kumar et al. (2013) observed that the lowest concentration of sewage sludge ($30\ g\ L^{-1}$) tested proved to be inadequate for effective SRB bioremediation. However, the measurable beneficial effects on SRB bioremediation efficiency tapered off when sewage sludge was added at the highest concentrations ($60\ g\ L^{-1}$). Their sewage sludge treatment microcosms maintained low dissolved oxygen (DO) until the end of their experiment, except for the lowest concentration ($30\ g\ L^{-1}$), which showed slight DO increases towards the end of the experiment. This DO trend in sewage sludge microcosms was expected as the organic material's initial role is to reduce DO concentrations for SRB activity (Kumar et al. 2013).

Because the AMD (control) had the low pH of 2.3, a 'trigger' for an initial substantial decrease in its acidity was necessary to activate the dormant SRB in the sludge. Prior to the establishment of an active SRB population, pH varied in proportion to time, sludge and sludge + ryegrass. Theng (1980) stated that liming soils with variable charge leads to an increase in their negative charge. Thus, because the sludge has a high cation exchange capacity (CEC) (Table 3) caused partly by a high clay content (140 %) and 2.5 % of fine $CaCO_3$ particles, it is likely that for the CS and CS/RY treatment regimes, Ca or $CaCO_3$ was a main neutralizer of acidity from the AD. Likewise, Strosnider and Nain (2010) found that limestone-rich sediments passively mitigated municipal waste water. Hence, it can be seen (Fig. 2) that within the first 10-day period ryegrass-only treatments had significantly increased AMD pH (by concurrently releasing organic acids into the AD) because of the introduction of significant amounts of Ca (9 % more than the control concentration) into the AMD. Although the ryegrass-only treatments also released Ca^{2+} into the AD, after day 10 the below-control-levels pH values were caused by the production of organic acids and HC from the decomposition of ryegrass. Additionally, this decomposition would have increased CO_2 levels, which would have further increased AD pH in the samples.

As previously stated, it can be seen that in contrast to the RY treatment regime's pH increases (which stopped at day 10), pH increases for the CS (with the exception of that of CS1, possibly because it was the lowest rate) and CS/RY regimes retained their pH increases until at least day 50 (Fig. 2). According to Ulrich (1991), the pH decreases over time of an AD occur in a series of steps, each representing the dissolution of specific buffering species present at that pH. The mineral species they concluded to be responsible for pH plateaus are:

1. calcium-based carbonate pH 5.5–6.4
2. aluminium hydroxide pH 4.3–5
3. iron hydroxide and jarosite pH 3–3.7
4. alumino-silicates pH < 3.

Thus the large amount of alumino-silicates (40 %), and $CaCO_3$ (2.5 %) placed in the sludge would have been major factors that resisted and delayed AD acidity increases, and were responsible for preventing the pH from falling far below a value of 3 in those samples that had a sludge component, as Strosnider and Nain (2010) also observed. The release of Al cations from clay particles can cause an increase in the pH of the surrounding liquid medium. Further, the proportion of clays and lime (that remained) to AD water increased as sampling continued, thereby depleting the actual amount of AD in the containers. This lowered ratio of AD to solids was greatest in the CS/RY samples (and most greatly in the CS3/RY), which had the highest solid:liquid ratio initially and, therefore, the highest solid:liquid ratio throughout the incubation. The increasing solid: liquid ratio in those samples would have further increased the net neutralization potential of the sludge-ryegrass samples, thereby resisting acidity more efficiently.

On the other hand, the steady reduction in pH for the RY samples during its decomposition under anoxic conditions, showed that even the Ca^{2+} introduced by the RY were not sufficient to counteract the organic acids from the RY. This highlights the effectiveness of the buffers in the sludge, whose absence in the RY caused pH to decrease after day 10, for the remainder of incubation. The brief pH decreases recorded for the CS2/RY and CS3/RY treatments between days 50 and 70 coincided with the precipitation of iron oxy-hydroxides in the samples, seen as orange-brown staining on the sides of the containers and as a thin (<0.2 mm) scum floating on top of the liquid. According to Bear (1968), precipitation of iron oxyhydroxides leads to the release from water of H^+, hence lowering the pH, as represented by the reaction

$$Fe^{2+} + 2H_2O = Fe(OH)_2 + 2H^+ \tag{1}$$

Beginning in the latter stages of the pre-SRB activity phase, the reddish staining of iron oxides persisted on the sides of the CS/RY containers. These precipitates could have caused changes in pH as mentioned above.

Because the AD used in the present study was not filtered or altered after its removal from the mine-site, it is very likely that it contained substantial amounts of iron oxides in suspension (hence its reddish colour) at the start of the experiment. Brief, sharp pH decreases (Fig. 2), recorded when reddish-brown staining was first seen for the CS/RY regime (particularly CS3/RY treatment on days 50–60), suggest that these pH decreases may have been cut short by the SRB carbonate production at this time because CS3/RY samples also showed black iron sulfide trapped in the substrate at this time, and recorded the highest sulphate reduction activity.

4.3 Acidity Levels After SRB Activation

Low pH values in the AD were responsible for the long period before SRB became active. Lyew et al. (1994) and Elliott et al. (1998) found that when AD dropped

<pH 4, SRB needed an adaptation period of up to 20 days for the resumption of active metabolism and hence continued population growth. Therefore, considering the extreme acidity (pH 2.3) of the AD in this experiment, and the fact that the SRB had not been previously primed with a less acidic medium, the long period of acclimatization (60 days) that elapsed before signs of SRB were observed in this study had not been totally unexpected.

Secondly, SRB colonization and metabolic activities are most successful where there are physical surfaces for the attachment of colonies. Thus, signs of SRB activity were eventually observed only within samples that throughout incubation had the highest ratio of solid to liquid material (that is, the CS3/RY). This is, in effect, comparable to the observation of Lyew et al. (1994) who found that when AD ($pH < 4$) was added in smaller increments over time, SRB acclimatization time was decreased. This effect was indirectly achieved in the CS3/RY samples of this study because the higher solid/AD ratio was achieved much earlier compared with the other treatments. Only comparatively small amounts of acidic liquid were initially present in the CS3/RY samples, thereby encouraging the growth of SRB. Additionally, this caused, of all the treatments, the closest physical contiguity of solid particles and SRB of this study.

In samples exhibiting sulphate reduction through bacterial activity, the signs of microbial activity started at the bottom of each cylindrical container in each case, where free sulfide (S^{2-}) reacted with the ferrous iron to form the black amorphous FeS. Substantial pH increases occurred either concurrently or shortly after that. Norton et al. (1990) suggested that the pH of a solution in intra-aggregate pores may be different from that of the bulk solution. This seems possible because the liquid in smaller pores is in immediate contact with or close to a source of solid acid-consuming materials. Given the evidence that signs of SRB activity were always seen starting at the bottom zone of the containers in this study, where the denser inorganic buffering colloids settled, under the stagnant incubation conditions of this experiment the pH was likely to have been higher in those lower zones of the substrate/sediment. Because Herhily and Mills (1985) concluded that within bottom sediment interstices, a higher pH within the sediments (compared with that of the surrounding water) activated SRB, localized zones of higher pH in the sediment interstices of this experiment could therefore have prevailed to initiate SRB mitigation of the AD. Such a pH differential would not necessarily need to have been great (an increase of <1 pH unit) because Elliott et al. (1998), using an organic substrate in fine sand, achieved SRB activation at a pH of just 3.25. In addition, the increasing anoxicity with depth could have similarly helped to activate the dormant SRB at the bottom zone. Herhily and Mills (1985) point out that sulphate adsorption onto oxy-hydroxides, which then sink in the water, can be a mechanism for the replenishment for SRB colonies in lower anoxic sediments. In the present study, the reddish iron oxy-hydroxide layer formed a scum at the top of the cylinders. As mentioned earlier, all samples were shaken every 10th day of the incubation. In the CS3/RY samples, this shaking action broke up the oxy-hydroxide scum layer into small flocs, which being positively charged could have then acted as adsorption sites for sulphates, with the flocs being then carried downwards by gravity to supply the SRB with sulphates in the lower zones. It is likely that the positively charged $Fe(OH)_3$ colloids adsorbed the sulfides by mutual neutralization, which probably helped to cause further extension of the black precipitate. The smallest amounts of water left in the liquid phase in the latter stages of the incubation were those found in the CS3/RY cylinders. This would have increased the ratio of oxy-hydroxides to water. Consequently, (1) a shortened downward distance, and (2) a higher concentration of flocs travelling during and after the shaking would have further intensified the effects of mixing during the shaking periods. This mechanism could have helped to maintain and enhance the SRB colonies seen in the CS3/RY samples.

5 Conclusion

A clay-lime treated sludge was capable of supplying the pH conditions needed to activate SRB in an AD of pH 2.3. The ryegrass-only treatments caused decreases in pH to levels significantly below those of the controls. However, at no time during this study did any of the pH decreases involving sludge-only, (or in any combination with ryegrass) show a decrease to levels below that of the controls. Thus sludge-only as a component had a greater positive effect on pH than ryegrass-only did. The steady reduction in pH for the RY samples during its decomposition under anoxic conditions showed that even the Ca^{2+} ions introduced by the decomposing ryegrass were not sufficient to counteract the organic acids from the RY. This highlights the effectiveness of the buffers in the sludge, whose absence in the RY caused pH to decrease after day 10 until the end of incubation. Therefore, decomposable ryegrass was effective in reducing Al concentration only when it was combined with sludge.

However, when sludge-only was added to the AD without an additional nutrient source, SRB populations did not increase up to a level high enough to significantly increase pH, and further precipitate metals. Consequently, the CS treatment regime failed to remove most of the toxic metals from the AD. The organic matter component inherent in the sludge-only thus failed to complex the Cu and other toxic metals inherent in the AD, and those introduced by the sewage sludge. Without the SRB, further decrease in the level of pollutants was therefore curtailed. During incubation, a sharp recovery of pH values was recorded for the sludge-ryegrass treatments, but the recoveries for the other treatment regimes were either slower or non-existent. Therefore, it is concluded that SRB can survive and become active in minute pores within buffering sediments in water below pH 2.5, if supplied with an energy source.

Throughout this experiment it was seen that the major improvements in the quality of the AD occurred only after SRB mediated a combination of sludge and ryegrass added to the AD. The greatest SRB activity occurred where the highest quantities of sludge were combined with the ryegrass. Since the ryegrass component was the same mass for each CS/RY treatment, higher quantities of sludge within the treatment regime meant a lowering of the ryegrass: sludge ratio. Yet, the speed, magnitude and effectiveness of SRB activation occurred in increasing order of their sludge masses: CS1/RY < CS2/RY < CS3/RY. Therefore, although it was the ryegrass that sustained SRB activity, it was the particular ratio of treatment ingredients that caused the greatest ameliorative effect on the AD. Under the conditions of the experiment, the ideal mix of ryegrass and sludge for SRB driven AD mitigation was in the ratio of 1:17:67 (w/w) for ryegrass, sludge and AD respectively (i.e. for CS3/RY). Higher quantities of sludge initially would have contained more SRB spores, and more buffering material. Therefore, it is reasonable to conclude that the greater the sludge (buffer): AD ratio, the greater the effectiveness of the SRB in the presence of an appropriate nutrient source. However, where there was no ryegrass, there was no SRB activation. Thus, had SRB been activated, and that activity sustained in the sludge-only treatments, similar toxicity decreases would have been expected in those treatments.

The long adaptation period needed for SRB in this study may not be of great practical value under constant flow conditions. Further work is needed to reduce the long acclimatization periods for SRB in highly acidic AD, probably with the incorporation of more highly efficient buffering additives that could reduce the hydraulic retention time (HRT) of a constant-flow reactor. In addition, after-treatment solid waste disposal problems would be lessened if the high solid treatment: AD ratio of 1:3.7 (w/w) of this experiment could be reduced.

6 Geobiotechnical Applications

6.1 Mosquitoes: Impacts of Acidity

When acidity builds up to high levels in water, it poisons plants in and around affected creeks and ponds. It can also kill fish and other aquatic creatures if they are unable to escape (Queensland Government 2015). Lower levels of acidity will simply make aquatic plants and animals weaker and more vulnerable to disease, and make it harder for young organisms to reach adulthood. Over time, sensitive species may be driven out and replaced by stronger, acid-tolerant invaders.

One example is mosquitoes, which can tolerate acidic water much more easily than the insects that prey on them. Acidified wetlands can therefore be a source of mosquito plagues. In addition, the current spread of the Chikungunya, Dengue, and Zika viruses increase risks of disease. Moreover, acidic water is unhealthy for drinking and can cause skin irritation.

Sulfuric acid can also attack concrete and steel, slowly destroying pipes, roads, bridges, and building foundations. In areas where acid sulphate soils are not treated properly before construction, (expensive) repairs may be required, or infrastructure may need to be replaced well before the end of its intended lifespan (Queensland Government 2015).

6.2 Dangers to Aquatic Life

If acid-mine drainage reaches the surface, very detrimental effects on aquatic organisms are likely. Because rock waste material is usually very coarse, the dump can exhibit high permeability (to water and gas) and low capacity for water retention by capillarity. As a result, when the reactive minerals are exposed to natural conditions, ideal conditions can be created for generating acid mine drainage. Once the critical reactions have been initiated, it has proven to be a difficult process to stop. The amount of AMD generated, depending on the dump conditions, can be significant. The process can last for centuries. One corrective is to improve the aggregation of these dumps. Harris (2009) showed that this can be achieved by (1) applying bio-solids preceded by shattering of the rocks (to fine powder) followed by (2) incubation under air-tight conditions at ambient temperatures (at a minimum) for several weeks to increase water retention of the remolded soil aggregates.

6.3 Guyana, Jamaica—Gold Mine Tailings

The most commonly used process for gold extraction is gold cyanidation, which extracts gold from low-grade ore by converting the gold to a water soluble coordination complex (Rubo et al. 2006). The heap leach and milling processes, which involve dewatering of gold-bearing ores, spraying of dilute cyanide solutions on extremely large heaps of ores containing low concentrations of gold, or the milling of ores with the use of cyanide and subsequent recovery of the gold–cyanide complex, have created a number of serious environmental problems affecting wildlife and water management (Eisler and Wiemeyer 2004). Sodium cyanide (NaCN—CAS No 143-33-9) is a highly soluble, white deliquescent crystalline powder forming a liquid solution when exposed to air. In Guyana and Jamaica, sodium cyanide is mainly used in the mining industry to recover gold from ore.

Bacterial mitigation can be applied in alkaline conditions which occur in gold mine tailings in the Guianas. Cyanide-resistant strains of microorganisms are now used routinely to degrade cyanide (Mosher and Figueroa 1996). Biological degradation of cyanide in which CN^- is converted to CO_2, NH_3, and OH^- by bacteria, when appropriate, is considered the most cost-effective method in cyanide detoxification and has been used in cyanide detoxification of heap leaches containing more than 1.2 million t (Mosher and Figueroa 1996).

Advantages of Anaerobic Treatment (after Sinha et al. 2009)

Anaerobic treatment has several attributes that are considered favorable in biological treatment

1. Less energy is required as mechanical aeration is avoided. On the contrary, a potential energy source methane (CH_4) is produced;
2. Significantly less land is required with smaller reactor volumes;

3. Extremely high destruction rates of wastes molecules can be achieved;
4. Solids (biological sludge) generation from the growth of biological cells (biomass production) is about 20 times less than the aerobic process, and therefore cost of disposal of solids are reduced;
5. Certain toxic organic wastes can only be destroyed by anaerobic digestion;
6. Many industrial waste-water lacks sufficient growth nutrients to support aerobic bacteria. Fewer nutrients are required to support the anaerobes;
7. There is complete elimination of 'off-gas' air pollution.

Disadvantages

1. There is a longer start-up time.
2. While it is in days for aerobic process, it may be in months for anaerobic process.
3. It may be more susceptible to upsets due to toxic substances.
4. It has potential for production of odors and corrosive gases.
5. It is not possible to remove biological nitrogen (N) and phosphorus (P).
6. May require alkalinity addition.

Acknowledgements The authors thank the following for their very helpful suggestions during the writing of this paper: Professor Martin Williams, Dr. P. Rengasamy and Dr. Vic Gostin of the University of Adelaide; Dr. Graham Taylor of the Commonwealth Scientific and Industrial Research Organisation (CSIRO), Glen Osmond, South Australia.

References

Ahlrichs JL (1962) Bonding of organic polymers to clay minerals. Diss Abstr 22:2121

Ahmad F, Tan KH (1985) Effect of Lime and organic matter on soybean seedlings grown in aluminium toxic soil. Soil Sci Soc Am J 50:656–661

Alexander M (1977a) Soil microbiology, 2nd edn. Wiley, Chichester, pp 362–370

Alexander M (1977b) Symbiotic nitrogen fixation. Introduction to soil microbiology. Wiley, New York, pp 305–330

Amato M (1983) Decomposition of plant material in Australian soils. Aust J Soil Res 21:563–570

Baldock JA, Aoyama M, Oades JM, Gran CD (1994) Structural amelioration of a South Australian red-brown earth, using calcium and organic amendments. Aust J Soil Res 32:571–594

Barrow NJ, Spencer K, McArthur WM (1969) Effects of rainfall and parent material on the ability of soils to adsorb sulphate. Soil Sci 108:120–126

Bear F (1968) Chemistry of the soil. In: Bear F (ed) American chemistry society monograph. Oxford University Press, New York

Bechard G, Rajan S, Gould WD (1993) Characterization of a microbiological process for the treatment of acidic drainage. In: Tarma AE, Apel ML, Brierley CL (eds) Biohydrometalurgical technologies. The Minerals, Metals & Materials Society, CANMET, Energy, Mines and Resources Canada Ottawa, Ontario, pp 277–286

Bechard G, Yamazaki H, Gould W, Bedard P (1994) Use of cellulosic substrates for the microbial treatment of acid mine drainage. J Environ Qual 23:111–116

Bell LC, Ward SC, Kabay ED, Jones CJ (1989) Mine tailings reclamation in Australia—an overview. In: Walker DG, Power CB, Pole MW (1989) (eds) In: Proceedings of the conference Reclamation, a global perspective, Calgary, Canada, vol 2. 27–31 August 1989, Rep no RRTAC 89-2, Alberta Land conservation and Reclamation Council, Edmonton, pp 769–778

Ben-Yaakow S (1973) pH buffering of pore water of recent anoxic marine sediments. Limnol Oceanogr 18:86–94

Berner RA, Scott MR, Thomlinson C (1970) Carbonate alkalinity in the pore waters of anoxic marine sediments. Limnol Oceanogr 15:544–549

Boyle M (1990) Biodegradation of land applied sludge. J Environ Qual 19:640–644

Bruschi M, Goulhen F (2006) New bioremediation technologies to remove heavy metals and radionuclides using Fe (III)-sulphate- and Sulfur reducing bacteria; In: Singh SN, Tripathi RD (eds) Environmental bioremediation technologies, Springer Publication, NY, pp 35–55

Chang IS, Shin PK, Kim BH (2000) Biological treatment of acid mine drainage under sulphate-reducing conditions with solid waste materials as substrate. Water Res 34:1269–1277

Charles D (1998) Wasteworld. New Sci 157(2119):32–35

Connell WE, Patrick WH Jr (1968) sulphate reduction in soil: effect of redox potential and pH. Science 159:86–87

Eisler R, Wiemeyer SN (2004) Cyanide hazards to plants and animals from gold mining and related water issues. Rev Environ Contam Toxicol 183:21–54

Elliott P, Ragusa S, Catcheside D (1998) Growth of sulphate reducing bacteria under acidic conditions in an upflow anaerobic bioreactor as a treatment system for acid mine drainage. Wat Res 00:1–7

Figueroa L, Seyler J, Wildeman T (2004) Characterization of organic substrates used for anaerobic bioremediation of mining impacted waters. In: Proceedings of the international mine water association conference, Newcastle, pp 43–52

Gibert O, de Pablo J, Cortina JL, Ayora C (2002) Treatment of acid mine drainage by sulphate-reducing bacteria using permeable reactive barriers: a review from laboratory to full-scale experiments. Rev Environ Sci Biotechnol 1:327–333

Gillman GP (1974) The influence of net charge on water dispersible clay and sorbed sulphate. Aust J Soil Res 12:173–176

Gillman GP (1985) Influence of organic matter and phosphate content on the point of zero charge of variable charge components in oxidic soils. Aust J Soil Res 23:643–646

Gonen N, Kabasakal OS, Ozdil G (2004) Recovery of cyanide in gold leach waste solution by volatilization and absorption. J Hazard Mater 113(1–3):231–236

Greenland DJ (1965) Interaction between clays and organic compounds in soils. Part II. Adsorption of soil organic compounds and its effect on soil properties. Soil Fert 28:521–532

Grindley P (1998) Chief mines engineer. Brukung a Mines, South Australia

Gyure RA, Konopka A, Brooks A, Doemel W (1990) Microbial sulphate reduction in acidic (pH 3) strip-mine lakes. FEMS Microbiol Ecol 73:193–202

Hard B, Friedrich CS, Babel W (1997) Bioremediation of acid mine water using facultatively methylotrophic metal-tolerant sulphate reducing bacteria. Microbiol Res 152:65–73

Harris MA (2009) Structural improvement of age-hardened gypsum-treated red mud wastes using readily decomposable phyto-organics. Environ Earth Sci 56(8):1517–1522

Harris MA, Ragusa S (2000) Bacterial mitigation of pollutants in acid drainage using decomposable plant material and sludge. Environ Geol 40(1±2):195

Havas M (1990) Recovery of acidified and metal-contaminated lakes in Canada. In: Norton SA, Lindberg SE, Page AL (eds) Acid precipitation, vol 4, pp 187–204

Hayes MHB (1980) The role of natural and synthetic polymers in stabilising soil aggregates. In: Berkeley RCW, Lynch JM, Melling J, Rutter PR, Ellis VB (eds) Microbial adhesion to surfaces. Ellis Horwood, Chichester, pp 262–294

Herhily AT, Mills AL (1985) sulphate reduction in freshwater sediments receiving acid mine drainage. Appl Environ Microbiol 49:179–186

Hinesly TD, Jones RL, Ziegler EL, Tyler JJ (1977) Effects of annual and accumulative applications of sewage sludge on assimilation of zinc and cadmium by corn (*Zea mays* L.). Environ Sci Technol 11:182–188

Horvath DJ, Koshut RA (1981) Proportion of several elements found in sewage effluent and sludge from several municipalities in West Virginia. J Environ Qual 10:491–497

Hue NV, Craddock GR, Adams F (1986) Effect of organic acids on aluminium toxicity in subsoils. Soil Sci Soc Am J 50:28–34

Kalin M, Fyson A, Smith MP (1993) ARUM acid reduction using microbiology. In: Biohydrometallurgical technologies

Koschorreck M, Frömmichen R, Herzsprung P, Tittel H, Wendt-Potthoff K, (2002) The function of straw for in situ remediation of acidic mining lakes: results from an enclosure experiment. Water Air Soil Pollut 2:97–109

Koschorreck M, Boehrer B, Friese K, Geller W, Schultze M, Wendt-Potthoff K (2011) Oxygen depletion induced by adding whey to an enclosure in an acidic mine pit lake Ecol. Eng. 37(12):1983–1989

Kobo K, Fujisawa T (1963) Studies on the clay-humus complex, 3. Adsorption of humus acid by clay. Soil Sci Plant Nutr 9:36–37

Kroth EM, Page JB (1946) Aggregate formation in soils with special reference to cementing substances. Proc Soil Sci Soc Am 11:27–34

Kumar RN, McCullough CD, Lund MA (2011) How does storage affect the quality and quantity of organic carbon in sewage for use in the bioremediation of acidic mine waters? Ecol Eng 37:1205–1213

Kumar RN, McCullough CD, Lund MA (2013) Upper and lower concentration thresholds for bulk organic substrates in bioremediation of acid mine drainage. Mine Water Environ 32:285–292. doi:10.1007/s10230-013-0242-8

Larsen VJ, Schierup H (1981) The use of straw for removal of heavy metal from waste water. J Environ Qual 10(2):188–193

Lottermoster BG (2003) ISBN: 3-540-00-526-9. Springer, Berlin, Heidelberg, New York, pp 3–11

Lyew D, Knowles R, Sheppard J (1994) The biological treatment of acid mine drainage under continuous flow conditions in a reactor. Trans I Chem E 72:42–47

McCullough CD, Lund MA (2011) Bioremediation of acidic and metalliferous drainage (AMD) through organic carbon amendment by municipal sewage and green waste. J Environ Manage 92:2419–2426

McCullough CD, Lund MA, JM May (2006) Microcosm testing of municipal sewage and green waste for full-scale remediation of an acid coal pit lake, in semi-arid tropical Australia. In: Proceedings of the 7th international conference on acid rock drainage

McCullough CD, Steenbergen J, te Beest C, Lund MA (2009) More than water quality: environmental limitations to a fishery in acid pit lakes of Collie, south-west Australia. In: Proceedings of the IMWA conference, Pretoria

Mills AL (1985) Acid mine waste drainage: microbial impact on the recovery of soil and water ecosystems. In: Tate RL, Klein DA (eds) Soil reclamation processes. Marcel Dekker, New York, pp 35–81

Mosher JB, Figueroa L (1996) Biological oxidation of cyanide: a viable treatment option for the minerals processing industry. Miner Eng 9:573–581

Muyzer G, Stams AJM (2008) The ecology and biotechnology of sulphate-reducing bacteria. Nat Rev Microbiol 6:441–454

Narwal RP, Singh BR, Panhwar AR (1983) Plant availability of heavy metals in a sludge-treated soil: effect of sewage sludge and soil pH on the yield and chemical composition of rape. J Environ Qual 12:358–365

Norton SA (1983) The chemical role of lake sediments during lake acidification and de-acidification. In: Proceedings of the 11th Nordic symposium on sediment, Universitet Oslo, pp 7–19

Norton SA, Kahl JS, Henriksen A, Wright RF (1990) Buffering of pH depressions by sediments in streams and lakes. In: Norton SA, Lindberg SE, Page AL (eds) Acidic precipitation, vol 4., Soils, aquatic processes and lake acidificationSpringer, Berlin, pp 132–157

Pashley RM (1985) electromobility of mica particles dispersed in aqueous solutions. Clay Clay Minerals 3:193–199

Peine A, Peiffer S (1998) In-lake neutralization of acid mine lakes. In: Geller W, Klapper H, Salomons W (eds) Acidic mining lakes. Springer, Berlin, pp 47–632

Pierce FJ, Dowdy RH, Grigal OF (1982) Concentrations of six trace metals in some major Minnesota soil series. J Environ Qual 11:416–422

QGRM (2002) Queensland government dept of natural resources and mines State Planning Policy 2/02 Guideline Acid sulphate Soils Version 2. http://www.dip.qld.gov.au/resources/policy/spp-guidelines-oct-02-v2.pdf

Queensland Government (2015) http://www.qld.gov.au/environment/land/soil/soil-testing/soil-terms/

Quirk JP (1960) Negative and positive adsorption of chloride by kaolinite. Nature 188:253–254

Ramasamy K, Kamaludeen, Sara PB (2006) Bioremediation of metals: microbial processes and techniques. In: Singh SN, Tripathi RD (eds) Environmental bioremediation technologies. Springer Publication, NY, pp 173–187

Robertson WK, Lutrick MC, Yuan TL (1982) Heavy applications of liquid-digested sludge on three ultisols: effects on soil chemistry. J Environ Qual 11:278–282

Romero FM, Prol-Ledesma RM, Canet C, Alvarez LN, Perez-Vazquez R (2010) Acid drainage at the inactive Santa Lucia mine, western Cuba: natural attenuation of arsenic, barium and lead, and geochemical behavior of rare earth elements. Applied Geochemistry 25(5):716–727

Rubo A, Raf K, Jay R, Norbert S, Wolfgang H (2006) Alkali Metal Cyanides. In: Ullmann's encyclopedia of industrial chemistry. Wiley-VCH, Weinheim, Germany. doi:10.1002/14356007.i01_i01

Schindler DW, Wagemann R, Cook RB, Ruszcynski T, Prokopowich J (1980) Experimental acidification of lake 223, experimental lakes area: background data and the first three years of acidification. Can J Fish Aquat Sci 37:342–354

Schnitzer M, Skinner SIM (1964) Organo-metallic interactions in soils. 3. Properties of iron- and aluminium-organic matter complexes, prepared in the laboratory and extracted from soil. Soil Sci 98:197–203

Schofield RK, Samson HR (1954) Flocculation of kaolinite due to the attraction of oppositely charged faces. Discuss Faraday Soc 18:135–145

Sen AM, Johnson B (1999) Acidophilic sulphate-reducing bacteria: candidates for bioremediation of acid mine drainage. Process Metallurgy 01/1999; 9:709-718. doi:10.1016/S1572-4409(99)80073-X

Siller H, Winter J (1998) Degradation of cyanide in agroindustrial or industrial wastewater in an acidification reactor or in a single-step methane reactor by bacteria enriched from soil and peels of cassava. Appl Microbiol Biotechnol. 50(3):384–389

Sinha RK, Dalsukh V, Sinha S, Singh S, Herat S (2009) Bioremediation of contaminated sites: a low-cost nature's biotechnology for environmental clean-up by versatile microbes, plants, and earthworms. In: Timo F, Johann H (eds) Solid waste management and environmental remediation. ISBN: 978-1-0741-761-3. Nova Science Publishers, Inc

Singh Y, Singh B, Khind CS (1992) Nutrient transformations in soils amended with green manures. Adv Soil Sci 20:238–305

Stark LR, Kolbash HJ, Webster HJ, Stevens SE, Dionis KI, Murphy ER (1988) The Simco #4 wetlands: biological patterns and performance of a wetland receiving mine drainage. In: Mine drainage and surface mine reclamation. IC 9183. Bureau of Mines, US Dept. of Interior, Pittsburgh, vol 1, pp 332–344

Strosnider WH, Nairn RW (2010) Effective passive treatment of high strength acid mine drainage and raw municipal wastewater in Potosı´, Bolivia using simple mutual incubations and limestone. J Geochem Explor 105:34–42

Swift MJ, Heal OW, Anderson JM (1979) Decomposition in Terrestrial Ecosystems. Blackwell, London

Theng BKG (1974) The chemistry of clay-organic reactions. Adam Hilger, London

Theng BKG (1980) Soils with variable charge. New Zealand Society of Soil Science

Torma AE, Apel ML, Brierly CL (1995) (eds) The minerals, metals and materials society. Boojum Research Ltd., Toronto, Canada, pp 312–327

Tuttle JH, Dugan PR, Randles CI (1969) Microbial sulphate reduction and its potential utility as an acid mine water pollutant abatement procedure. Appl Microbiol 17:297–302

Ulrich B (1991) An ecosystem approach to soil acidification. In: Ulrich B, Sumner ME (eds) Soil acidity. Springer, Berlin

USP Technologies (2016) Cyanide Treatment with H_2O_2. www.h2o2.com/industrial/applications.aspx?pid=106&. Accessed May 9, 2016

Waybrant KR, Blowes DW, Ptacek CJ (1998) Selection of reactive mixtures for use in permeable reactive walls for treatment of mine drainage. Environ Sci Tech 32:1972–1979

Yeoh NS, Oades JM (1981) Properties of soils and clays after acid treatment. 1 Clay minerals. Aust J Soil Res 19:147–158

Zehnder AJB, Stumm W (1988) Geochemistry and biogeochemistry of anaerobic habitats. In: Zehnder AJB (ed) Biology of anaerobic microorganisms. Wiley Interscience, New York, pp 1–38

Detoxification of Acid Drainage Using Inorganic pH Buffers in a Constant Flow Bioreactor

Mark Anglin Harris and Santo Ragusa

Abstract To substantially reduce the acclimatization period for SRB in a highly acidic medium, we studied some effective pH buffers for flowing AMD. Silicate buffers increased pH before SRB was introduced. A mixture of AMD and sewage-sludge-impregnated ryegrass from a previous incubation contained an enriched strain of *Desulfovibrio vulgaris,* one of two major SRBs. They were active at a pH above 5 at 30 °C for six days, producing a black coloration probably because of the precipitation of sulfide, thereby initially suggesting the concentrated presence of viable SRB. The in situ pH was 6.3. The odor of H_2S was detected from the inoculants. An up-flow bioreactor made from rigid PVC tubing with sampling ports placed 5 cm apart in the column was plugged with rubber seals. The treatment consisted of ground ryegrass mixed thoroughly with a mollisol soil. Freshly cut ryegrass ground to <1 mm diameter was hand-mixed with the mollisol in 1:5 ratio (w/w) and the mixture placed on the sand in the column to occupy the column volume, except for a 10 cm wide space at the top of the column. An electrically powered peristaltic pump slowly pumped the AMD into the column from the bottom, until the top of the treatment material in the column was just covered with the influent. The pH was measured after 1 day. At the end of 20 days without any added lime, the oxide achieved the highest pH increase of all the buffers. The bioreactor supported sulphate reduction at continuous flow rates of 30, 60, 90, and 180 ml day^{-1}. The results of this study show that it is possible to establish an SRB population in a constant flow reactor, using readily decomposable material and a fast-acting soil-based buffer. It was concluded that cation exchange capacity (CEC), and not merely free lime, was the crucial buffer characteristic at the low pH level of <3.0 units. The CEC of the substrate used in this study was three times that of a sludge buffer used by Harris and Ragusa (2000). The only effective substrate difference (excluding the influent flow rate) between the present study and that of Harris and Ragusa (2000) is that of the buffers used. Hence, it was concluded that for pH increases, and decreases of soluble cations in AMD, CEC is

Original article: **Bioremediation of acid mine drainage using decomposable plant material in a constant flow bioreactor**. *Environmental Geology* Volume 40, Issue 10, pp. 1192–1204.

M.A. Harris, *Geobiotechnological Solutions to Anthropogenic Disturbances*,
Environmental Earth Sciences, DOI 10.1007/978-3-319-30465-6_8

a major determining factor among the buffering materials. Where sources of lime are insufficient, the use of medium to high organic C soil-based buffers with small quantities of lime can be an effective alternative for activating SRB in a reactor.

Keywords Acid-drainage · Bioremediation · Sewage-sludge · Sulphate-reducing-bacteria

Highlights

- Inorganic buffers increase pH for rapid SRB detoxification of heavy metals in acid mine drainage
- SRB activated in flowing AMD.

1 Introduction

1.1 Effects of Sulfide Exposure

At metal mines, the target ore (like gold, silver, copper, etc.) is often rich in sulfide minerals. When the mining process exposes the sulfides such as Pyritic minerals (FeS_2) to water and air, oxidization to sulfuric acid occurs as reactions produce sulfuric acid. This acid can dissolve other harmful metals and metalloids (like arsenic) from the surrounding rock (Maiti 2003). According to Whitbread-Abrutat (2008), the most serious environmental concerns at orphaned and abandoned mines are acid rock drainage from underground workings, open pit mine faces, waste rock piles, and tailings storage areas that were left exposed to the elements or inadequately reclaimed, resulting in the contamination of water with dissolved metals and acidity. Sheoran et al. (2010) state that the pH of a given mine soil can change rapidly as the rock fragments weather. They observed that with pH levels as low as two, the sulfuric acid mobilizes aluminium, heavy metals and uranium from the rock, and oxidize, producing iron oxide as a by-product (hence the characteristic reddish-brown colour of AMD). Unweathered (or un-oxidized) mine soils which contain a significant amount of pyritic-S in excess of their neutralizers (carbonates) will rapidly drop the pH to a range of 2.2–3.5 after exposure to water and oxygen. When such toxic fluids flow into groundwater, streams and rivers, they destroy ecosystems and taint water in regions where freshwater already may be in short supply.

Vegetation achieves optimal growth in soil at a neutral pH, whilst carbonate ($Ca/MgCO_3$) bearing minerals and rocks tend to increase the pH as they weather and dissolve (Sheoran et al. 2010). When the soil pH drops to below 5.5, reduced legume and forage growth occur due to metal toxicities such as aluminium or manganese, phosphorus fixation, and reduced population of N-fixing bacteria (Sheoran et al. 2010).

1.2 Active or Passive Treatments? Justification

Treatment of acid mine drainage (AMD) and municipal wastewater (MWW) is central to maintaining global water resource quality. Untreated AMD causes water quality degradation in coal and metal mining regions worldwide (Bell and Donnelly 2006). Discharges of untreated MWW degrade water resources in many developing nations (Gadgil 1998; Kivaisi 2001). In developed nations, where MWW is generally addressed actively, treatment consumes considerable financial, material and energy resources (Muga and Mihelcic 2008). Compared to passive methods, conventional MWW and AMD treatment are energy-intensive with higher operational and maintenance costs (Younger et al. 2002; Muga and Mihelcic 2008).

Several methods have been applied to reduce the toxicity of drainage water from disused mine-sites. Using a gentle spray, James and Mrost (1965) have attempted to leach out soluble salts and oxidation products from pyritic material to relatively low water table, where they are fixed onto clay particles. However, such a measure does not always solve the problem of metal toxicity to plant life (Hore-Lacy 1978; Strosnider et al. 2009) especially where water tables rise in wet seasons. Most techniques involve the application of chemical precipitators such as calcium compounds. Precipitation by lime and limestone is one of the oldest and cheapest methods for removal of heavy metals from acid drainage solutions (Strosnider et al. 2009). Such measures have not been highly successful. Research has shown (Dean et al. 1972) that Cd, Cu, Pb and Zn will begin to precipitate from dilute solution at pH values exceeding 5.3, 6.0, 6.7, and 7, respectively. However, lime additions have not been highly successful for pyritic materials as they soon tend to form pans. The longevity of the oxidation process is calculated in decades rather than years (Elliott et al. 1998) and sources of lime are not ubiquitous. Hore-Lacy (1978) observed that in one United States case where 2500 t ha^{-1} was added to the tailings at Bingham (the "Iron Mountain") the pH was back to a value of 3.5 after 3 months. Elliott et al. (1998) concluded that the management of AMD should be focused on containment and prevention rather than the chemical treatment of effluents. The present methods utilizing lime are not cost-effective in the long term. Biological mitigation using sulphate reducing bacteria (SRB) has been reported (Lyew et al. 1994; Elliott et al. 1998). Because substrates are needed for the reduction process, costs may be prohibitive. For a microbial process to be economically feasible, the carbon and energy source should be cheap, widely available, and highly effective (Hard et al. 1997; McCullough and Lund 2011).

Citing the high cost of commercial available organic acids as an energy source, Elliott et al. (1998) suggested further research into the use of alternate carbon source in bioreactors for potential large scale use. In South Australia organic waste comprises almost 50 % of all household rubbish, a total of 97, 546 tons (Handreck 1996). A large fraction of this is cut grass that is wasted through the process of decomposition. Readily decomposed and finely ground rye grass could be used as a cheap and effective energy source for the mitigation of AMD (Harris and Ragusa 2000).

They found that under stagnant conditions a lime-stabilized sewage sludge in combination with decomposable plant material and SRB substantially decreased the acidity of AMD that had a prior pH of <3.

1.3 Character of Substrate

Frömmichen et al. (2003) found that whilst pure and complex carbon sources can serve as suitable substrates for stimulating microbial reductive processes in (coal) pit lake sediment for alkalinity generation, complex substrates (i.e. straw, wood chips) were inefficient at remediation as acidic waters often lack the micro-flora that can degrade lignin. Sulphate-reducing bacteria are thus unable to use complex organic substrates such as starch, cellulose, proteins, and fats. Chang et al. (2000) tested different organic waste materials as electron donors for sulphate reducers and found that cellulose polysaccharides were the main components of the waste materials consumed in their reactors. SRB are dependent on other microbes that degrade these complex substrates and ferment them to products that can serve as substrates for SRB (Figueroa et al. 2004; Muyzer and Stams 2008). Thus Dvorak et al. (1992), Nagpal et al. (2000), and Tsukamoto et al. (2004) note that SRB use the easily degradable fraction of organic matter that contained short-chain low-molecular weight organic compounds (e.g. methanol, ethanol, lactate).

Lactate is the best substrate for SRB (Postgate 1984) and a superior electron donor compared to others (e.g. ethanol) in terms of moles of bicarbonate produced per mole of substrate consumed (3 vs. 2, respectively), since it is better at neutralizing the acidity in the treated effluent (Kaksonen et al. 2003). In Cuba, some detailed studies about the use of some types of biotechnological barriers such as bacteria or mushrooms that neutralize the acidic environmental contamination during the metal exploitation works in the area have been planned (Díaz de Villalvilla et al. 2003). The main drawback is that not all SRB species are capable of oxidizing lactate and ethanol to CO_2. Some SRB can only partially oxidize simple organic molecules to acetate, and very few can use acetate alone (Nagpal et al. 2000).

SRB activity rates are thus dependent on the nature of organic waste used, and, in particular, the bioavailability of organic carbon (Gibert et al. 2002). As naturally refractory organic substrates release carbon and other nutrients slowly, it would seem beneficial if combinations of labile and refractory substrates are used, as this decomposing mixture can continue to provide carbon after the initial labile carbon fractions are exhausted (Koschorreck et al. 2002; McCullough and Lund 2011). They observed that a good organic carbon source must both initiate and sustain SRB-based bioremediation.

The rationale for selecting 30 $g\ L^{-1}$ of sewage sludge as the lowest concentration evaluated was based on previous studies that indicated that this is the minimum concentration necessary to initiate bioremediation of AMD affected pit lakes (McCullough and Lund 2011; McCullough et al. 2006).

Thus the availability of carbon from plant matter depends on decomposition, and decomposition is extremely limited in acidic and anoxic conditions (Kalin et al. 1993), which is typical of acid sulphate mine-spoils. Gibert et al. (2002), in a review, found that the specific composition of the organic matter was a primary determinant of the efficacy of the passive SRB-based treatment system; in particular, the lability and biological availability of the material. However, the ability of the organic matter to be decomposed over a prolonged period is likely more important to bioremediation of AMD than initial lability and bioavailability. Harris and Ragusa (2001) also noted that the availability of carbon from plant matter was dependent on decomposition, which can be extremely limited in acidic and anoxic conditions.

Further, Gyure et al. (1990) showed that concentration of organic acids (from decomposition of organic matter) greater than 5 mM completely inhibited SRB activity in sediment at pH 3.8. Harris and Ragusa (2000) found that readily decomposable finely ground plant material placed in an AMD of pH 2.30 released some organic acids, but at such a high potency as to have decreased the AMD pH to a value of 1.89. Subsequent inoculation of these samples with SRB did not result in establishment of viable colonies in this medium. This was probably one factor that delayed the activation of SRB for their sewage sludge/ryegrass experiment, where, SRB was not activated until a period of >50 days had elapsed after incubation had commenced, even for the most favourable sludge + ryegrass treatment. For the sewage sludge/ryegrass experiment (Harris and Ragusa 2000), readily decomposable material by itself did not substantially change the concentration of heavy metals in AMD. Neither did the sludge-only treatments. Yet, Waybrant et al. (1998) observed that sewage sludge encouraged the highest levels of sulphate reduction over seven other organic carbon sources. Moreover, multiple studies have noted that mixtures of sewage sludge with multiple organic carbon sources generally promote higher sulphate reduction rates than single sources (Waybrant et al. 1998; Harris and Ragusa 2000; McCullough et al. 2006).

Van Breemen et al. (1983) concluded that alkalinity produced during sulphate reduction, but not removed from the soil as acid neutralizing capability (ANC; aq) causes ANC (s) to increase. Using a mixture of sewage sludge and green waste (leaf mulch, woodchips and sawdust). Kumar et al. (2013) reduced 4500 mg L^{-1} of sulphate to <25 mg L^{-1} in only 35 days. A laboratory experiment by Harris and Ragusa (2000) also found that (under stagnant conditions), a mixture of sewage sludge and plant material (fresh rye grass) was effective in initiating amelioration of acidity and metal concentrations of acid mine waters through sulphate reduction. They were able to increase pH levels from 2.3 to >3 within 30 days.

The above-mentioned "stagnant treatment" mitigation system (Harris and Ragusa 2000) is not always useful for field applications because flowing effluent can remove suspended acid neutralizing components. Therefore the high acidity (pH < 3.0 for many acidic mine waters) must first be appreciably decreased before biological sulphate reduction can occur (Alexander 1977; McCullough and Lund 2011). SRB are known to have their optimum pH at around neutral to slightly alkaline (Church et al. 2007; Gyure et al. 1990). Harris and Ragusa

(2000) observed that SRB can operate in waters at significantly lower pH values when such waters were previously provided with effective pH buffers. The buffer they utilized (a mixture of decomposable ryegrass and aerobically treated sewage sludge) took 50 days to increase the AMD pH values to 3.0 from its initial 2.3. With a continuous flow of new quantities of AMD influent into a bioreactor, it would seem that the above-mentioned acclimatization period would be longer. Substantially reducing this acclimatization period would require a more efficient pH buffer for the AMD than that utilized by the above-mentioned authors. In a column experiment, Larsen and Schierup (1981) showed that the efficiency of straw to adsorb heavy metals from an AMD could be improved by increasing the pH with NaOH. They also point out that one disadvantage of using NaOH is that some organic constituents of the straw are more soluble under basic conditions.

When soil pH falls below a value of 4, clay minerals, and Al/Fe hydr(oxides) become the most effective proton sinks in the ground (Bruggenwert et al. 1991). At 6 h after lowering the pH of a stream with sulfuric acid, Norton et al. (1990) found that below pH 5, neutralization was dominated by the release of Al. Upon progressive acidification of the soil, large amounts of Al ions are eventually liberated (Bolt and Bruggenwert 1976). This acidity can be used up as ANC when the Al cations enter the adsorption sites of the clay minerals or oxides (Bruggenwert et al. 1991). For example, after testing the adsorptive power of Fe_2O_3 by adding it to trace elements mobilized in fermented (anaerobic) plant material, up to 59 % of the Cu, Zn, Cu and Ni was removed from solution at pH 7 (Toth 1968), although adsorption decreased with lower pH values. The same effect was observed for a sand coated with ferric oxide. The dissolution of Fe_2O_3, Al_2O_3 and MnO_2 is negligible at pH of >5 (Van Breemen et al. 1983). However, they are strong buffers at pH < 5 because oxides are the chief materials accepting protons (Bolt and Bruggenwert 1976; Barrow and Bowden 1987). Therefore, they may be very important in buffering against hydrogen ions. The ability of these and other buffers to increase the pH of AMD needs to be investigated. It would seem from the foregoing that clay minerals or oxides added to AMD may increase the pH of such waters up to a level which could trigger the activation of SRB.

1.4 Cuba: Camagüey Sulfide System

Because of its vast number of ore deposits, the Camagüey mining region (Fig. 1) is the most important in Cuba, constituting a district of epithermal deposits (Díaz de Villalvilla et al. 2003). They found that the nature of the rocks of the volcanic arc in Camagüey allows correlation with those of The Dominican Republic. There are several types of mineral deposits in Central Cuba including high sulfide (Golden Hill) and low sulfide systems (Beatriz, Florencia), Cu–Mo–Au porphyry (Palo Seco y La Purísima) and Au–Ag–Zn Skarns (La Purísima) deposits (2003; Simon et al. 1999).

Fig. 1 Physiography of Cuba. Note the high density of natural waterways in the heavily mined Camaguey sulfidic region. Acidic mine drainage threatens the integrity of surface water and ground water aquifers

The presence of several types of mineral deposits genetically distinctive in these formations are characterized by pyrite presence as main mineral, either disseminated or accompanied by quartz in veins (Díaz de Villalvilla et al. 2003). The high prices of gold and silver at the present time, and a variety of minerals such as gold, silver, lead, bismuth and nickel belonging to the telluride group suggests at least several decades of future mining operations. In Golden Hill, Florencia and Beatriz, the corresponding environmental studies have been defined in order to take the systematic measures to avoid or decrease damage as possible (López et al. 1998).

The deposits are small and located near each other, standing out as small elevations on the surface, with ores formed by pyrite as the main mineral (Díaz de Villalvilla et al. 2003). Typical oxidation zones reach 10–20 m depth with quartz, carbonates, and mineral-loamy products of the wall-rock alteration. The reserves are approximately 5–10 tons of gold. The veins usually have high gold contents and erratic distributions could be present. Small mining operations are needed to open pit facilities and benefit chord in the studied resources (Díaz de Villalvilla et al. 2003).

The above workers concluded that the oxidation of these sulfurs could generate acid drainage, which can potentially affect the superficial drainages of the area of influence in the mine and consequently enter in the food-chain. However, the presence of carbonates in the mineral bodies and associated with the wall rocks could directly limit the mobility of the metals and the acid drainage product of the water circulation and the pyrite presence during the mining activity (open pit) (Díaz de Villalvilla et al. 2003). Thus samples collected by Du Bray (2004), of stream sediment using the open-pit method and heap leaching show results of <5–12 silver ppm, 0.04–0.4 ppm of gold, <5 ppm cadmium, 18–81 ppm cooper, 1.4–4.2 % of iron 200–2.020 ppm manganese, 7–197 ppm lead, and 70–1.131 ppm zinc (Du Bray 2004).

1.5 Dominican Republic: Pueblo Viejo Sulfides

1.5.1 Physiography and Climate

The Dominican Republic occupies the eastern portion of the Caribbean island of Espanola (the nation of Haiti occupies the western portion). The Dominican Republic was estimated to be the world's 11th ranked producer of nickel in 2008 (Wacaster 2010). The Pueblo Viejo gold deposit ranks as one of the largest high-sulfidation acid epithermal deposits in the world (Mueller et al. 2008; Kuck 2009; Redwood 2009).

The central region of the Dominican Republic is dominated by the Cordillera Central Mountain Range, which runs from the Haitian border to the Caribbean Sea (Fig. 2). The highest point in the Cordillera Central is Pico Duarte at 3175 m. Pueblo Viejo is located in the eastern portion of the Cordillera Central (Fig. 2) where local topography ranges from 565 m at Loma Cuaba to approximately 65 m at the Hatillo Reservoir (Smith et al. 2008).

The mine was inadequately closed and reclaimed, resulting in the extensive formation of acid mine drainage (AMD), leading to a legacy of contaminated infrastructure, soil, and water at the site and surrounding area. Untreated and uncontrolled AMD resulted in the transport of high metal concentrations and acidity into local waterways (Kuck 2009). Other environmental, health, and safety hazards identified at the site included rusting machinery, oil and gasoline-contaminated soils, mercury-contaminated materials, asbestos, and local groundwater contamination (Kuck 2009). Two rivers run through the concession, the Margajita and the Maguaca. The Margajita drains into the Yuna River upstream from the Hatillo Reservoir while the Maguaca joins the Yuna below the Hatillo Reservoir. The flows of both rivers vary substantially during rainstorms (Smith and Stephenson 2011).

The Dominican Republic has a tropical climate with little fluctuation in seasonal temperatures, although August is generally the hottest month, and January and February are the coolest. Temperatures at the project site range from daytime highs of 32 °C to night-time lows of 18 °C. Annual rainfall is approximately 1800 cm, with May through October typically being the wettest months. The Dominican Republic is in a hurricane channel; the hurricane season is typically

Fig. 2 The eastern section of the Central Cordillera contains Pueblo Viejo, one of the world's largest sulfidic gold mines, headquartered at Cotuí, within 150 km of Santo Domingo. It therefore can be seen that, as in Camaguey (Cuba), several river basins (at least 4) are at risk from toxic heavy metal pollution from exposed sulfidic wastes

August to November. As a result of previous mining and agriculture, there is little primary vegetation on the Pueblo Viejo site and surrounding concessions, but secondary vegetation is abundant outside of the excavated areas and can be quite dense (Smith and Stephenson 2011).

1.6 Aim and Hypothesis

Given adequate substrates and a long enough acclimatization time (Harris and Ragusa 2000), dormant SRB colonies could become extremely active in acid drainage (AD) conditions ($pH < 3$). The SRB placed in the harsher environment of a similar (in character) but flowing AMD would require a longer acclimatization period (i.e. >50 days) than that in the above study. The aim of the present study was to decrease that time period. A more efficient pH buffer than that used previously is required. One objective of the present experiment was to identify materials that may more efficiently improve the quality of AMD. Some of these materials include oxides and other silicates. It is hypothesized that (1) clay or Fe/Al (hydr)oxides released in AMD of pH < 2.5 may increase the pH of the AMD to a point where sulphate reduction by SRB is begun, and (2) SRB will mitigate SRB of pH < 3 in a constant flow bioreactor energized by decomposable plant matter, but initially activated by an inorganic pH buffer.

2 Materials and Methods

Properties of silicate buffers and whole soil buffers.

A description of silicate buffers used in this study is shown in Table 1. The physical and chemical properties of the two whole soils are shown in Table 2. The first was an alfisol (illitic-kaolinitic) or red-brown earth (RB) and the other a mollisol (MC; illitic-kaolinitic minerals). To increase its clay fraction, the RB was dug from the B-horizon (10–20 cm depth) of a structurally degraded continuous wheat plot from the Waite Long-term Trial, Waite Campus, University of Adelaide. The mollisol was a black earth (hence having a high clay content) from 1–4 cm of a natural follow plot.

2.1 Processing of Whole Soils

After air-drying for 5 days at 30 °C, the clods were broken, gently crushed, and thoroughly mixed. During mixing, any obvious organic matter and stones observed were removed from the samples. The soil was then passed through a 2 mm sieve, thereafter being stored in air-tight containers. The properties of the soils together with the test methods are listed in Table 2.

Table 1 Some properties of clay and soil buffers

Name	Abbreviation	pH	CEC (cmol (+)/kgcmol)	CaO	SiO_2	Fe_2O_3	Al_2O_3
Active bond	AB	10	75	1.0	56.0	4.5	16.2
Acid drainage	AD	2.3	–	–	–	–	–
Active gel	AG	10	75	0.9	56.0	4.5	16.2
Ball clay	BCL	4.7	n.d.[a]	0.2	58.2	1.0	26.3
Trufeed bentonite	TRP	7.0	80	0.02	61	2.9	14.8
China clay	CLH	8.4	n.d.	0.1	50.8	0.9	33.1
Oxide	OX	6.5	n.d.	0.3	5.0	28.0	14.5
Mollisol	MC	7.2	59.1	n.d.	n.d.	n.d.	n.d.
Illite	IL	7.1	6.8	n.d.	n.d.	n.d.	n.d.
Pure kaolinite	PURK	8.2	6.7	0.1	53.7	0.9	35.1

[a]No data

Table 2 Some chemical and physical characteristics of two soils

Soil property	Alfisol	Mollisol	Test method
pH (1:5 sludge:water ratio)	6.5	7.2	Watson-Victor glass electrode
Electrical conductivity (dS m^{-1})	0.12	0.23	Ionode glass electrode
Organic C (%)	1.24	4.6	Leco C analyzer
Total C (%)	1.4	4.61	Leco C analyzer
Total N (%)	0.11	0.69	Leco N analyzer
Inorganic C (%)	<0.01	<0.01	Difference
CEC (cmolc (+)/kg)	7.3	59.1	1N NH_4OAC (buffered at pH 7)
Illite (total)	D	D	X-ray diffractometry
Kaolinite (total)	p	p	X-ray diffractometry
Randomly interstratified minerals (total)	p	p	X-ray diffractometry
Clay (%)	19	40.3	Pipette (Day 1965)
Silt: 2–20 μm (%)	31.7	36.1	Difference
Fine sand: 20–200 μm (%)	44.8	17.2	Pipette
Coarse sand: 0.2–2.0 mm (%)	3.9	6.4	Pipette

D Dominant; *p* clay minerals present

2.2 Inoculation

Inoculation was not necessary in order to initiate the sulphate reducing process (Christensen et al. 1996; Harris and Ragusa 2000; Koschorreck 2008; Koschorreck et al. 2011; Kumar et al. 2011), but was applied to shorten the initial lag phase (Christensen et al. 1996). A mixture of AMD (Table 3) and sewage-sludge impregnated ryegrass form a previous incubation containing an enriched strain of *Desulfovibrio vulgaris* one of two major SRBs active at a pH above 5 (Alexander 1976) at 30 °C for six days, Alexander (1976) found that the bacteria grew, and the medium turned black in colour. He had successfully used his enriched culture as an inoculum in a subsequent experiment. This effected a 4-fold decrease in sulphates and the production of sulfides after only 3 days. The medium in the present study had turned black probably because of the precipitation of sulfide, thereby initially suggesting the concentrated presence of SRB. The in situ pH was measured at 6.3. The presence of viable SRB was strongly suggested. In addition to the above-mentioned colour change, the smell of H_2S was detected from the inoculants. The presence of H_2S was confirmed by the lead acetate test: a filter paper dampened with lead (II) acetate4 solution was placed inside the tube. Lead (II) ions reacted with the H_2S gas to from brown lead (II) sulfide.

Aliquots of 20 g of the AMD-suffused sewage sludge and ground decomposable ryegrass containing this culture were removed for use as an inoculant for the bioreactor. The inoculants were thoroughly mixed with the mixture already placed in the column. Continuous flow of the AMD influent (pH 2.3) through the column was started two days after inoculation.

Table 3 Concentration of cations in solution in an AMD at Brukunga Mines, South Australia

Aluminium	46
Arsenic	0.2
Boron	<0.1
Cadmium	<0.1
Cobalt	0.1
Chromium	0.01
Copper	0.1
Iron	2100
Magnesium	290
Manganese	110
Nickel	0.03
Sulfur	2700
Selenium	<0.01
Zinc	5.7
Molybdenum	<0.1
Sodium	220
Phosphorus	<1
Acidity as $CaCO_3$ (kg)	2006–2602

2.3 The Reactor

The reactor was made from rigid PVC tubing 500 mm high with an 80 mm internal diameter (Fig. 3). This up-flow bioreactor was designed by Elliott et al. (1998), who achieved growth of SRB at pH 4, using sodium lactate as the energy source. Eight sampling ports placed 5 cm apart in the column were plugged with rubber seals (Fig. 3). The treatment consisted of ground ryegrass mixed thoroughly with a mollisol soil. To avoid backflow of the mixture with possible clogging of the influent inlet, a 6 cm thick gravel bed of 1 cm quartz gravel was laid at the bottom of the column, which was overlain by a 4 cm thick pure washed quartz (250 mμ) layer. Freshly cut ryegrass ground to <1 mm diameter was hand-mixed with the mollisol in 1:5 ratio (w/w) and the mixture placed on the sand in the column to occupy the column volume, except for a 10 cm wide space at the top of the column. The AMD was slowly pumped into the column from the bottom, using an electrically powered peristaltic pump, until the top of the treatment material in the column was just covered with the influent. The pH was measured after 1 day.

2.4 Sampling Methods: Bioreactor

The interstitial water in the column was sampled with a syringe and needle, after removing the rubber seals in turn from the holes. The sampling regime was as follows: 5 ml were taken every day from the top surface of the water in the column (i.e., at 40 cm height) and at 5 cm (the base of the column). Water samples were taken at

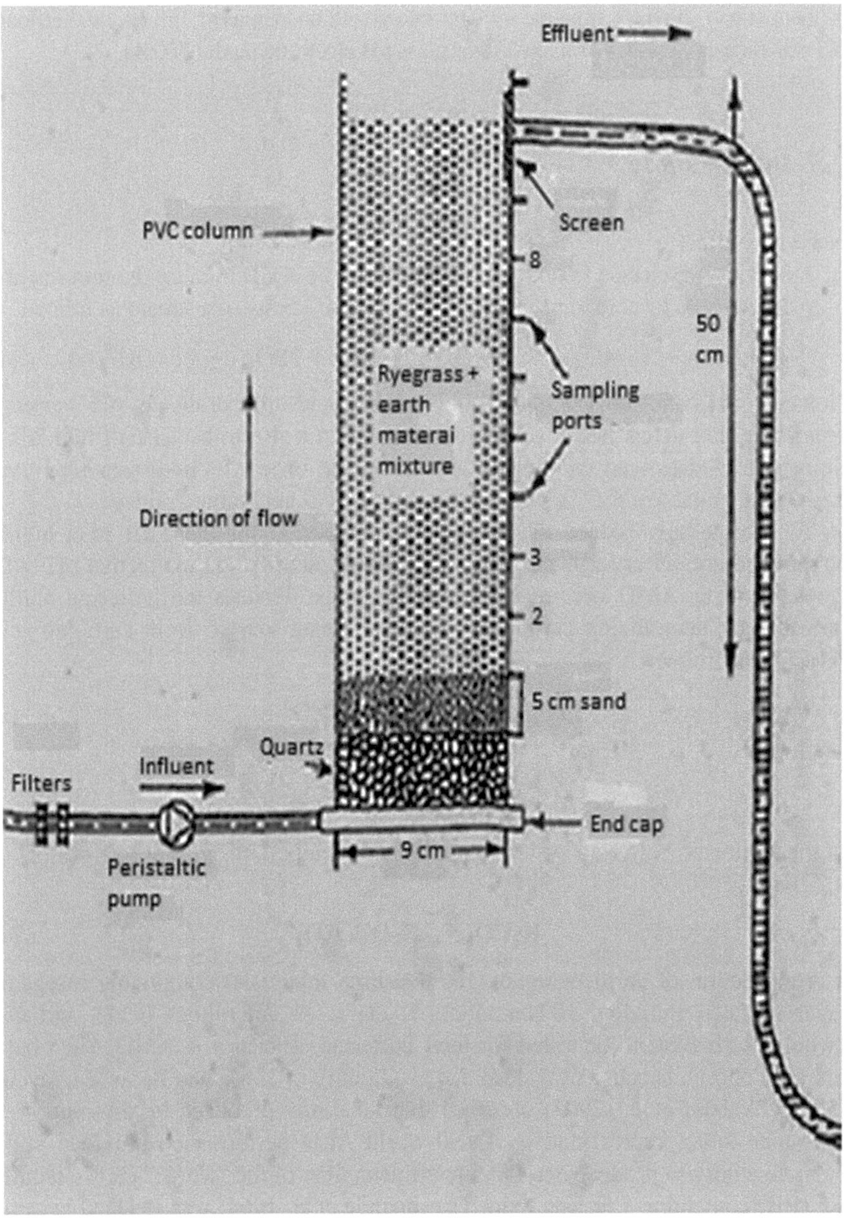

Fig. 3 Bioreactor containing nutrient source and a mollisol as a pH buffer to accelerate the proliferation of SRB (after Elliott et al. 1998)

15-, 25-, 35 cm (hereafter in this study referred to as ports 4, 7, and 10) on the 1st, 3rd, and 5th day after each increase in AMD flow rate of the column incubation. The sampling plan was to continue every 3 days until a pH reading of at least 5.5 was simultaneously obtained from all ports above port 3 (i.e., at least 15 cm above the point of influent entry), at which time the influent rate would be increased. Influent and effluent pH was measured with a Watson-Victor glass pH electrode model 5003.

2.5 Buffer Study

Need for buffers

A buffer was selected to remove H^+ ions from the AMD because the introduction of hydrogen ions by acid mine drainage changes the acidity of a stream as follows:

$$FeS_2(s) + 15/4O_2 + 7/2H_2O \longleftrightarrow 4H^+ + 2SO_4^- + Fe\ (OH)_3(s) \quad (1)$$

However, pH is not always a good indicator of acid mine drainage; pH measurements may not detect heavy acid mine drainage in a stream because of high alkalinity due to dissolved carbonates. Assessing the excess of hydrogen ions over basic ions, "total acidity," is a better measurement of acid mine drainage.

To substantially reduce the acclimatization period for the SRB in a highly acidic medium, an experiment was conducted to identify some effective pH buffers for flowing AMD because acid mine drainage depletes the buffering ability of water by neutralizing carbonate and bicarbonate ions to form carbonic acid (H_2CO_3) as follows:

$$H^+ + CO_3{}^{2-} \longleftrightarrow HCO^{3-} \quad (2)$$

$$H^+ + HCO^{3-} \longleftrightarrow H_2CO_3 \quad (3)$$

As exposure to acid mine drainage depletes the acid neutralizing capacity (ANC) of the carbonate buffering system, thereby completely destroying it below a pH of 4.2 (Eq. 4),

$$H_2CO_3 \rightarrow H_2O + CO_2 \quad (4)$$

a cross-section of earth materials (14 non-lime materials) comprising inorganic layer silicates including 10 bentonites, 1: clays, an amorphous oxide, and two "whole" soils were to be tested for their buffering efficiency in AMD. Such buffers were chosen because of widespread availability of clays, and because Cravotta (2003), Watzlaf et al. (2003) observed that alkalinity produced by dissolution in limestone zones representative of real-world ALD performance provided sufficient alkalinity to produce net-alkaline effluent. One of the "whole" soils—an alfisol (RB) containing 2 % iron oxide (Deshpande et al. 1964) was selected because addition of iron oxide to a Ferralli-Haplic acrisol leads to an increase in buffering capacity against hydrogen ions (Yu 1985). The other whole soil was selected because of its high level of organic matter and clay content. The materials are listed in Table 1. Pulverized $Ca(OH)_2$ (<1 mm diameter) at rates of 0-, 2.5-, 5-, 7 % (w/w), was mixed with each clay or soil. The samples were each submerged in stagnant AMD (pH < 5) for 30 days. The pH of the AMD was recorded every day for thirty days, or until no further changes occurred for consecutive readings.

2.6 Bioreactor Experiment

The SRB bioreactor experiment was carried out under anaerobic conditions at mesophilic temperatures (30–35 °C) because the formation of sulfide by sulphate reduction is enhanced by increasing water levels and rising temperatures (Alexander 1977; Tsukamoto et al. 2004). The pH was recorded daily throughout the incubation. Cation concentration of metals in the samples was monitored by ICPAES. The passage of AMD effluent in the proposed column was to occur through the material which was to be shown in experiment 2 as the most effective buffer.

3 Results

3.1 Buffer Study

The effects of the various buffers on the pH of the AMD are shown in Fig. 4, where the relationship between pH and different rates of lime application can be observed. Incrementally higher lime rates did not necessarily cause an increase in the pH of the AMD. Although most buffers did not respond well to lime addition, even at the highest liming rate of 7.5 %, it was, in general, always the same ones that caused the pH increases at various rates of added lime (Fig. 4). Of the fifteen buffers studied, even at the highest application rate (7.5 %)—only four—the mollisol whole soil (MC), the active bond clay (AB), the active gel clay (AB), and the oxide (OX) decreased the AMD acidity to pH > 5. At the 5 % added $Ca(OH)_2$ level, only three earth materials—the MC, OX, and AG—accomplished AMD neutralization at pH > 5 (Fig. 4).

At the lowest lime application rate (2.5 %), only the mollisol caused the AMD pH to increase to >5 (Fig. 4). Although the ANC of the AMD is comparable to that of the mollisol and the oxide, the mollisol is seen as the most significantly

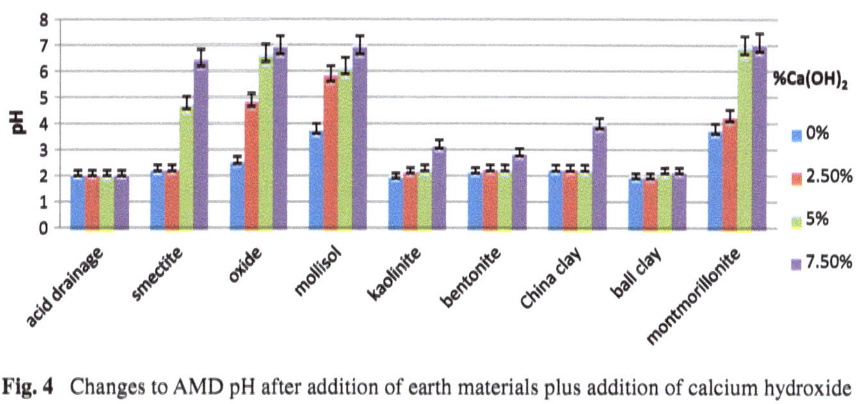

Fig. 4 Changes to AMD pH after addition of earth materials plus addition of calcium hydroxide

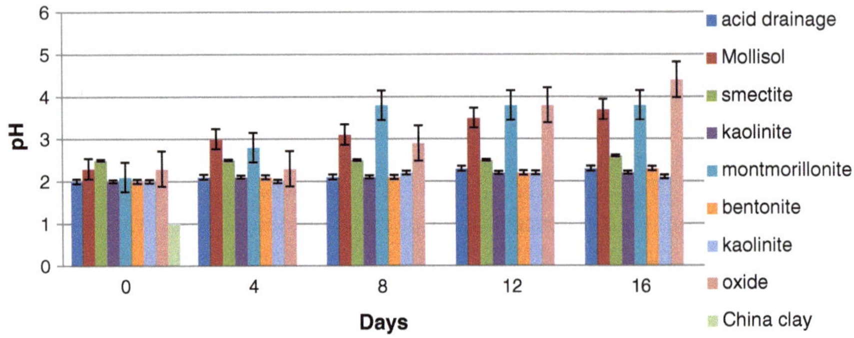

Fig. 5 Changes to pH of acid drainage after addition of non-calcic earth materials

responsive to low-level lime spiking (Fig. 4). Thus at just 2.55 of added lime, the pH of the mollisol-treated AMD rose dramatically from 2.2 to 5.9 units, whereas the oxide increased AMD pH from 2.2 to a value of under 5.0. This represents (after 2.5 % added lime) proportional pH increases of 38-, 33-, and 28 % for MC, OX, and AG respectively. The control pH (AMD) did not change significantly even at the highest lime rate because its pH, even after a 7.5 % addition of lime, still remained unchanged at <2.5 % units after 5 days and up to 25 days (Figs. 4 and 5). This shows that without an adequate buffer (in combination with lime) in AMD, the quantities of lime applied becomes increasingly large. With increasing rates of added lime, the pH increases became less substantial for all the above-mentioned three highly responsive buffers (i.e., MC, OX, and AG). Whereas there are significant pH increases up to the 5 % lime addition for all three responsive buffers, after that, at the 7.5 % lime treatment, pH increases for them are small or negligible. The AB clay had previously not increased the pH of AMD, even when 2.5 % of $Ca(OH)_2$ was added, but increased markedly when 5 % $Ca(OH)_2$ was added (Fig. 4). In general, the more responsive the buffer, the earlier it peaks (i.e., stabilizes; Fig. 4). While the MC-treated AMD peaked at just 5 % added lime, the less responsive AB increased the pH of the AMD with 5 % added lime and did not peak, even at 7.5 % added lime. This increase (for AB) occurred from pH 4.5 to >6 units when 7.5 % lime was added, with a pH improvement of 33 %.

At the end of 20 days without any added lime, the oxide achieved the highest pH increase of all the buffers (Fig. 5), being at least equally as responsive as the MC without lime during the 20 day period. In the first 5 days the pH of the MC-treated AMD exceeded that of the oxide-treated AMD (Figs. 4 and 5). Because of its quicker action, the mollisol and not the oxide was chosen as a potential buffer for use in the planned bioreactor. The oxide was, in addition to the above reasons, found to have inhibited the activation of the SRB. No evidence of SRB was detected after a 30-day period following a subsequent trial inoculation of the oxide-treated AMD in stagnant conditions (not illustrated here). However, for the mollisol, the waiting period for SRB was only 5 days. Even without added lime, the MC, OX, and AG substantially increased the AMD pH from 2.3 to >3.0 units at 5 days. These materials therefore exhibited a high ANC (Fig. 4). However, the AG, although a reasonable good buffer at 0- or 2.5 % added lime, changed the physical characteristics of the AMD from watery to a highly viscous consistency because of its swelling. At the end of 21 days, the AG-AMD container was observed to be fully occupied with a swelling, sticky mass which made it difficult to visually demarcate the boundary between the active gel and liquid AMD. If used in the bioreactor at that stage, this viscous AMD would have clogged the tubing in the pumping system. A pre-filtering would have been required, which might have led to potential waste disposal problems. Because the objective of this experiment was to improve the water quality and generally expedite the remediation process for AMD, the above-mentioned uncertainties associated with the AG resulted in its rejection as a buffer for the bioreactor experiment. Although the active bond clay (AB) did not swell appreciably, a pH peak of 5.0 units was not even approached until 5.0 % $Ca(OH)_2$ had been added to the AMD (Fig. 4). The buffering response of the AB without lime was therefore very slow or non- existent because it did not, after 21 days, increase the AMD to a value of even 3.0 units (Fig. 5). Such a small increase in acidity was insufficient, and was even less than that of the buffer used by Harris and Ragusa (2000) for SRB activation in the less drastic environment of a no-flow reactor system (compared with the upward-through-flow system). On the basis of all the above-mentioned results, the MC was chosen to buffer fluids in the following investigations.

3.2 *Bioreactor Experiment*

3.2.1 Appearance

The bioreactor supported sulphate reduction at continuous flow rates of 30, 60, 90, and 180 ml/day. Initially, a test flow rate of 60 ml/day was applied. The pH values rose from 2.3 at port 1 (near the basal entry of AMD influent), to a value of 5.0 in the surface of the water (45 cm upwards from the base) in the column (port 10) in just 4 days. The usual signs of SRB were not detected for a further 20 days. The water became cloudier. The red coloration was gradually lost. The suspended matter of a water sample settled out after 20–40 days. Dissection of the column on the 20th day did not detect an odour of H_2S or the characteristic blackening coloration by iron sulfides, although the influent had a very high measured concentration of Fe ions. This suggested that while the buffer (MC) worked effectively to reduce AMD acidity, high flow conditions inhibited SRB respiration. The flow rate was consequently reduced, and the experiment was conducted at half the influent test rate (i.e., 30 ml day^{-1}). At day 5 the column began to darken in the middle with a strong odour of H_2S when any of the ports 3 through 8 were slightly opened. Hydrogen sulfide occurs as a respiration product of sulfide reduction. However, adaptation periods for the reactor to adjust increased with each higher influent flow rate, which was suggested by an immediate drop or a steadying drop in pH values form all ports on the day after the flow rate increased.

Table 4 Changes in pH with distance from an influent AMD after 10 days in an SRB bioreactor

Location	Distance (cm)	Average pH	pH change (units)
Port 2	5	2.3	+0.1
Port 3	10	2.7	+0.4
Port 4	15	5.8	+3.4
Port 5	20	6.8	+4.5
Port 6	25	6.5	+4.1

3.2.2 Changes in pH and Adaptation of SRB to Acidic Conditions

The response of the reactor to low pH conditions was detected by sampling at column ports on the second day and every third day after the influent flow rate was increased. Figure 4 and Table 4 show the changes in pH at three different levels in the bioreactor. The pH changes were less marked in the upper zone of the reactor. The zone of highest pH readings was not, as had been expected, furthest away from the influent inlet at the water surface at port 10 (45 cm), but was located about three quarters of the distance up from the base (Fig. 4; Table 4). As the influent flow period lengthened after the first day of each flow rate increase, the front of pH increase advanced downwards towards the inlet valve at the base of the column. The zone of the maximum pH values never at any time reached the surface of the AMD (Fig. 4).

After 3 days of operation at 30 ml/day, the pH at 15 cm above the influent inlet (port 4) was 5.4 (Fig. 4). The pH had risen to 5.7 one day after the influent flow rate was increased to 60 ml/day (day 5). Thereafter, it took 4 additional days to recover to the pH of 5.9 cm day 10 (Table 4).

When the influent flow rate was again increased to 90 ml/day, the pH did not drop after 1 day (as it did when the flow rate had been increased from 30- to 60 ml/day). Instead, the pH leveled off, and after 3 days, began to rise to its highest values of the study prior to day 18, at 15 cm (Fig. 4). In general, as long as the influent flow rate through the reactor increases over time by small increments of 30 ml day^{-1}, pH at all ports also increases incrementally at least until day 15. This suggests that the SRB colony was becoming acclimatized to the slowly increasing acidic conditions in the column. However, between days 15 and 18 during the initial period after the influent rate was doubled from 90- to 180 ml day^{-1}, the pH began to fall at most ports (Fig. 4). This suggested an inability of the SRB colony to cope with such a large, sudden increase in acidity. Even though pH dropped at a larger increment at port 7 than at port 10 at day 18, the absolute value was still the highest at port 7. In addition, port 7 (30 cm) was the only position at which pH did not continue to fall at day 18 (Figs. 4 and 5). The subsequent pH recovery for 180 ml day^{-1} did not begin in the zone furthest (45 cm) above the influent inlet, but was observed at 30 cm (Fig. 4). This not only occurred at day 21, just 5 days after the largest influent rate increase, but while pH values were still dropping at the other ports above and below (Figs. 4 and 5).

A laboratory experiment by Harris and Ragusa (2000) also found that a mixture of sewage sludge and plant material (fresh rye grass) was effective in initiating amelioration of acidity and metal concentrations of acid mine waters through sulphate reduction. They were able to increase pH levels from 2.3 to 3 within 30 days. With the organic materials we have tested, 60 days is usually sufficient to achieve a pH of 7, as also shown by Kumar et al. (2011), McCullough and Lund (2011), McCullough et al. (2006). This reflects the sigmoidal response of pH (Fig. 2) where early treatment phases can be slow, most likely due to limited initial supplies of labile organic carbon, which then rapidly accelerate up to pH 6 before reaching an asymptote near pH 7. This does not appear to be an artefact of completing experiments once pH neutrality is acquired, because microcosm pH does not seem to increase above 7–7.5 regardless of how long the experiment had been allowed to continue (Harris and Ragusa 2001; McCullough and Lund 2011). McCullough and Lund (2011) further note that this threshold pH does not appear to be due to limited sulphate or labile organic carbon and further research is required to understand the cause of this phenomenon.

With respect to pH, the interstitial waters in the quartz sand-gravel zone at the bottom were markedly different from the interstitial water in the substrate level above, which contained viable SRB. At day 24, pH in the quartz-gravel-sand layer at the bottom was still unchanged at 2.3, as opposed to pH of 4.0 at 15 cm above the surface of the sand layer, i.e., at the base of the lowest one-third of the column substrate interstices (Fig. 4). Throughout incubation, the quartz-gravel-sand layer never exceeded a pH of 2.4 (Fig. 4). Acidity at a distance of 10 cm up from the surface of the sand bed was also monitored. Even at this height up inside the substrate, pH remained at <3 at day 10 (Table 4), while at just 5 cm further up, the value was 5.8 units. This suggested that SRB were not as active at 10 cm and below, as they were further up the column. Continuous fresh additions of AMD below 10 cm would have been a more demanding environment for SRB.

3.2.3 Cation Concentration

The concentration of the cations Al, Cu, Fe, Mg, Mn and sulphate was monitored by ICPAES. During this study, even the most efficient zone in the column experienced marked decreases in cations and sulphate concentration. After just 6 days of operation the following decreases had occurred at just 15 cm above the influent inlet (Fig. 7):

1. Fe from 1800- to 70 mg L^{-1}
2. Sulphate down from 2700- to 1330 mg L^{-1}
3. Although Al concentration had doubled from 32- to 62- by day 6, it had decreased to 1.8 mg L^{-1} by day 9.

In contrast the metal concentrations were comparable in the quartz-gravel-sand interstices (port 1) to that for the raw AMD (Tables 3 and 4). Changes in the AMD pH with cation concentration after the first 10 days at the lowest influent flow rates

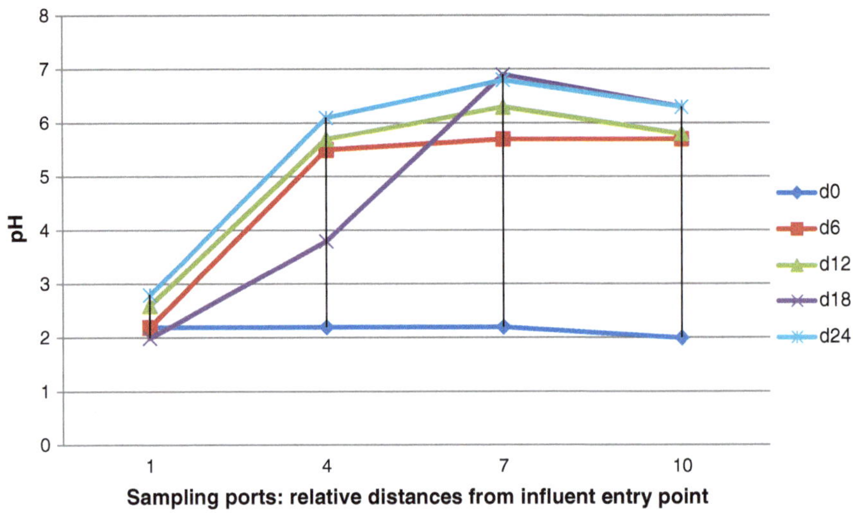

Fig. 6 Changes to pH with distance from AMD entry point in an SRB-driven bioreactor

(30–60 ml day^{-1}) are also shown (Table 4). These and other data (such as Fig. 5) highlight four observations. Firstly, as the incubation progressed, the concentration of heavy metal cations in the AMD appeared to decrease, but by increasingly small amounts. Secondly, as the distance away from the influent entry point increased, metal cation and sulphate concentration in the AMD decreased. Thirdly, with increasing distance up the column from the influent entry point, the cation concentration perturbations following increased flow rates in the AMD decreased in magnitude (Fig. 7). This indicates that effects of an increasing influent flow rate on the metal and sulphate concentration became less marked with distance from the influent entry point. The pH changes inversely with toxic metal concentration over time (Fig. 6).

These observations, the fourth one in particular, suggest the presence of a growing SRB colony despite the increasingly acidic environment (caused by increasing flow rates), even though the nutrient substrate was being used up. An example is shown in Fig. 7a, at the 15 cm port (the 3rd port nearest the influent inlet) where the largest influent increase (on day 16, i.e., from 90- to 180 ml day^{-1}) caused the sulphate concentration to rise most sharply at 18 days from 1300- to 2200 mg L^{-1}(a 59 % increase). It was brought back down to 1750 mg L^{-1} 3 days later by day 21. The general trend describes a gradual lowering of sulphate concentration over time. At the lower column levels (and to some extent at the highest column levels) the largest sudden change to the sulphate and Fe regime coincided with the largest increase in influent flow rates (Fig. 7). At three quarters of the way up the column, sudden changes in influent flow rates caused the smallest perturbations in sulphate and other toxic ion concentrations (Fig. 7).

Fig. 7 Cation concentration at distances from AMD influent entry. **a** 15 cm above influent inlet, **b** 30 cm above influent inlet, and **c** 45 cm above influent inlet

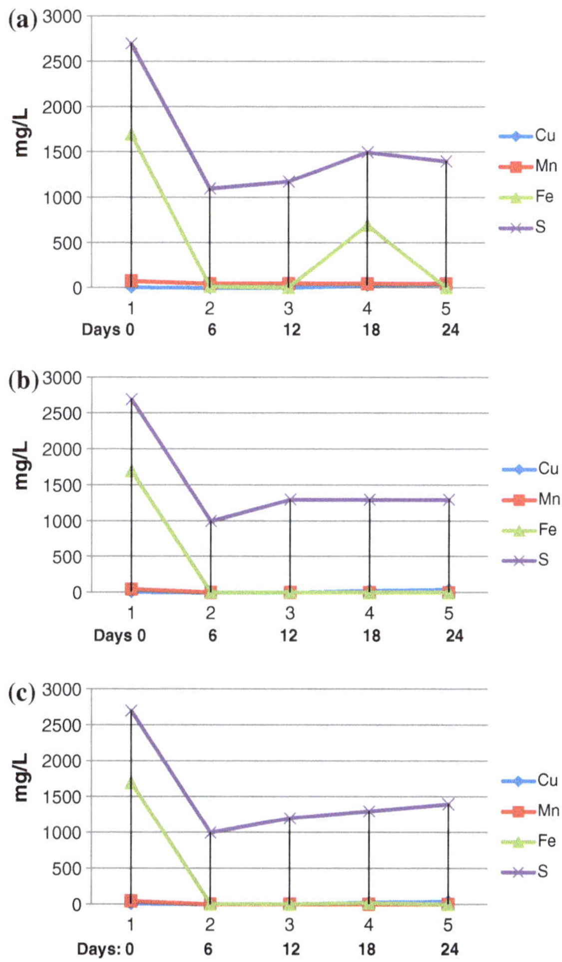

4 Discussion

4.1 Buffer Study

The MC soil of this study contained 40 % clay, whereas the other whole soil (RB) used in the buffer study contained only 19 % clay (Table 2). Throughout the buffer study, the pH of the RB-treated AMD always remained at <4 pH units. Only when the added lime was increased to 7.5 % did the pH of the RB-treated AMD exceed 3.0 units. A previous study indicates that a kaolinite sludge increased the pH of an AMD from 2.3 to 2.6 over a 40-day period (Harris and Ragusa 2000). This is a small increase, and is not dissimilar to that of the RB of the present study. The RB clay fraction was dominated by illite, which is generally unreactive. Because one of the aims of this experiment was to shorten the acclimatization period for SRB activation, it was not necessary to prolong the experimental period merely to identify the peak characteristics of the less efficient buffers. The buffering study was, effectively, stopped at 20 days, when some of the materials had begun to markedly increase the AMD pH to a far greater degree than kaolinites and illites did. Yu (1985) who observed the buffering capacity of three soils consisting of 30-, 40-, and 65 % clay, and Singh et al. (2003) found that buffering capacity increased in ascending order of clay content. The two pure clays that were among the three most effective buffers were the active bond and active gel, both being smectites (montmorillonites) with high CEC's. The third material, a whole soil, i.e., the mollisol, contained high fractions of illite and kaolinite. The native organic matter was 4.6 % (w/w), and on a w/w basis organic matter influences CEC far more greatly than any other factor in the soil (Tisdall and Oades 1982).

4.2 Bioreactor Experiment

The pattern of AMD cation concentration and AMD pH during this incubation is seen to be, in general, an inverse relationship. As pH increases, the concentration of soluble cations decreases, and vice versa (Fig. 6). The rise in AMD pH began before the detection of any evidence of SRB activities in the column. Although evidence from the prior buffer experiment indicates that the MC by itself would have caused a rise in the AMD pH, the pH was at all times below a value of 4.0 units. The second stage of this study must have been the combination of MC submerged with rapidly decomposable plant material that initially caused the rise to >4.0 units in the pH of the AMD. Qualitatively, this is in agreement with the findings of Ponnamperuma (1972), who found that green manures added to soils caused a remarkable rise in pH two weeks after submergence. Singh et al. (1992) found that pH increases after submergence of an acid soil (pH 5.6) with and without green manure went up to 6.0 and 7.1 in pH values, respectively. Their explanation was that organic reducing substances formed during the decomposition of green manures may reduce Fe and Mn oxides. Protons are consumed in the course of oxide reduction, which causes soil pH to rise. The process of reduction could have accounted for the decrease in concentration of the soluble form of those metals 3 days after incubation began, but prior to the activation of SRB. This is corroborated by the findings of Katyal (1977). Using *Sesbania aculeata* as green manure under flooded conditions, he recorded a decrease in (metal) toxicity and favourable changes to pH.

Strong evidence for SRB activity was observed, including production of H_2S, the blackening of the substrate buffer, and the lead acetate test. The sulphate reduction was not as great at the water surface as it was at port 7. Large temporal and spatial differences in the pH and soluble cations in a uniformly blended material were detected.

1. The front of pH increase advanced downwards in the column towards the influent as the incubation proceeded.
2. The sulphate reduction in the influent increased with distance away from the zone of fastest flow (i.e., from where it entered the column).
3. A general increase in pH values towards the top of the column always occurred in this study. The highest pH values were never at any time detected at the water surface, or near the surface. The highest pH values always existed at some distance below the surface zone (but always within the top half of the column).
4. The major inflection point for sulphate and other cations appear in the above-mentioned upper region between ports 3 and 5, where the pH values begin to rise markedly, i.e., about a third of the distance down the column from the water surface.

However, the pH values, although lower at the surface than at 5–8 cm below the surface, never actually fall below the threshold for SRB growth (pH 5.5) in either of these zones. In this uppermost zone, sulfide reduction would not have been greatly hindered after day 3 as a result of pH changes. Even at day 10, sulphate, although less concentrated at port 7, was concentrated at 1300 mg L^{-1} at the water surface in the reactor. The two cations most affected by distance from the inlet port are S and Fe (Fig. 7). As distance increases further up the column from 30- to 45 cm, the differences in sulphate reduction become smaller (Fig. 7a–c). SRB activity did not increase as much between 20- and 45 cm (the water surface) as it did between 15- and 30 cm above the inlet port. A change in sulphate concentration from a decreasing one in the column to an increasing trend at the water surface can be seen (Table 4). Alexander (1977) concluded that even though SRB are active at pH > 5.5, they are most active at a neutral pH. The area around port 7 showed consistently greatest sulphate reduction. Nevertheless, such pH changes in this case are a symptom, not the cause of decreasing SRB activity at the water surface.

The answer may lie in the increasing consumption of oxygen in the upper layers of the AMD. Anoxic to anaerobic ORP is critical for initiation and continued SRB activity for AMD bioremediation (Strosnider and Nairn 2010) and can be, in fact, indicative of SRB activity. Anaerobic bacteria begin to use other electron acceptors in decreasing order of their electrochemical reduction potential once dissolved oxygen (DO) is removed (Castro and Moore 2000). SRB start reducing sulphate after most of the oxidizing species have been consumed, at an anaerobic redox potential between −75 and −200 mV (Connell and Patrick 1968).

In this experiment the AMD did not completely fill the column. In colder months, the top few centimeters of sediments in a lake bed may be oxidized (Herhily and Mills 1985), and SRB is inhibited above the oxidation depth. This is because the process of sulphate reduction requires low oxidation-reduction potentials (Eh) characteristic of anaerobic habitats. Herhily et al. (1987) found that maximal rates of SR occurred at the surface of shallow lake sediments in summer, but at 3–7 cm in other seasons. They explained that this was because the biological oxygen consumption (by aerobic organisms) increased in summer, leading to anoxic conditions at the surface. In this study, the top of the reactor was left open to the air to simulate field conditions. Because the AMD was acidic, few aerobic organisms would have survived at the surface. At the higher column levels near the water surface, more dissolved oxygen would be present, causing an oxygen surplus in the upper zones of the water. The alkalinity may be consumed as the reduced sulfur species diffuse into oxic regions and are oxidized (Herhily et al. 1987). In the present study, oxygen was always present in the column. During its setting up, no attempt was made to avoid exposure to air. There was an air-filled space above the water surface. The dissolved oxygen in the top layer of the water would have inhibited SRB and microbial sulphate reduction. The difference between the sulphate reduction rate in the AMD at 30- and 45 cm could thereby be resolved. Compared with that of a previous study using aerobically treated sewage sludge as a pre-SRB-stage buffer (Harris and Ragusa 2000), the lag phase (i.e., pre-SRB activation time) of this study was shorter. The toxic cations were removed earlier and more efficiently than in that study. Both studies utilized buffers containing similar proportions of clay (40 %), but the sludge in the previous study contained 2.5 % $CaCO_3$ (MA Harris, unpublished data), as opposed to <0.01 $CaCO_3$ for the mollisol used in this study (Table 2). Even without the high proportion of $CaCO_3$ of the sludge, the mollisol is a significantly more efficient pH buffer. The CECs of the two buffers differ markedly differ markedly at 18.4 and 59.1 cmol (+) kg^{-1} respectively. The CEC of the mollisol Ca is four times as high as that of the sludge Ca. These significant CEC differences must have caused the differing adsorptive capacities, and may have resulted from the organic C fractions of the two buffers. CEC can be reduced by as much as 20–50 % by removal of the organic phase, which normally constitutes just 3–5 % of the soil mass. The organic C fractions for the mollisol and the sludge were 4.6- and 1.4 % respectively. They both contained the same proportion of clay. It is reasonable to conclude that the inherent influence of organic matter on the CEC of the mollisol was three times that of its effect on the sludge (Harris and Ragusa 2000). This would be the case because the clay fraction of both soils were dominantly kaolinite-illites. By further increasing the organic C fraction, the addition of decomposable plant material would have increased the CEC in this experiment, and also of the sludge in a previous study (Harris and Ragusa 2000). However, during the incubation with green manure, the mollisol, with the greater native organic fraction, would have exhibited greater increases in CEC.

As the capacity of the bioreactor exceeded 2 L, the 20 g L^{-1} inoculation rate of sewage sludges was seemingly less than half the minimum rate of 30 g L^{-1} minimum required according to Kumar et al. (2013). Thus Kumar et al. (2013) selected 30 g L^{-1} of sewage sludge as the lowest concentration evaluated based on previous studies that indicated that this is the minimum concentration necessary to initiate bioremediation of AMD affected pit lakes (McCullough and Lund 2011; McCullough et al. 2006). Therefore, the explanation of the success of this low rate of sewage sludge used in this study may be the fact that it had been already been SRB-activated, having been aliquots from a prior SRB-producing experiment by

Fig. 8 Variations of pH with AMD concentration of soluble cations in an SRB-driven bioreactor

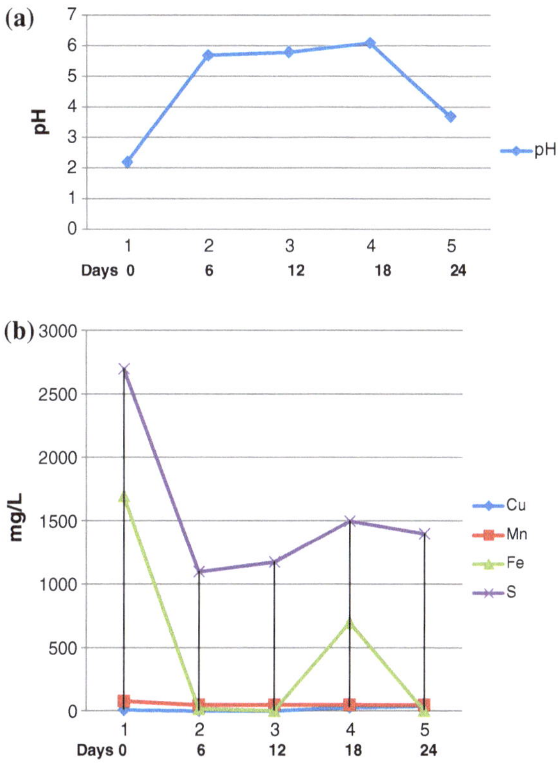

Harris and Ragusa (2000). Such priming of the substrate therefore promises a possible lowering of the 30 mg L^{-1} minimum concentration found for sewage sludge which had not been previously exposed to SRB (Fig. 8).

5 Conclusions

The results of this study show that it is possible to establish an SRB population in a constant flow reactor, using rapidly decomposable material and a fast-acting soil-based buffer. Under such conditions, the SRB can become acclimatized to the high AMD acidity, even when the hydraulic retention time (HRT) is shortened. Had the above-mentioned buffer of a previous study been used in this experiment, it would have been expected to have been significantly less effective because of the more drastic conditions of a moving acidic effluent. Yet, the buffer/substrate materials of the present study proved to be far more effective than those of Harris and Ragusa (2000). It seems CEC, and not merely free lime, was the crucial buffer characteristic at the low pH level of <3.0 units. The CEC of the substrate used in this study was three times that of a sludge buffer used by Harris and Ragusa

(2000). The only effective substrate difference (excluding the influent flow rate) between the present study and that of Harris and Ragusa (2000) is that of the buffers used. Hence, it can be concluded that for pH increases of the soluble cations in the AMD, CEC is a major determining factor among the buffering materials. Where sources of lime are unavailable or insufficient, the use of medium to high organic C soil-based buffers with small quantities of lime can be an effective alternative for activating SRB in a reactor.

6 Geobiotechnological Applications

6.1 A Procedure for the Hydrogeological Confinement of Acid Mine Drainage

By sealing off the source of contaminants, hydrogeological barriers can reduce or prevent the spread of AMD. The objective is to "buy time" required for detoxifying before re-introducing the liquid into the environment. To this end Navarro et al. (2000) dug a perimetrical ditch to catch and drain surface runoff. To determine the dimensions and positioning of the ditch, they identified the existence and propagation of leakages by noting the surge of contaminated water through contiguous aquifers and by analytical data from piezometers.

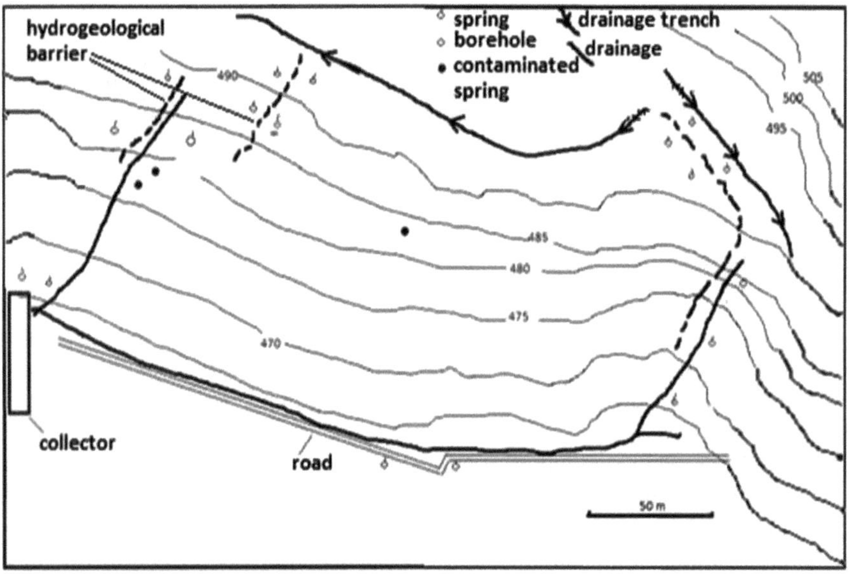

Fig. 9 Hydrogeological containment ground for aqueous contaminants emanating from mine waste tailings heap (after Navarro et al. 2000)

Their method includes control of contamination by preventing water from entering the waste-disposal site, and catching the inevitable outputs of contaminated water for subsequent treatment. The advantage of impermeable screens is the avoidance of high energy costs as pumping then becomes unnecessary. Initial costs of the impermeable barriers however can be extremely high, depending on the areal extent of the waste. The above workers however by taking advantage of natural hydrogeological barriers, reduced construction costs (Fig. 9).

References

Alexander M (1976) Symbiotic nitrogen fixation. In: Alexander M (ed) Introduction to soil microbiology. Wiley, New York, pp 305–330

Alexander M (1977) Soil microbiology, 2nd edn. Wiley, Chichester, pp 362–370

Barrow NJ, Bowden JW (1987) A comparison of models for describing the adsorption of anions A on a variable charge mineral surface. J Colloid Interface Sci 119(1):236–250. doi.org/10.1016/0021-9797(87)90263-3

Bell FG, Donnelly LJ (2006) Mining and its impact on the environment. Taylor & Francis, New York

Bolt GH, Bruggenwert MGM (eds) (1976) Soil Chemistry. Part A. Basic Elements, Elsevier, Amsterdam, The Netherlands, p 281

Bruggenwert MGM, Hiemstra T and Bolt GH (1991) Proton sink in soil controlling soil acidification. In: Ulrich B, Sumner ME (eds) Soil acidity. Springer, Verlag

Castro JM, Moore JN (2000) Pit lakes: their characteristics and the potential for their remediation. Environ Geol 39:1254–1260

Chang IS, Shin PK, Kim BH (2000) Biological treatment of acid mine drainage under sulphate-reducing conditions with solid waste materials as substrate. Water Res 34:1269–1277

Christensen NL, et al. (1996) The report of the ecological society of America committee on the scientific basis for ecosystem management. Ecol Appl (6)3: 665–691

Church CD, Wilkin RT, Alpers CN, Rye RO, McCleskey RB (2007) Microbial sulfate reduction and metal attenuation in pH 4 acid mine water. Geochem Trans 8(10):297–300. http://www.imwa.info/docs/imwa_2008/IMWA2008_076_Martins.pdf

Connel WE, Patrick WH Jr (1968) sulphate reduction in soil: effects of redox potential and pH. Science 159:86–87

Cravotta CA (2003) Ecosystem restoration: size and performance of anoxic limestone drains to neutralize acidic mine drainage. J Environ Qual 32:1277–1289

Dean JG, Bosqui FL, Lanouetle KH (1972) Removing heavy metal from waste waters. Environ Sci Technol 6:518–522

Deshpande TL, Greenland DJ, Quirk JP (1964) Role of iron oxides in the bonding of soil particles. Nature 201:107–108

Díaz de Villalvilla L, Milia I, Santa-Cruz M, Aguirre G (2003) Formación Los Pasos: Geología, Geoquímica y su comparación con el Caribe. In: Estudios sobre los Arcos Volcánicos de Cuba. Centro Nacional de Información Geológica. I.G.P. CD, Cuba, pp. 54–61

Du Bray E (2004) Preliminary compilation of descriptive geoenvironmental mineral deposit models. U.S Geological Survey. Open File Report. 95–831

Dvorak DH, Hedin RS, Edenborn HM, McIntire PB (1992) Treatment of metal-contaminated water using bacterial sulphate reduction: results from pilot-scale reactors. Biotechnology and Bioengineering 40(5):609–616

Elliott P, Ragusa S, Catcheside D (1998) Growth of sulphate reducing bacteria under acidic conditions in an upflow anaerobic bioreactor as a treatment system for acid mine drainage. Water Res 1–7

Figueroa L, Seyler J, Wildeman T (2004) Characterization of organic substrates used for anaerobic bioremediation of mining impacted waters. In: Proceedings of the international mine water association conference, Newcastle, pp 43–52

Frömmichen R, Kellner S, Friese K (2003) Sediment conditioning with organic and/or inorganic carbon sources as a first step in alkalinity generation of acid mine pit lake water (pH 2–3). Environ Sci Tech 37:1414–1421

Gadgil A (1998) Drinking water in developing countries annual review of energy and the environment, vol 23 pp 253–286 (Volume publication date November 1998) doi:10.1146/annurev.energy.23.1.253

Gibert O, de Pablo J, Cortina JL, Ayora C (2002) Treatment of acid mine drainage by sulphate-reducing bacteria using permeable reactive barriers: a review from laboratory to full-scale experiments. Rev Environ Sci Biotechnol 1:327–333

Gyure RA, Konopka A, Brooks A, Doemel W (1990) Microbial sulphate reduction in acidic (pH 3) strip-mine lakes. FEMS Microbiol Ecol 73:193–202

Handreck K (1996) Losses of iron in leachates from organic components of potting media. Communications in Soil Science and Plant Analysis 27(9–10):1996. doi:10.1080/00103629609369688

Hard B, Friedrich CS, Babel W (1997) Bioremediation of acid mine water using facultatively methylotrophic metal-tolerant sulphate reducing bacteria. Microbiol Res 152:65–73

Harris MA, Ragusa S (2000) Bioremediation of acid drainage using decomposable plant material and sludge. Environ Geol 40(1/2):195–215

Harris MA, Ragusa S (2001) Bioremediation of acid mine drainage using decomposable organic matter in a constant-flow bioreactor. Environ Geol 40:1192–1240

Herhily AT, Mills AL (1985) sulphate reduction in freshwater sediments receiving acid mine drainage. Appl Environ Microbiol 49:179–186

Herhily AT, Mills AL, Hornberger GM, Bruckner AE (1987) The importance of sediment sulphate reduction to the sulphate budget of an impoundment receiving acid mine drainage. Water Resources Research 23(2):287–292. doi:10.1029/WR023i002p00287.

Hore-Lacy I (1978) Sulphide ores: an overview of the effect of mining. In: Rummery RA, Howes KMW (eds) Management of lands affected by mining. Division of Land Resources Management, CSIRO, Australia, pp 111–119

James AL, Mrost MJ (1965) Control of acidity of tailings dams and dumps as a precursor to stabilization by vegetation. Journal of the South African Institute of Mining and Metallurgy 65:488–495

Kaksonen AH, Franzmann PD, Puhakaka JA (2003) Performance and ethanol oxidation kinetics of a sulphate-reducing fluidized-bed reactor treating acidic metal-containing wastewater. Biodegradion 14:207–217

Kalin M, Fyson A, Smith MP (1993) ARUM acid reduction using microbiology. In: Biohydrometallurgical technologies. http://origin-ars.els-cdn.com/content/image/1-s2.0-S0883292710000533-si1.gif. Accessed 1 June 2015

Katyal JC (1977) Influence of organic matter on the chemical and electro-chemical properties of some flooded soils. Soil Biol Biochem 9:259–266

Kivaisi AK (2001) The potential for constructed wetlands for wastewater treatment and reuse in developing countries: a review. Ecol Eng 16:545–60

Koschorreck M (2008) Microbial sulphate reduction at a low pH. FEMS Microbiol Ecol 64:329–342

Koschorreck, Boehrer B, Friese K, Geller W, Schultze M, Wendt-Potthoff K (2011) Oxygen depletion induced by adding whey to an enclosure in an acidic mine pit lake. Ecol Eng 37(12):1983–1989

Koschorreck M, Frömmichen R, Herzsprung P, Tittel H, Wendt-Potthoff K (2002) The function of straw for in situ remediation of acidic mining lakes: results from an enclosure experiment. Water Air Soil Pollut 2:97–109

Kuck PH (2009) Nickel: U.S. geological survey mineral commodity summaries 2009, pp 110–111

Kumar RN, McCullough CD, Lund MA (2011) How does storage affect the quality and quantity of organic carbon in sewage for use in the bioremediation of acidic mine waters? Ecol Eng 37:1205–1213

Kumar N, Clint R, McCullough D, Lund MA (2013) Upper and lower Concentration thresholds for Bulk Organic Substrates in Bioremediation of Acid Mine Drainage. Mine Water Environ 32:285–292. doi:10.1007/s10230-013-0242-8

Larsen VJ, Schierup H (1981) The use of straw for removal of heavy metal from waste water. J Environ Qual 10(2):188–193

López JM, Moreira J, Pantaleon GJ, Lavandero RM, Montano J, Cruz-Martín J (1998) Tipos mineralógicos de algunos yacimientos auríferos de Cuba. III Congreso Cubano de Geología y Minería (GEOMIN 98). Geología y Minería 98. La Habana, pp 371–374

Lyew D, Knowles R, Sheppard J (1994) The biological treatment of acid mine drainage under continuous flow conditions in a reactor. Trans Inst Chem Eng 72:42–47

Maiti SK (2003) Moef report, an assessment of overburden dump rehabilitation technologies adopted in CCL, NCL, MCL, and SECL mines (Grant no. J-15012/38/98-IA IIM)

McCullough CD, Lund MA (2011) Bioremediation of acidic and metalliferous drainage (AMD) through organic carbon amendment by municipal sewage and green waste. J Environ Manage 92:2419–2426

McCullough CD, Lund MA, May JM (2006) Microcosm testing of municipal sewage and green waste for full-scale remediation of an acid coal pit lake, in semi-arid tropical Australia. In: Proceedings of the 7th international conference on acid rock drainage (ICARD). American society of mining and reclamation (ASMR), St Louis, MO

Mueller AG, Hall GC, Nemchin AA, O'Brien Darren (2008) Chronology of the Pueblo Viejo epithermal gold-silver deposit, Dominican Republic—Formation in an Early Cretaceous intra-oceanic island arc and burial under ophiolite: Minera Deposita, Nov 2008, vol 43(8), pp 873–889

Muga HE, Mihelcic JR (2008) Sustainability of wastewater treatment technologies. J. Environ. Manage. 88:437–447

Muyzer G, Stams AJM (2008) The ecology and biotechnology of sulphate-reducing bacteria. Nat Rev Microbiol 6:441–454

Nagpal S, Chuichulcherm S, Livingston A, Peeva L (2000) Ethanol utilization by sulphate-reducing bacteria: an experimental and modeling study. Biotechnol Bioeng 70:533–543

Navarro JA, Lopez Caloma, Perez Garcia A (2000) Characterization of groundwater flow in the Bailin hazardous waste-disposal site (Huesca, Spain). Environ Geol 40(1, 2):217–222

Norton SA, Kahl JS, Henriksen A, Wright RF (1990) Buffering of pH depressions by sediments in streams and lakes. In: Norton SA, Lindberg SE, Page AL (eds) Acidic precipitation, vol 4., Soils, aquatic processes and lake acidificationSpringer, Berlin, pp 132–157

Ponnamperuma FM (1972) The chemistry of submerged soils. Adv Agron 24:29–96

Postgate J (1984) The Sulphate-Reducing Bacteria, 2nd edn. Cambridge University Press, Cambridge

Primelles L, Alvarado B, Torres M (2003) Campo Mineral Maclama. Breve Caracterización Geológica y Mineralógica de la mineralización Oro-Telurídica. Tipos esperados de depósitos. In: Estudios sobre los Arcos Volcánicos de Cuba. CD. Centro Nacional de Información Geológica. I.G.P, Cuba, pp 33–42

Primelles L, Barroso A, Lugo R, Escobar E (1998) Geología y Metalogenia del Campo Mineral Guaimaro. Enfoque actual. III Congreso Cubano de Geología y Minería (GEOMIN 98). La Habana. pp 390–393

Redwood SD (2009) Dominican Republic packs a punch: Mining Journal, Jan 23, p. 4

Sheoran V, Sheoran AS, Poonia P (2010) Soil reclamation of abandoned mine land by revegetation: a review. Int J Soil Sediment Water 3(2), Article 13, ISSN 1940-3259

Simon G, Kessler SE, Russell N, Hall CM, Bell D, Piñero E (1999) Epithermal gold mineralization in an old volcanic arc: the Jacinto deposit, Camagüey District. Cuba. Economic Geology. 94:487–506

Singh B, Odeh IOA, McBratney AB (2003) Acid buffering capacity and potential acidification of cotton soils in northern New South Wales. Aust J Soil Res 41:875–888

Singh Y, Singh B, Khind CS (1992) Nutrient transformations in soils amended with green manures. Adv Soil Sci 20:238–305

Smith HA, Stephenson PR (2011) Pueblo Viejo gold project Dominican Republic Technical Report for Pueblo Viejo Dominicana Corporation, Barrick gold corporation. Goldcorp Inc, AMC Mining Consultants (Canada) Ltd

Smith HA, Stephenson PR, Butcher MG, Carr CA (2008) Pueblo Viejo Gold Project: Dominican Republic Technical Report. Canadian Securities Administration. https://www.sec.gov/.../o40716exv99w1.htm

Strosnider WH, Nairn RW (2010) Effective passive treatment of high strength acid mine drainage and raw municipal wastewater in Potosí, Bolivia using simple mutual incubations and limestone. J Geochem Explor 105:34–42

Strosnider WH, Winfrey BK, Nairn RW (2009) Performance of an ecologically-engineered multi-stage acid mine drainage and municipal wastewater passive co-treatment system. Paper was presented at the 2009 National Meeting of the American Society of Mining and Reclamation, Billings, MT. In: Barnhisel RI (ed) Revitalizing the environment: proven solutions and innovative approaches, 30 May–5 June 2009. Published by ASMR, Lexington

Tisdall JM, Oades JM (1982) Organic matter and water-stable aggregates in soils. J Soil Sci 33:141–163

Toth J (1968) Deposition of submarine crusts rich in manganese and iron. Geological Society of America Special Paper 521(91)1, pp. 44–54, doi:10.1130/0016-7606(1980)9

Tsukamoto TK, Killion HA, Miller GC (2004) Column experiments for microbiological treatment of acid mine drainage: low-temperature, low-pH and matrix investigations. Water Res 38:1405–1418

Van Breemen N, Mulder J, Driscoll CT (1983) Acidification and alkalization of soils. Plant & Soil 75(3):283–308

Wacaster S (2010) The mineral industries of the islands of the Caribbean Aruba, The Bahamas, Barbados, Dominican Republic, Jamaica, Trinidad and Tobago, and Other Islands. U.S. Department of the Interior, U.S. Geological Survey. August 2010

Watzlaf GR, Kairies CL and Schroeder KT (2003) Flushing of metals from reducing and alkaline producing systems. Paper presented at the 2003 National Meeting of the American Society of Mining and Reclamation and the 9th Billings Land Reclamation Symposium, Billings, MT, June 3–6, 2003. Published by ASMR, 3134 Montavesta Rd., Lexington, KY 40502

Waybrant KR, Blowes DW, Ptacek CJ (1998) Selection of reactive mixtures for use in permeable reactive walls for treatment of mine drainage. Environ Sci Tech 32:1972–1979

Whitbread-Abrutat PH (2008) Mining legacy survey: informing the background paper [cited 2012 August 7]. Post-Mining Alliance, Eden Project. Available from: http://www.postmining.org/documents/surveysummaryreportFINAL25Feb08.pdf

Younger PL, Banwart SA, Hedin RS (2002) Mine Water: Hydrology, Pollution. Kluwer Academic Publishers, Dordrecht, Netherlands, Remediation

Yu Tian-ren (1985) Physical chemistry of paddy soils. Springer, Berlin Heidelberg New York

Part IV
Mine-Spoils and Mine Tailings: Microbial Remediation

Aggregation of Sodic Subsoils Using Gypsum and Decomposable Phyto-organics: Interactions and Implications for Bauxite Red Muds

Mark Anglin Harris and Pichu Rengasamy

Abstract The ability of gypsum with or without the addition of N-rich decomposable phyto-organics to increase soil aggregation in sodic subsoils was examined. Before flowering, young leaves from *Vicia sativa (common vetch)* were finely chopped to <2 mm particles and incubated into the subsoil at 80 % field capacity for 12 weeks. Experiments were carried out on two alkaline sodium affected sub-soils (15–30 cm) from Strathalbyn, and Two Wells in South Australia, under glasshouse conditions in pots containing 1 kg of soil. To a sodic subsoil, rapid encrustation was induced to a depth of 8 mm. The average size of dispersed materials in control soils was < 5 μm, whereas after green manuring, the average particle size increased up to 30 μm. Such particles were insufficient to resist translocation by light to moderate winds. The products of the released Ca^{2+} from native lime aggregated the clay particles and stabilized them into masses which exceeded 100 μm in diameter. The results of this study promise the use of gypsum as an ameliorant for sodic alkaline environments such as the red muds of bauxite mine-spoils. Where no organic matter had been added, it is concluded that the binding action of Ca^{2+} released in solution replaced the Na^{+1} which had been dispersing the particles prior to the treatment.

Keywords Air-pollution · Bauxite dust · Bauxite wastes · Caribbean red muds · Gypsum-amendments

---

Original article: **Sodium affected subsoils, gypsum, and green-manure: Inter- actions and implications for amelioration of toxic red mud wastes**. *Environmental Geology* Volume 45, Issue 8, pp. 1118–1130.

M.A. Harris, *Geobiotechnological Solutions to Anthropogenic Disturbances*, Environmental Earth Sciences, DOI 10.1007/978-3-319-30465-6_9

1 Introduction

1.1 Effects of Particle Size

The extraction of aluminium oxide from bauxite creates an environmental problem of large lakes containing toxic red mud. In addition to polluting underground water supplies (Zhang et al. 2001), the caustic soda applied (involving the dumping of millions of tons of finely ground ore into tanks of hot caustic soda under high pressure) in alumina extraction completely breaks down the soil fabric by disrupting its clay microstructure and inhibiting re-alignment of clay particles. The resultant sodium clays are eventually characterized by thick, tough surface crusts in dry conditions, and water logging under even gentle rain. Exchangeable sodium in concentrations above 15 % stifles plant growth (Fig. 1), severely restricting the capability of roots to absorb nutrients, regardless of the amounts of inorganic fertilizer added. With the added problems of high pH and the accompanying toxicity, revegetation programs in such environments are doomed. Furthermore, the salinity risk in surrounding soils and ground water is vastly increased.

Having been derived from highly weathered humid tropical materials producing mostly insoluble secondary clay minerals, red mud tailings from the Bayer process exhibit a particle diameter of between 0.8 and 50 μm with an average value of 14.8 μm (Ping and Dong-Yan 2012). Nevertheless, the prevalence of the fine fractions has been cited as a major constraint limiting red mud reclamation efforts (Courtney et al. 2013). Thus, in an unamended state, mine residues exhibit

Fig. 1 Surface crusting of a sodic soil. The soil attains high strength which restricts root penetration. *Source* usgs.gov

degraded soil structure as well as other chemical and biological characteristics which are inhibitory to long-term plant growth. For example, at a bauxite residue disposal area in South-west Ireland, a series of restoration treatments were investigated for evidence of soil development (Courtney et al. 2013). Unamended residue had high pH, sodicity, salinity, dominance of fine fraction and poor aggregate stability as well as low carbon content (Courtney et al. 2013).

Soil aggregates have been divided into two broad categories based on their size: macro-aggregates >250 μm diameter and micro-aggregates <250 μm diameter. For macro-aggregates >2 mm in diameter, the most effective binding agents are plant roots and fungal hyphae (Tisdall and Oades 1982). With decreasing size of macro-aggregates, however, physically smaller entities, such as polysaccharides of plant and microbial origin become important stabilizing agents. Micro-aggregate stability depends on the dispersion and coagulation of individual or packets of clay particles, including the composition and concentration of electrolyte in the soil solution.

Deflocculation occurs because unlike the polyvalent cations of calcium and aluminium, sodium is monovalent. When sodium is adsorbed onto the negatively charged particles it can only attach to one particle and hence unlike with calcium and aluminium it cannot form a bridge between the one it is attached and any nearby. This means the two particles will repel each other and create a dispersed soil condition (Figs. 2 and 3).

Sekhon and Bajwa (1993) observed that the enhancement of the release of Ca^{2+} from $CaCO_3$ into the soil solution can be achieved through the incorporation of rapidly decomposing organic materials. The Ca^{2+} reduces dispersion of clay particles from aggregates, by becoming cation bridges between negatively charged clay particles, and between organic anionic groups.

1.2 Chemical Conditions

Due to the existence of sodium carbonate in red muds, a very important barrier to remediation, re-use and long term sustainability of bauxite residue management is difficult due to its high alkalinity. Although high pH of sodic soils has no direct adverse effect on plant growth per se, it lowers the availability of microorganisms with the exception of the alkalophilic ones. High pH, low organic matter content and presence of calcium carbonate strongly modify the ability of microorganism to exist in the soil medium.

1.3 Gypsum Justification

Gypsum is chemically $CaSO_4{\cdot}2H_2O$ and is a white mineral that occurs extensively in natural deposits. It is most effective when it is ground before it is applied to the

Fig. 2 Dispersive behaviour of sodic soil crumb in rain water compared to that of non-sodic soil crumb (*top left*)

soil. Gypsum is soluble in water to the extent of about 0.25 %, and is, therefore, a direct source of soluble calcium. Gypsum reacts with both the Na_2CO_3, and the adsorbed sodium as follows:

$$Na_2CO_3 + CaSO_4 \rightarrow CaSO_3 + Na_2SO_4 \text{ (leachable)} \quad (1)$$

Calcium sulphate is a salt, but unlike sodium chloride it is not toxic to plants. Gypsum slightly increases the salinity of the soil solution, and hence reduces swelling and dispersion without detrimental effect on plants through two mechanisms. Though the same effect can be seen when using saline bore water, such water often contains high levels of sodium and chlorine that are toxic to plants. Having added gypsum to soil/red mud mixtures Lehoux et al. (2013) found that even at relatively low concentrations (1 % w/w), the gypsum was sufficient to buffer experimental pH to 7.5–8.5. This effect was attributed to the reaction of (1) Ca^{2+} supplied by the gypsum with OH^- and (2) carbonate from the red mud to precipitate calcite (by removing carbonate, calcite indirectly decreases production of OH^-, as stated above).

Fig. 3 Bayer-processed red mud waste behaves in a similar manner to the sodic sub-soil in Fig. 2 (*right beaker*). *Source* Jamaica Bauxite Institute

Gypsum addition could therefore provide a cheaper alternative to the "dig and dump" method for the treatment of red mud-affected soils (Lehoux et al. 2013). The observed inhibition of trace metal release within red mud-affected soils was relatively insensitive to either the percentage of red mud or gypsum present, thereby lowering the risk of the treatment. The lowered pH enhanced trace element sorption and largely inhibited the release of Al, As and V (Lehoux et al. 2013). However, they concluded that there is the risk that excessive application of gypsum could lead to detrimental long-term increases in soil salinity.

1.4 Aggregation

Soil aggregates have been divided into two broad categories based on their size: macro-aggregates >250 μm diameter and micro-aggregates <250 μm diameter. For macro-aggregates >2 mm in diameter, the most effective binding agents are plant roots and fungal hyphae (Tisdall and Oades 1982). With decreasing size of macro-aggregates, however, physically smaller entities, such as polysaccharides of plant and microbial origin become important stabilizing agents. Micro-aggregate stability depends on the dispersion and coagulation of individual or packets of clay particles, including the composition and concentration of electrolyte in the

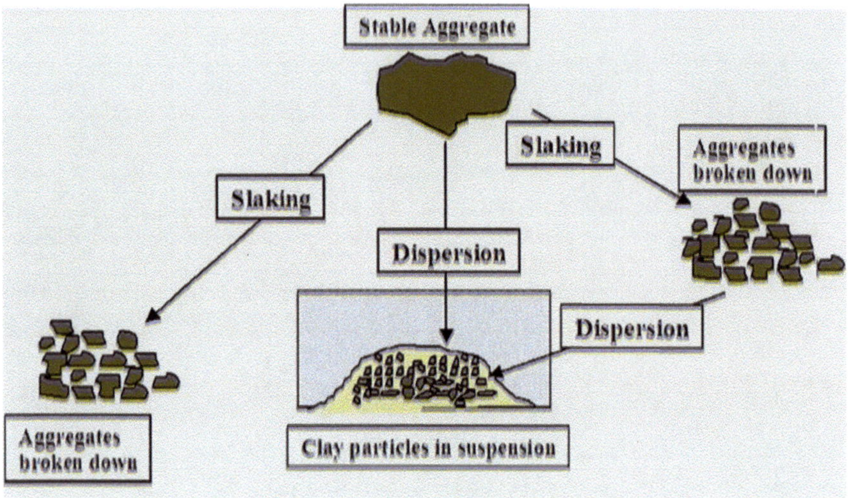

Fig. 4 The various soil physical conditions. Adapted from nrcs.usda.gov

soil solution. High sodicity causes clay to swell excessively when wet. The clay particles move so far apart that they separate (disperse) (Fig. 4). This weakens the aggregates in the soil, causing structural collapse and closing-off of pores.

1.5 Role of Decomposable Organics

Applying sub-treatments of six different soil amendments (filter cake, mill ash, compost, gypsum, dolomite and rice husk) and a control with no amendment, Khodphuwiang et al. (2007) found that minimum tillage with any of the soil amendments gave higher sugar cane yield and number of millable canes, both in plant and ratoon cane. However, none of these amendments were readily decomposable, and no differences in cane yield were found among the treatments with different types of soil amendments except that the application of 15.0 t ha^{-1} of filter cake (dry weight) tended to give higher yield and quality of cane. This similarity of yield results among decomposed or recalcitrant organic additives suggest that changes in soil physical condition of the soil caused the increases in yield. Further, Nelson et al. (2001) did not find significant differences between various soil amendment applications, though soil chemical properties SAR improved by the minimum tillage management. As fresh, readily decomposable organic matter had not been used with gypsum, it is postulated here that such additives could have stronger effects on the soil physical properties, and, by extension, on crop yields.

1.6 Hypothesis

The effectiveness of green manure with or without gypsum to ameliorate and restore the structural integrity of sodic subsoils of a high pH status will be investigated. The hypothesis for this study is two-fold: (1) when used in sodic subsoils, green manures can increase micro-aggregation of dispersed clays (2) Gypsum acting alone can increase the micro-aggregation in sodic subsoils.

2 Methods and Materials

Table 1 shows the main features of two sodic subsoils obtained from a 15–30 cm depth of a sodic red-brown earth in South Australia from (1) Strathalbyn, and (2) Two Wells. The pH (>9) readings were obtained from 1:5 (v/v) soil/water extracts.

2.1 Soil Pre-treatment

The sodic subsoils were sifted through a 2-mm sieve and thoroughly hand mixed with gypsum during air drying, where the equivalent of 20 t ha^{-1} of gypsum was added to the soil in half the number of pots, the other half being "no gypsum" controls. The pots were lined on the inside with a fine meshed nylon cloth to

Table 1 Properties of calcareous, sodic sub-soils (15–30 cm)

Sub-soil property	Strathalbyn	Two Wells
Soil pH (soil:water dilution 1:5)	9.1	9.5
EC (1:5) sludge:water extract (dS m^{-1})	0.25	0.52
Organic carbon (%)		
CEC (mmolc kg^{-1})	39	16
SAR	16	15
Sand	45	67.1
$CaCO_3$	45	12.9
Particle size distribution		
Sand		
Silt	0.25	0.52
Clay (%)		
Clay mineralogy (% of clay fraction)		
Illite	28	32
Kaolinite	14	18
Quartz	15	4
Randomly interstratified minerals	43	45

reduce losses of clay particles during the drainage. The soil was placed in pots of dimensions 12 cm deep, 50 cm wide and 80 cm long. The soils were then wet to 80 % saturation. Six weekly leaching cycles followed, where the water was slowly added from a dropper onto a layer of inert plastic granules resting on a nylon cloth placed on the soil surface. This mechanism reduced the mechanical impact of the water drops on the soil. At the end of the six week leaching period, soils were air-dried and plants were grown in them.

2.2 Incubations

For the incubations of green manure and gypsum in the sodic soils, the following treatments were applied:

1. Gypsum 20 t ha^{-1} (G)
2. Gypsum 20 t ha^{-1} green manure 20 t ha^{-1} (GMGY)
3. Green manure 20 t ha^{-1} (GM)
4. Control (no gypsum, no green manure).

Phyto-organics were prepared after Harris and Rengasamy (2004b). The following techniques were used to assess the soil structural stability as affected by the various treatments.

2.3 Water Stable Macro-Aggregation, Hydraulic Conductivity

The method of Kemper and Rosenau (1986) was adopted and saturated hydraulic conductivity after Klute (1986).

2.4 Spontaneous Dispersion

The major factor affecting hydraulic conductivity in sodic soils is dispersed clay. Even as little as 1 % of the total clay when dispersed affects the hydraulic conductivity by blocking micro-channels in the soil mass. Spontaneous dispersion was calculated as follows. Distilled water was slowly and carefully added down the side of a flask, to 25 g of air-dry soil aggregates without disturbing the soil. The flask was then left to stand for 24 h. The water above the soil was then stirred for 15 s and 10 ml of clay suspension removed using a pipette. The clay obtained by spontaneous dispersion of the soil was calculated as a percentage of the total soil weight on an oven-dry basis.

2.5 Mechanical Dispersion

For the determination of mechanically dispersible clay, the samples prepared as previously were placed in an "end over end" mechanical shaker for 24 h. After 8 h of standing, 10 ml of the suspension was removed and then oven-dried for 24 h. The mechanically dispersed clay was calculated as a percentage of total oven-dried soil.

Particle size analysis of dispersed material
Particle size analysis of the suspension measures the average size of soil particles after mechanical dispersion. Clay particles can be bound together into floccules by cements (Warkentin 1982; Emerson 1983). The size of these are therefore an indication of the level of aggregation of clay particles caused by treatments. After soils were mechanically dispersed (1:5 soil water), particle size distribution was measured in suspensions by using a NICOMP model C370 submicron particle sizer version 5. The standardization was done using 0.09-lm latex rubber spheres. The values were within 5 % error.

2.6 Electrophoretic Mobility

Because dispersion is reduced by lowering the repulsive charges of the clay particles, the measurement of particle charge and their mobility reveals the effect of the treatments on clay dispersion. This is because organo-mineral complexing neutralizes such charges, thereby lowering the electrophoretic mobility of particles. The following Smoluchowski equation is related to the electrophoretic mobility of the particles (Pashley 1985):

$$\text{Zeta} = 4\pi n\mu e/\acute{E} \quad (2)$$

where zeta is the potential in mV, n is the electrophoretic mobility (mm s^{-1} V^{-1} cm^{-1}), μe is the viscosity of the medium, and $\acute{E}$ is the dielectric constant. For aqueous media at 20 °C, the relation between electrophoretic mobility and the zeta potential is 12.8 mV/mobility unit. The zeta potential was measured using a Malvern Zetamaster, "Zetamaster particle electrophoresis analyzer". All zeta potential measurements were at constant temperature (20 °C) and with a constant field strength of 80 V cm^{-1}. Measurements were carried out over a 30-s run time and the final values used were prepared in reverse osmosis (RO) water. The zeta potential measurements in this study were all negative, indicative of the anionic charge on the clay surfaces causing the clay to migrate towards the cathode. The pH, electrical conductivity (EC), total cation concentration (TCC) and sodium adsorption ratio (SAR) were measured by standard methods (Rengasamy et al. 1984).

3 Results and Discussion

3.1 Soil Chemical Analyses

A decreasing pH trend occurred for all treatments at 4, 8, and 12 weeks for 1:5 soil: water extracts (Table 2; Fig. 5), but the untreated original control samples (no gypsum, no green manure) fluctuated above values recorded before the incubation. For both subsoils, decreases in pH by GY were comparable to those of the GM treatments (Table 2). At week 12, the pH readings were, in order of effectiveness: GMGY Strath 7.86 >GMGY Two Wells 8.12 >GY Strath 8.21 >GM Strath 8.32 >GY Two Wells 8.34. Thus, the most effective treatment was GMGY, especially for the Strathalbyn samples (from pH 9.65 to 7.86), which contained more native lime (Table 2; Fig. 5). However the pH value of 7.86 occurred only for the more calcareous subsoil, and was not achieved until week 9. This is discussed later. Ponnamperuma (1972) reported small but significant and steady pH decreases in submerged alkaline soils with or without additives while for very acidic soils (pH 3.5 soils) under submerged conditions, Harris and Megharaj (2001) after 12 weeks observed a slight but increasing trend in pH values (to pH 4.2) without additives, and from pH 3.5 to pH 5.6 with a GM additive. This corroborates Ponnamperuma (1972) who also stated that regardless of its initial pH, a soil under submerged conditions eventually achieves pH neutrality, a trend enhanced by decaying organic matter in the soil. In the present study, the control samples (no GM, no GY), though kept continually moist, were not submerged, and moved slightly

Table 2 **a** Effects of green manure on the pH of two sodic subsoils (no added gypsum). **b** Effects of gypsum on the pH of two sodic subsoils (no added gypsum)

Weeks of incubation				
Green manure (t ha^{-1})	Week 0	Week 4	Week 8	Week 12
(A)				
Two Wells				
0	9.2^{g}	9.7^{a}	9.6^{i}	9.6^{i}
20	9.2^{g}	8.9^{e}	8.8^{d}	8.2^{b}
Strathalbyn				
0	9.1^{f}	9.3^{h}	9.1^{f}	9.3^{h}
20	9.1^{f}	8.7^{c}	8.2^{b}	8.1^{a}
(B)				
Two Wells				
0	9.3^{g}	9.8^{a}	9.7^{i}	9.7^{i}
20	9.3^{g}	8.9^{e}	8.8^{d}	8.3^{b}
Strathalbyn				
0	9.1^{f}	9.2^{h}	9.1^{f}	9.2^{h}
20	9.1^{f}	8.6^{c}	8.2^{b}	8.1^{a}

Significance ($P < F$ from ANOVA) = 0.001. Column values followed by different letters are significantly different ($P < 0.05$) as determined by Tukey's HSD

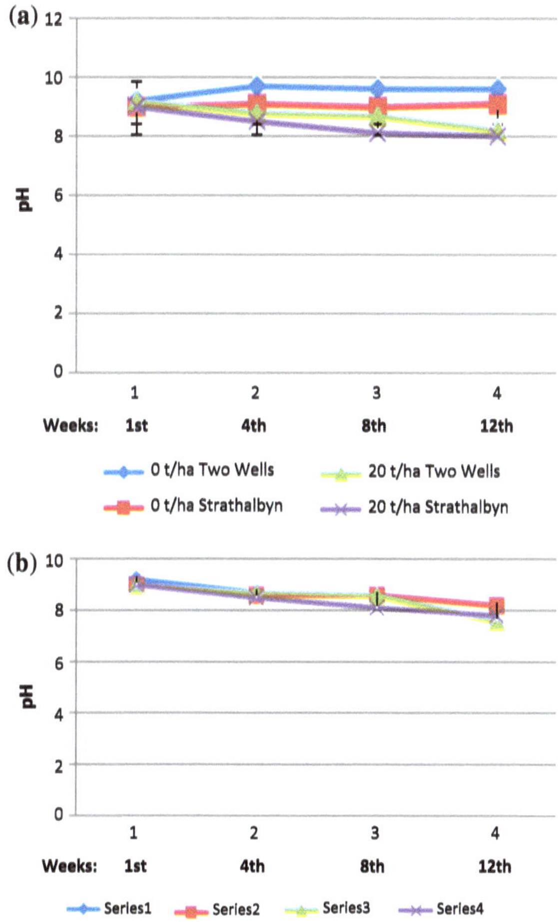

Fig. 5 **a** Effects of green manure without gypsum on the pH of two sodic soils. **b** Effects of green manure with gypsum on the pH of two sodic soils

away from pH neutrality. Two possible factors could have caused this movement away from pH neutrality in the controls of this study: (1) though continually moist, the samples were not submerged, and (2) there was insufficient organic matter (approximately 0.5 %, Table 1) in the control samples concerned, to furnish protons which could have induced a reduction in pH. Thus the decreasing pH trend (from pH 9) exhibited by the GM-impregnated samples (0.97 pH units for Strathalbyn, 1.19 for Two Wells), and continued throughout incubation (Table 2; Fig. 5), may have been even greater had the samples been submerged.

3.2 Gypsum and Green Manure as Separate Treatments

GM almost doubled EC when the samples were first treated with gypsum (Fig. 6). The Strathalbyn GMGY samples at 12 weeks were 30 % more conductive than

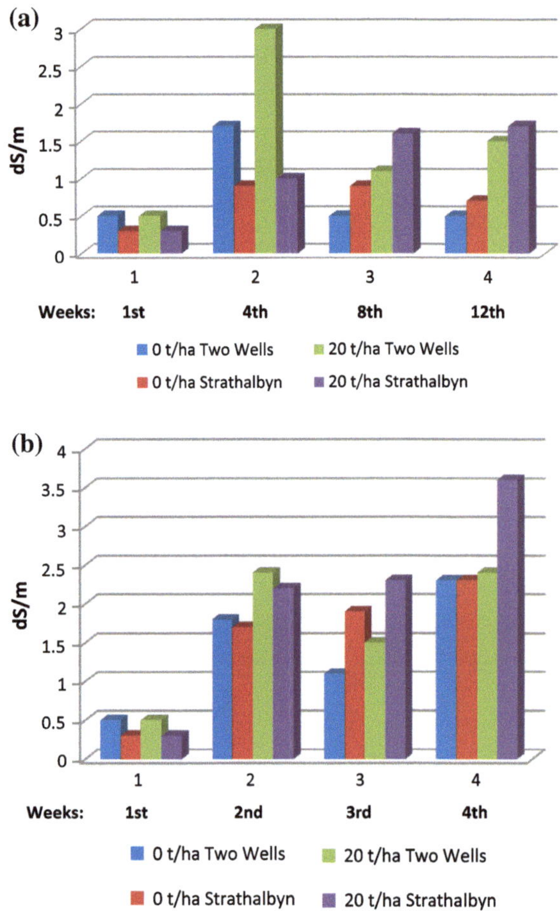

Fig. 6 **a** Effect of green manure without gypsum on the EC of two sodic soils. **b** Effect of phyto-organics with gypsum during incubation, on the EC of two sodic soils

the corresponding Two Wells samples (Fig. 6). Both subsoils were tested for the presence of dissolved cations against controls in 1:5 (v/v) soil:water suspensions. Analysis revealed that the Ca^{2+} cation count at weeks 8 and 12 for the treated Strathalbyn subsoils was significantly higher than that for any other time or treatment in this study (Fig. 7). Because the rate of applied gypsum had been the same in all soils, the extra Ca^{2+} found in the Strathalbyn soil at 12 weeks must have come from the indigenous $CaCO_3$ in those samples. The higher conductivity for the GY samples (compared to the non-gypsum treated calcareous Strathalbyn samples) can be explained by elevated Ca^{2+} concentrations consistent with its greater solubility compared to that of native lime in both subsoil controls. This probably led to the strong increase at weeks 8 through 12 for the Strathalbyn samples, as its native $CaCO_3$, which was not as soluble as gypsum, would not have yielded as many cations in the earlier stages of incubation. Over 12 weeks of incubation, SAR decreases occurred in the following descending order:

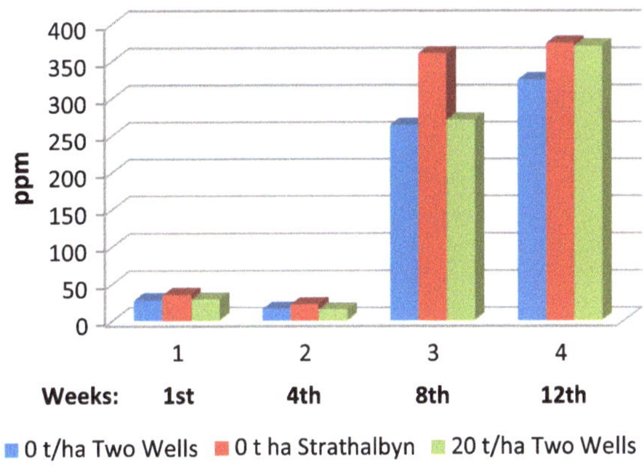

Fig. 7 The effect of phyto-organics, sampling site, and weeks of incubation, on the soluble calcium (ppm) of two sodic subsoils

GMGY Two Wells >GMGY Strathalbyn >GY Two Wells >GY Strathalbyn >GM Strathalbyn >GM Two Wells (Fig. 7).

Thus, GM with gypsum was more effective than either gypsum or green manure acting alone. Green manure without gypsum was more effective in Strathalbyn samples compared to those of Two Wells, halving the SAR of the Two Wells samples and decreasing by approximately 2/3 that of the Strathalbyn samples (Fig. 7). However, gypsum treatment and Two Wells samples, respectively, were observed (Fig. 7). The combined treatment of GM with GY was the most effective for either subsoil, as it caused 7- and 9-fold SAR decreases, respectively, for Two Wells and Strathalbyn samples. When values for both subsoils are combined, all these substantial SAR decreases correlate very strongly with Ksat increases (Tables 3 and 4), confirming that a decrease in the number of sodium cations attached to the clay particles decrease dislocation of, and pore blockage by those clay particles. High correlations of 0.96 and 0.68 for the Strathalbyn and Two Wells samples, respectively, were found between SAR and EC, and EC and K_{sat}, especially for the GMGY samples, where EC values doubled those of the GM treatments for either soil (Tables 3 and 4). Clay dispersion is reduced by lowering SAR and increasing EC of the soil solution (Quirk and Schofield 1955; Rengasamy and Olsson 1991; Sumner 1993). Acid production during green manure decomposition (as exhibited by falling pH values), either as organic acids or as H_2CO_3 from CO_2 evolution, resulted in the dissolution of some of the native $CaCO_3$ present in the soil. This process led to the increase of Ca^{2+} in solution, which simultaneously reduced SAR and increased EC.

Thus, the beneficial reaction of protons from green manuring depended on the action of products of green manure decomposition and on indigenous $CaCO_3$ in the soil. Because the Strathalbyn soil contained more indigenous $CaCO_3$ (14 %) than the Two Wells soil (4 % $CaCO_3$), more Ca^{2+} was released from the Strathalbyn soil, hence causing a lower SAR (Table 1; Fig. 8). However, SAR

Table 3 Linear correlation matrix of measured parameters for Strathalbyn subsoil

	pH	EC	TCC	spdis (%)	mcdis (%)	zeta	z mob	ps (nm)	SAR	cm/h	% wsa
pH	1										
EC	−0.92	1.00									
TCC	−0.36	0.16	1.00								
spdis (%)	0.32	−0.31	−0.34	1.00							
mcdis (%)	0.70	−0.56	−0.45	0.23	1.00						
zeta	0.30	0.39	0.38	−0.26	−0.25	1.00					
z mob	−0.42	0.41	0.45	−0.56	−0.58	0.74	1.00				
ps (nm)	−0.64	0.75	0.26	−0.37	−0.39	0.47	0.58	1.00			
SAR	0.84	−0.86	−0.23	0.37	0.63	−0.30	−0.29	−0.58	1.00		
cm/h	0.93	0.96	0.29	−0.31	−0.64	0.33	0.35	0.70	−0.94	1.00	
% wsa	0.14	−0.15	0.24	−0.25	0.32	0.53	0.10	−0.24	0.06	−0.19	1.00

PH pH; *EC* electrical conductivity; *TCC* total cation capacity; *spdis %* spontaneous dispersion; *modis %* mechanical dispersion; *zeta* zeta potential; *z mob* electrophoretic mobility; *ps (nm)* particle size; *SAR* sodium adsorption ratio; *cm:h* Ks; *% wsa %* waterstable macro-aggregates alone was more effective than GM, in that 3- and 5-fold SAR decreases for the Strathalbyn

Table 4 Linear correlation matrix of measured parameters for Two Wells subsoil

	pH	EC	TCC	spdis (%)	mcdis (%)	zeta	z mob	ps (nm)	SAR	cm/h	% wsa
pH	1										
EC	−0.35	1.00									
TCC	−0.67	0.02	1.00								
spdis (%)	0.61	−0.27	−0.60	1.00							
mcdis (%)	0.67	−0.25	−0.26	0.48	1.00						
zeta	−0.52	0.68	0.03	−0.27	−0.50	1.00					
z mob	−0.55	0.63	0.05	−0.47	−0.70	0.94	1.00				
ps (nm)	−0.42	0.22	0.48	−0.49	−0.49	0.45	0.56	1.00			
SAR	0.72	−0.42	−0.22	0.37	0.84	−0.89	−0.87	−0.54	1.00		
cm/h	0.57	0.34	0.29	−0.29	−0.58	0.76	0.69	0.76	−0.85	1.00	
% wsa	0.48	−0.46	0.18	−0.45	0.34	0.53	0.48	−0.34	0.41	−0.33	1.00

PH pH; *EC* electrical conductivity; *TCC* total cation capacity; *spdis %* spontaneous dispersion; *modis %* mechanical dispersion; *zeta* zeta potential; *z mob* electrophoretic mobility; *ps (nm)* particle size; *SAR* sodium adsorption ratio; *cm:h* Ks; *% wsa* % waterstable macro-aggregates

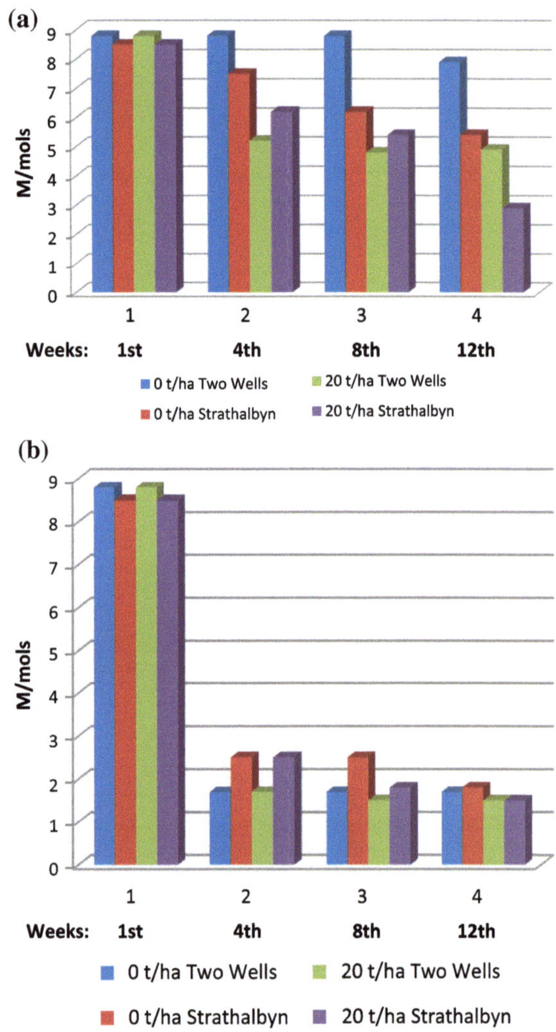

Fig. 8 **a** The effect of phyto-organics without gypsum on the sodium absorption ratio after incubation of two sodic subsoils. **b** The effect of phyto-organics + gypsum on the sodium absorption ratio of two sodic subsoils after incubation

was reduced in both treated soils while they remained undisturbed under incubation. This is similar to a minimum tillage field regime. Thus Khodphuwiang et al. (2007) decreased SAR by applying minimum tillage after adding gypsum. They concluded that this could be due to less soil disturbance by minimum tillage, which may have reduced movement and dispersion of calcium and magnesium into the soil solution. The two ions, Ca^{2+} and Mg^{2+}, are also easily oxidized when there is abundant oxygen in the pore space which is increased by the tillage. With minimum tillage, large amounts of calcium and magnesium remain adsorbed at soil particle surfaces, causing lower SAR which were similar to the recommendation of Franzen (2003).

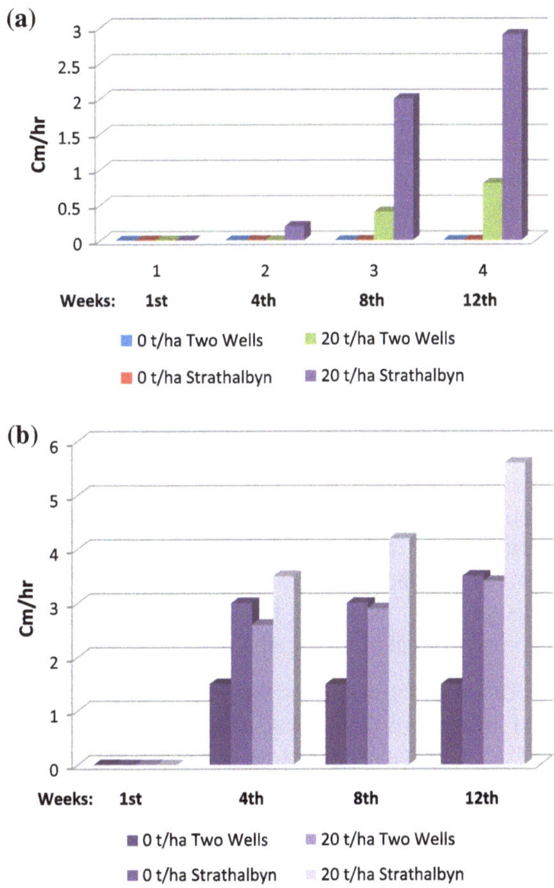

Fig. 9 **a** The effect of phyto-organics without gypsum on the hydraulic conductivity after incubation of two sodic subsoils. **b** Effects of phyto-organics with gypsum during incubation on the hydraulic conductivity of two sodic sub-soils

3.3 Hydraulic Conductivity

Green manure increased K_{sat} in all samples (Fig. 9), thereby signifying its ability to improve the permeability of sodic subsoils. However, when green manure was incubated in gypsum-treated subsoils, a minimum of 16 % higher K_{sat} than that of the green manure only treatment (GM) was observed throughout incubation (Fig. 9). Though the increase in K_{sat} attained during the gypsum pre-treatment (GY) had been maintained throughout the incubation, they were not significantly improved upon over the 12-week incubation period (Fig. 9). Even though initially not as responsive as the GY treatments early in incubation, the GM treatments (Strathalbyn only) exhibited a K_{sat} result approximately up to that of the GY treatment after 12 weeks (Fig. 9). Thus, whereas the effects of GY on K_{sat} ended at 4 weeks, those of GM continued on to the end of the 12-week incubation. The absence of decomposing GM in the gypsum-only treatment apparently prevented a higher level of, and a continued increase in, permeability.

The improvement in K_{sat} by the addition of green manure + gypsum is mainly due to the prevention of clay dispersion. As macro-aggregation in sodic soils was not changed substantially by these treatments, flocculation of particles and the prevention of clay dispersion seemed to be the major mechanism for the increases in K_{sat}. This is in agreement with Warkentin (1982) who found that wet sieving to measure size distribution of stable macro-aggregates >0.5 mm is not a valid test for the stability of clayey soil structure, but that the significant aspect of stability of clayey soil is the "permanence of fissures or pores." This is clearly shown by the significant negative correlations between mechanically dispersed clay and K_{sat} (Tables 3 and 4). The above-mentioned factors resulted in a decreasing magnitude of K_{sat} as follows: GMGY > GM > GY > controls for combined samples (Fig. 9). With subsoils separated, the sequence is:

GMGY Strathalbyn >GMGY Two Wells >GM Strathalbyn >GM Two Wells (Fig. 9).

Thus, the increase in K_{sat} for Strathalbyn samples was higher than that of Two Wells samples for all treatments. The higher $CaCO_3$ content 14 % ($CaCO_3$) of Strathalbyn (Table 1) probably contributed to a higher Ca^{2+} in soil solution towards the end of the incubation than that of Two Wells subsoil (4 % $CaCO_3$). This probably accounts for the K_{sat} of the Strathalbyn GM samples having approached that of the Strathalbyn GY treatment by week 12 of the incubation.

This did not occur for the Two Wells samples. Not only was the Strathalbyn GM Ksat more than three times that of the Two Wells GM, but the Two Wells GY actually attained almost twice the K_{sat} of the Two Wells GM (Fig. 9). For K_{sat} therefore, the Strathalbyn subsoil was substantially more responsive to green manure than was the Two Wells subsoil. After Na^+ ions have been flushed from soils, the need for the presence of divalent cations in such environments for indigenous clay materials to become flocculated has been recognized by Muneer and Oades (1989), Baldock et al. (1994), and Shanmuganathan and Oades (1984). In this study, H_2CO_3 had been formed as an indirect result of the increased CO_2 under continuously moist soil conditions, which in turn was caused by the increased respiration of microorganisms in response to green manure as a substrate in both subsoils. By week 12, the native $CaCO_3$ had probably become sufficiently solubilized under the continuous action of organic acids and H2CO3. It is thus possible that because considerable time (even under the prevailing conditions of adequate heat and moisture) was needed for such reactions to substantially occur, a dramatic rise in K_{sat} was observed only towards the end of the incubation period (Fig. 9).

3.4 Clay Dispersion

Clay dispersion was decreased by all treatments. Inherently, both subsoils were readily dispersible and a small amount of energy was required to further disperse the clays (Fig. 10).

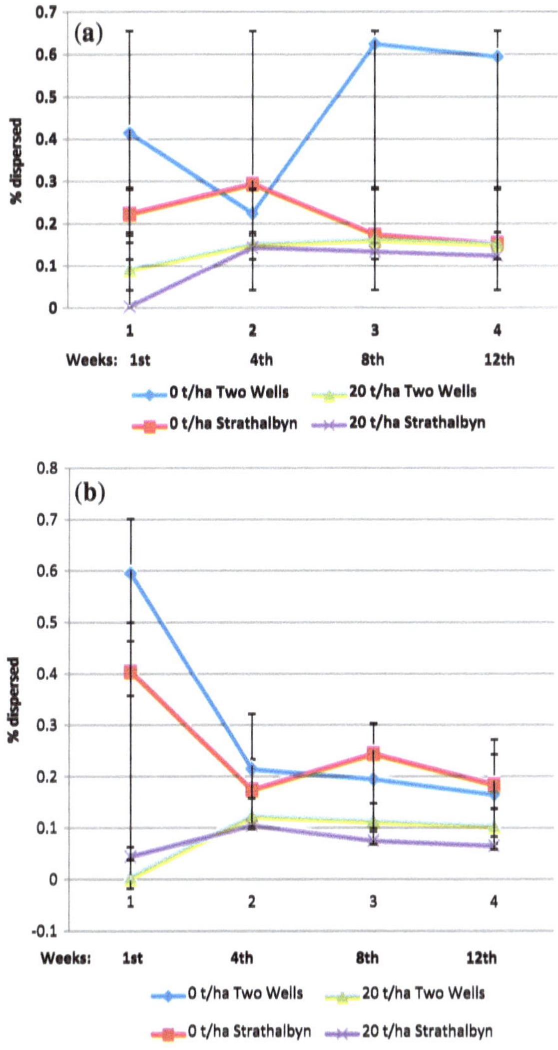

Fig. 10 **a** The effect of phyto-organics without gypsum on the spontaneous dispersion of two sodic sub-soils during incubation. **b** The effect of phyto-organics with gypsum on the spontaneous dispersion of two sodic sub-soils during incubation

However, after 4 weeks incubation without additives of gypsum or green manure (i.e. GM controls), the amount of clay spontaneously dispersed from the Two Wells GM controls was twice that of the Strathalbyn GM controls (Fig. 10), but for mechanical dispersion under the same conditions, Two Wells was only 25 % more dispersive (Fig. 11). Thus, the same amount of energy applied to both original subsoils resulted in far less dispersion of the former. A similar pattern emerged for the gypsum-treated controls (GY) up to week 8, but by week 12 the effect of GY had decreased below that of the GMGY. For the GM and GMGY samples, decreases in spontaneous and mechanical dispersion occurred more substantially than those by GY, throughout incubation. Though GM treatment

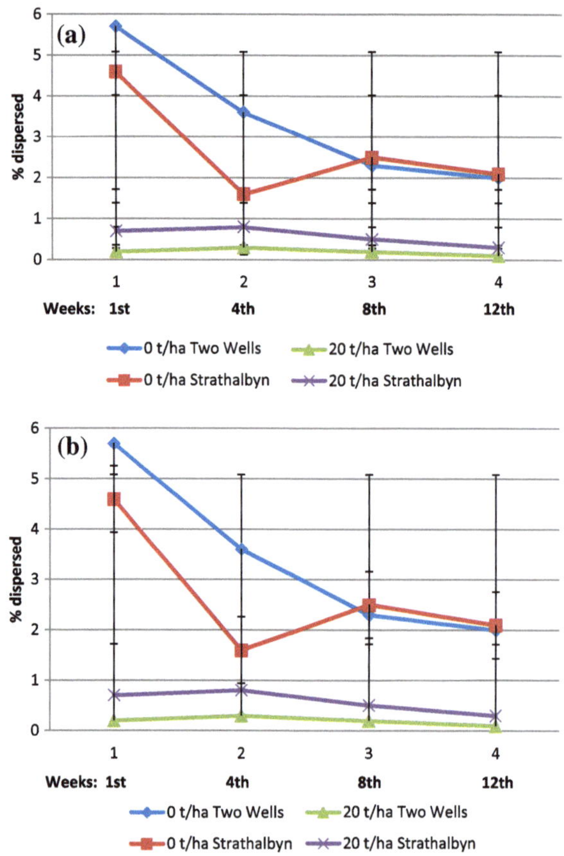

Fig. 11 a Effects of phyto-organics without gypsum on the mechanical disperison of two sodic subsoils after incubation. **b** Effects of phyto-organics (+gypsum) on the mechanical dispersion of two sodic subsoils after incubation

of itself drastically reduced mechanical dispersion for both soils, its effect on the Strathalbyn samples was twice as effective as that on the Two Wells samples (Fig. 11). Under GM treatment, the clays in the Strathalbyn samples thus became twice as tightly bonded together as those of Two Wells.

The configuration of the clays was probably also altered. The clay particles, it is reasonable to suggest, rather than having been dispersed downwards [as had been the case in the wheat straw experiment of Baldock et al. (1994)], required double the energy to sustain similar levels of dispersion.

The extra energy required by both subsoils to overcome their natural dispersiveness was therefore supplied by the GM, particularly in the case of the Strathalbyn samples. Baldock et al. (1994) found that high rates of wheat straw, when incorporated with gypsum, caused dispersion and pore blockage by mobilizing clay particles downwards. However, they referred to wheat straw as "readily decomposable", yet such is not the case compared to fresh leafy material, which decomposes at 60–80 % p.a. (Altieri 1989). Wheat straw (with a higher lignin content) decomposes slowly, at about 30 % p.a. (Sarkanen and Ludwig 1971;

Mason 1977). In the present study, the opposite result (i.e. decreasing dispersion) was caused by green manure, probably as a result of its faster rate of decomposition compared to that of wheat straw. Thus, in the case of Baldock et al. (1994), wheat straw was added over a five-year period (with increases of dispersion occurring not until after the third year), whereas in this study, green manure was incubated and measured for only a 12-week period. Susanto (1992) found that mobile organic colloidal materials, ranging in size from a few hundred molecules to a diameter in the order of microns, move around in the soil solution, and Goldberg and Glaubig (1987) found that the positive edge charges of clay particles are offset by their specific adsorption of organic anions when those anions are no longer than the clay particle edges. It is noteworthy that Baldock et al. (1994) found that dispersion did not increase until after a 3-year period of wheat straw in the soil. Therefore, the anions most likely had reached a state of advanced polymerization, and therefore small in size. Placed against clay particles, such small anions could have therefore acted as de-flocculants for the soil in the manner described above by Goldberg and Glaubig (1987). On the other hand, the "younger" products of decomposition from the present study were composed of organic anions larger than clay particles, which would have inhibited dispersion of clay particles.

3.5 Particle Size Analysis

In this study, the above explanation regarding large anionic sizes for "younger organic" materials is supported by particle size analysis of the treated samples. Thus, although gypsum acting alone dramatically increased particle sizes (5-fold) for the Two Wells samples (which contained 4 % of native lime), when combined with the young organic matter from green manure, the size increase for the same subsoil was 10-fold at week 4 and 20-fold at week 8 of incubation (Fig. 11).

Nevertheless, spectacular as these increases were, the anions may not have been composed entirely of organic matter. Green manure (combined values) without gypsum (GM) caused less than 30 % increases in particle size at 4 or 8 weeks, and only a 2-fold size increase for the same subsoil, at week 12. Clearly, the huge size increases occurring for gypsum in combination with the green manure is greater than the additive values for the separate gypsum and green manure treatments. The un-accounted for size differences could be largely the result of decreasing dispersion of clay by the products of green manure and gypsum. As clay dispersion was decreased by the GMGY treatment, the configuration of the clays was probably also altered. The clay particles, it is reasonable to suggest, rather than having been dispersed downwards (as had been the case in the wheat straw experiment of Baldock et al. (1994), not only remained stuck together, but became linked up by Ca^{2+} cations with the newly created large anions from the green manure. In other words, the larger particles were possibly composed not just of organic material from the green manure, but partly of undispersed clay particles. Hence, the source of the unaccounted-for size increases.

3.6 Aggregation

3.6.1 Gypsum and Crusting

For no treatment was there a significant difference in macro-aggregates at any stage of incubation, though GM showed an increasing trend. Gypsum acting alone does not seem to improve the macro-aggregation of soils (Baldock et al. 1994). Yet, Amezketa et al. (2005) found gypsum effective in reducing soil crusting in laboratory experiments using both sodic and non-sodic soil. Hence the role of gypsum is to flocculate the clay particles. As has been found by other workers (Baldock et al. 1994; Muneer and Oades 1989), Ca^{2+} in soil solution do not improve macro-aggregation. In this study, the addition of green manure to the gypsum-treated soils also caused no significant difference in macro-aggregate stability. Decomposing green manure was unable to bind micro-aggregates effectively into larger water-stable aggregates in the sodic subsoils. Even though macro-aggregation in the size range of >250 μm did not occur (Tables 5 and 6), particle size analysis shows that the clay particles had been aggregated by the green manure into a size range of 30 μm.

Warkentin (1982) stated that clay minerals must remain in a flocculated state to maintain soil aggregate stability. The average size of dispersed materials in control soils are <0.5 μm, whereas, after green manuring, the average particle size had gone up to 30 μm. This clearly shows that the products of decomposition, organic molecules and Ca^{2+} interacted with clay particles to form stable clusters, or domains. A synergistic result was achieved. The zeta potential and negative electrophoretic mobility of the clay particles (Tables 3 and 4) seem to support the above conclusions on clay aggregation. Because the zeta potential is a measure of the level of negative charge (which causes mutual repulsion) on clay particles, the greater the negative charges of adjacent particles, the greater their mutual repulsion and hence, greater mobility. Thus, the zeta mobility, (or electrophoretic mobility) varies directly with the zeta potential. As clay micro-aggregation increased (as was evident by the particle size or "p.s." increases), the zeta potential and mobility decreased (Tables 3 and 4). This is because an increase in clustering of negatively charged clay crystals results in the decrease of negative charge.

It appears that the role of green manure is in providing organic acids to dissolve $CaCO_3$ and flocculate clay particles rather than aggregation of clay domains to form macro-aggregates. This is supported by significant negative correlations between pH and EC and positive correlations between pH and SAR (Tables 3 and 4). The reduction of pH from 9.36 to 8.32 in Strathalbyn samples and from 9.65 to 8.12 in Two Wells samples after 12 weeks of addition of green manure only, correlated highly with increases in K_{sat} from 0.1 to 2.9 and 0 to 2.9 cm/h, respectively. The decrease of pH thus seems to be a major prerequisite in the reclamation of sodic soils.

3.7 Analysis of Flocculation

After macro-aggregates broke down under simulated rain, the disintegrated particles were dried and subjected to particle size analysis. It can be seen (Fig. 11) that the particle sizes for the gypsum treated samples ranged from 50–230 µm, with the average size being 140 µm. Yet for the controls the particle sizes after simulated rain were all under 50 µm. Further, no particle (trapped as dust) removed by wind pressure from the control RMW did exceed 50 µm in diameter (Fig. 11). Thus gypsum substantially increased the average weight of particles in the RMW by >100 % (Fig. 11). Therefore, for the gypsum-treated RMW, though the simulated rain destroyed the larger agglomerates, further breakdown of particles was more difficult under continued rain. The sub 250 µm particles were therefore more resilient than those of a larger size. In fact, it is possible that the larger agglomerates

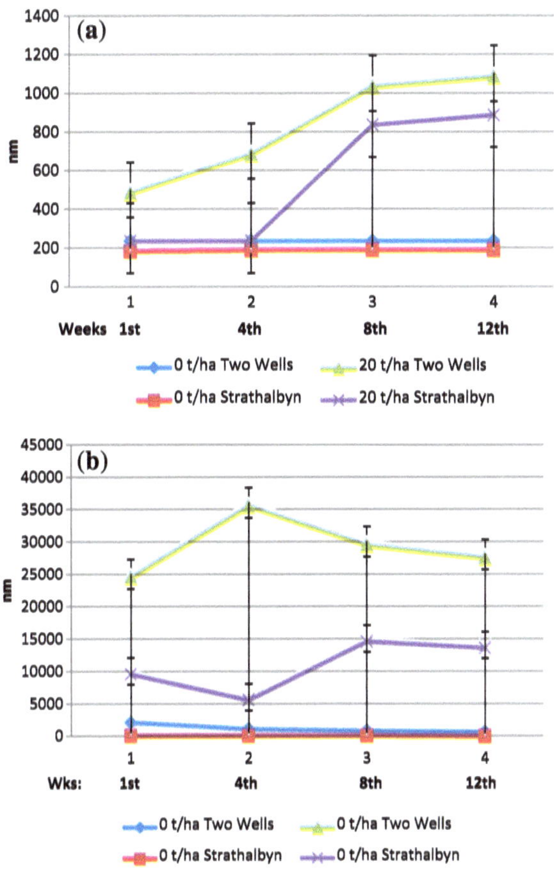

Fig. 12 **a** Sub-micron particle size (nm) of two sodic subsoils after incubation with phyto-organics without gypsum. **b** Sub-micron particle size (nm) of two sodic subsoils after incubation with phyto-organics with gypsum

had not been caused in fact by the gypsum but by weak thixotropic forces (or "age hardening") caused by the wetting and drying treatment cycles observed under similar conditions (Seng and Tanaka 2012). Thus in this study, the addition of gypsum apparently made no difference in macro-aggregate stability under simulated rain. This is consistent with the findings by other workers such as Harris and Rengasamy (2004a, b). Thus Baldock (1994), and Muneer and Oades (1989), found that Ca^{2+} in soil solution do not markedly improve macro-aggregation.

Nevertheless, lithification by atmospheric carbonation did not occur following the addition of gypsum to sodic subsoils. Under moist conditions followed by repeated water flushing during a 3-month incubation, Harris and Megharaj (2001), and Harris and Rengasamy (2004a, b) observed no subsequent hardening in the ensuing 12 months. On the other hand, a similar treatment applied to alkaline bauxite red mud waste without flushing (O'Callaghan and others 1998) resulted in the lithification of the alkaline red muds. Gypsum increased water retention in sodic subsoils (Fig. 12) but not in the gypsum-treated red muds (O'Callaghan 1998; Harris 2008). Therefore, whereas it is safe, and beneficial, physically and chemically, to add gypsum in large quantities to soil and subsoils, the physical results of the similar treatments to alkaline wastes such as red muds cause not only hard-setting but permanently lithified non-porous concrete-like entities. In this respect gypsum above certain thresholds can be edaphically counter-productive when added to red mud wastes.

4 Conclusion

The effect of green manure in sodic subsoils was examined with and without the addition of gypsum. The most important change occurring in the sodic subsoils is that of substantially increased hydraulic conductivity. This is due to the formation of tight clusters of organo-clay particles and the accompanying reduction in the amount of dispersible clay particles in the soil. Thus, the chief site for these effects is the microstructure of the soils, by the formation of clay particle clusters. The addition of green manure together with gypsum, improved permeability of the sodic subsoils more than that due to incorporation of gypsum. The additive effect of green manure and gypsum on Ks was considered to result from the influence on soil pH which resulted in the release of protons, solubilizing of the native $CaCO_3$ and leaching of Na_2CO_3. As a result of these extra protons in the soils, flocculation of the clay particles increased, along with a concomitant decrease in dispersion, leading to the creation of clay clusters and an improvement in the soil porosity. It was seen that the beneficial effects of green manure were increased by gypsum (and vice versa) in every instance, and that these effects were more substantial and longer lasting for the more calcareous samples. Therefore, rather than using green manure, or gypsum separately, consideration should be given to the incorporation of gypsum with green manure, along with a source of "slow-release" Ca^{2+}

to maintain high levels of amelioration for more extended periods. As dried out bauxite red mud lakes are alkaline-sodic, thereby presenting similar chemical and physical conditions to those of the subsoils in this study, including high pH, hard-setting, low permeability, and water logging, it is likely that treatments similar to the above ones may similarly improve plant growth conditions in such red mud environments. Further research is required to test this hypothesis.

5 Geobiotechnological Applications

5.1 Expressing Sodicity

Sodicity is expressed as exchangeable sodium percentage (ESP). Every soil has a definite capacity to adsorb the positively charged constituents of dissolved salts, such as calcium, magnesium, potassium, sodium, etc. This is termed the cation exchange capacity. The various adsorbed cations can be exchanged one for another and the extent of exchange depends upon their relative concentrations in the soil solution, the valency and size of the cation involved, nature and amounts of other cations present in the solution or on the exchange complex, etc. Exchangeable sodium percentage (ESP) is, accordingly, the amount of adsorbed sodium on the soil exchange complex expressed in percent of the cation exchange capacity in milli-equivalents per 100 g of soil. Thus, exchangeable sodium percentage (ESP):

$$= \frac{\text{Exchangeable NA (me/100 g soil)}}{\text{Cation exchange capacity (me/100 g soil)}} \times 100 \tag{3}$$

5.2 Sodicity Determinations

Experimental determination of exchangeable sodium percentage is tedious, time consuming and subject to errors. Incomplete removal of index salt solution during the washing step of CEC determinations can lead to high CEC values and therefore low ESP estimates. Similarly, hydrolysis of exchangeable cations during the removal of the index salt solution, fixing of ammonium ions from the index or replacement solution by the soil minerals and the dissolution of calcium carbonate or gypsum in the index or replacing solutions can all lead to low values of cation exchange capacity and therefore to high ESP estimates. Problems of CEC and ESP determinations are also encountered in soils of high pH containing zeolite minerals. These minerals, e.g. analcime, contain replaceable monovalent cations in their lattice which are readily replaced by monovalent cations used as the index or replacement cation resulting in unusually high values of ESP (Gupta et al. 1984). To overcome some of these difficulties several workers prefer to obtain an estimate of the exchangeable sodium percentage from an analysis of the saturated soil extract. Workers at the US Salinity Laboratory (Richards 1954) proposed that the sodium adsorption ratio (SAR) of the soil solution adequately defines the soil sodicity problem and is quantitatively related to the exchangeable sodium percentage of the soils. Sodium adsorption ratio, SAR, is defined by the equation:

$$SAR = Na^{+}/\sqrt{(Ca + Mg)^{++}/2} \qquad (4)$$

where all concentrations are in mmol (+)/litre.

Although some studies have shown that grouping calcium and magnesium in the above equation is not strictly valid, there appears only little loss of accuracy when this is done.

5.3 Soil Salinity in Cuba

Substantial soil salinity exists in the entire southern stretch of Guantánamo (at the eastern tip of the island), because of drought and because it's very low-lying land, and the seawater penetrates through the aquifers. The entire southern portion of eastern Cuba is dry and thus tends to have high levels of salinity. The region north of Santa Clara (east of the Cuban capital has a kind of clay in the soil that creates lumps, which make it difficult for plants to absorb water and nutrients (Simpson 2013).

Use of gypsum with indigenous sources of leguminous green manure such as Acacia & Leucaena:

Cuba is home to several types of acacia, including the *Acacia zapatensis* that is only found in Cuba. These are leguminous, providing large sources of N. Abundant sources and large stands of Acacia proliferate invasively and abundantly on idle lands in the Jamaica parishes of St. Catherine and Clarendon, where some sodic soils exist. For dried bauxite red muds the abundance of such invasive, rapidly growing acacia species could provide sufficient organic matter. The soft shoots are preferable, being more easily decomposed, and supplying the N required for sustaining a population of bacterial decomposers in the soil (Harris 2008). With a large source of gypsum at a distance of 150 km to the northeast in Rockfort, Kingston, the materials are available for the amelioration of sodic soils, including red mud wastes.

5.4 Kinds of Amendments

Chemical amendments for sodic soil reclamation can be broadly grouped into four categories:

(a) Soluble calcium salts, e.g. gypsum, calcium chloride (Amezketa et al. 2005).
(b) Acids or acid forming substances, e.g. sulfuric acid, iron sulphate, aluminium sulphate, lime-sulfur, sulfur, pyrite, etc.

(c) Calcium salts of low solubility, e.g. ground limestone.
(d) Minimum tillage, which decreases SAR after adding gypsum (Khodphuwiang et al. 2007). With minimum tillage, large amounts of calcium and magnesium remain adsorbed at soil particle surfaces, causing lower sodium adsorption ratio (SAR) (Frantzen et al. 2006).

The suitability of one or another amendment for sodic soil reclamation will largely depend on the nature of the soil and cost considerations. Ground limestone, $CaCO_3$, is an effective amendment only in soils having pH below about 7.0 because its solubility rapidly decreases as the soil pH increases (Table 22). It is apparent that the effectiveness of limestone as an amendment is markedly decreased at pH values above 7.0. Some soils that contain excess exchangeable sodium also contain appreciable quantities of exchangeable hydrogen and therefore have an acidic reaction, e.g. degraded sodic soils. Lime reacts in such soils according to the reaction:

$$\mathrm{Na, H - clay\ micelle} + CaCO_3 = \mathrm{Ca - clay\ micelle} + NaHCO_3$$

However, lime is not an effective amendment for most sodic soils with pH values. In fact, sodic soils contain measurable to appreciable quantities of sodium carbonate which imparts to these soils a high pH, always more than 8.2 when measured on a saturated soil paste, and up to 10.8 or so when appreciable quantities of free sodium carbonate are present. In such soils only amendments comprising soluble calcium salts or acids or acid-forming substances are beneficial. The following chemical equations illustrate the manner in which some of the amendments react in these soils.

Studies of Chawla and Abrol (1982) with a highly sodic soil containing free sodium carbonate showed that treatment of soil with very finely ground gypsum resulted in high initial hydraulic conductivity which decreased sharply with time. On the other hand, treatment with gypsum passed through 2 mm mesh and having a range of particle size distribution helped in maintaining permeability at higher level and for a longer period. Their results showed that higher solubility of finer particles caused them to react with free sodium carbonate, inactivating the soluble calcium due to formation of insoluble calcium carbonate.

5.5 General Gypsum Applications

Regardless of pH of the treated medium, additive Ca^{2+} in compounds added to soil are prone to atmospheric carbonation. The carbonate ion acts as a base in aqueous solution, readily absorbing hydrogen ions in solutions regardless of pH. In acidic solution the reaction is:

$$CO_3{}^{2-}(aq) + H^+(aq) \leftrightarrow HCO_3{}^-(aq) \quad (5)$$

while in alkaline or neutral solution it is:

$$CO_3{}^{2-}(aq) + H_2O(l) \leftrightarrow HCO_3{}^-(aq) + OH^-(aq) \quad (6)$$

The carbonate ion will react with various species. Where hydrogen ions are abundant (acidic solution) the hydrogen ion sticks to the carbonate ion to produce the hydrogen carbonate ion. With a paucity of hydrogen ions, carbonate in effect removes a hydrogen ion from water, leaving a hydroxide ion behind and producing an alkaline solution.

References

Altieri MA (1989) Agroecology. Westview Press Inc., Colorado, pp 354–355

Amezketa E, Aragues R, Gazol R (2005) Efficiency of sulfuric acid, mined gypsum, and two gypsum by-products in soil crusting prevention and sodic soil reclamation. Agron J. 97:983–989

Baldock JA, Aoyama M, Oades JM, Susanto, Grant CD (1994) Structural amelioration of a South Australian red-brown earth, using calcium and organic amendments. Aust J Soil Res 32:571

Chawla KL, Abrol IP (1982) Effect of gypsum fineness on the reclamation of sodic soils. Ag Water Manage 5:41–50

Courtney R, Harrington T, Byrne KA (2013) Indicators of soil formation in restored bauxite residues. Ecol Eng 58:63–68

Frantzen D, George R, James G (2006) Effectiveness of gypsum in the north central region of the U.S. North Dakota State University Extension Service 2006 SF-1321

Emerson WW (1983) Interparticle bonding. Soils: an Australian viewpoint CSIRO, Melbourne, Australia. Academic Press, London, UK, pp 477–497

EPA (2012) Rare Earth elements: a review of production, processing, recycling, and associated environmental issues EPA 600/R-12/572. Dec 2012. www.epa.gov/ord

Goldberg S, Glaubig RA (1987) Effect of saturating cation pH, and aluminum and iron oxides on the flocculation of kaolinite and montmorillonite. Clay Clay Miner 35:220–227

Gupta RK, Bhumbla DK, Abrol ID (1984) Sodium-calcium exchange equilibria in soils as affected by calcium carbonate and organic matter. Soil Sci 138: 109–114

Harris MA (2008) Structural improvement of age-hardened gypsum-treated bauxite red mud waste using readily decomposable phyto-organics. Environ Geol 56: 1517–1522. DOI: 10.1007/s00254-008-1249-5

Harris MA, Megharaj M (2001) The effects of sludge and green manure on hydraulic conductivity and aggregation in pyritic mine tailings materials. Environ Geol 41:285–296

Harris MA, Rengasamy P (2004a) Sodium affected subsoils, gypsum, and green-manure: Inter-actions and implications for amelioration of toxic red mud wastes. Environ Geol 45(8):1118–1130

Harris MA, Rengasamy P (2004b) Treatment of a potential soil capping for open-cut mines (Part 1): influence of prior management on responses to a phyto-organic amendment. Environ Geol 46(1):15–21

Kavamura VN, Esposito E (2010) Biotechnological strategies applied to the decontamination of soil polluted with heavy metals. Biotechnol Adv 28:61–69

Kemper WD, Roseneau RC (1986) Aggregate stability and size distribution. In: Klute A (ed) Methods of soil analysis, 2nd edn. American Society of Agronomy, Science Society of America, Madison, WI, pp 425–442

Khodphuwiang C, Kongyoo P, Prammanee P, Weerathaworn khodphuwiang P (2007) Sodic soil management of the sugarcane area in western Thailand. In: Proceedings of international society sugar cane technologists, vol 26

Klute A (1986) Methods of soil analysis, Part 1, physical and mineralogical methods, 2nd ed (Klute A editor), vol 9(1). American Society of Agronomy, Agronomy Monographs, Madison, Wisconsin. doi:10.1002/gea.3340050110

Lehoux AP, Lockwood CL, Mayes WM, Stewart DI, Mortimer RJG, Gruiz K, Burke IT (2013) Gypsum addition to soils contaminated by red mud: implications for aluminium, arsenic, molybdenum and vanadium solubility. Environ Geochem Health 35:643–656

Lone MI, He ZL, Stoffella PJ, Yang X (2008) Phytoremediation of heavy metal polluted soils and water: Progress and perspectives. J Zhejiang Univ Sci B 9(3):210–220

Mason CF (1977) Decomposition studies. In: Institute of biology, no 14. Edwin Arnold Publishers, London

Moynahan OS, Zabinski CA, Gannon JE (2002) Microbial community structure and carbon-utilization diversity in a mine tailings revegetation study. Restor Ecol 10:77–87

Muneer M, Oades JM (1989) The role of Ca-organic interactions in soil aggregate stability. I. Laboratory studies with 14C-glucose, $CaCO_3$ and $CaSO_42H_2O$. Aust J Soil Res 27:389–399

Nelson P, Ham G, Kingston G, Burgess D, Lawer A, Wood A, Christianos N, Wilson P, Grundy M, Smith D, Hardy S (2001) Diagnosis and management of sodic soils under sugarcane. CRC Sugar Technical Publication, Townsville, Australia, p 64

O'Callaghan WB, McDonald SC, Richards DM Reid RE (1998) Development of a topsoil-free vegetative cover on a former red mud disposal site. Alcan Jamaica Rehabilitaion Project Paper

Pashley RM (1985) Electromobility of mica particles dispersed in aqueous solutions. Clay Clay Miner 3:193–199

Ping Wang and Dong-Yan Liu (2012) Physical and chemical properties of sintering red mud and bayer red mud and the implications for beneficial utilization. Materials. 5:1800–1810. doi:10.3390/ma5101800. ISSN 1996-1944

Ponnamperuma FM (1972) The chemistry of submerged soils. Adv Agron 24:29–96

Quirk JP, Schofield RK (1955) The effect of electrolyte concentration on soil permeability. J Soil Sci 6:163–178

Rengasamy P, Olsson K (1991) Sodicity and soil structure. Aust J Soil Res 29:935–952

Rengasamy P, Greene RSB, Ford GW, Mehanni AH (1984) Identification of dispersive behaviour and the management of redbrown soils. Aust J Soil Res 22:413–431

Sarkanen KV, Ludwig CH (1971) Lignins. Wiley, New York, pp 57–60

Sekhon BS, Bajwa BS (1993) Effect of organic matter and gypsum in controlling soil sodicity in rice and wheat-maize system irrigated with sodic waters. Agri Water Manage 24:15–25

Seng S, Tanaka H (2012) Properties of very soft clays: A study of thixotropic hardening and behavior under low consolidation pressure. Soils and Foundation 52: 335–345 doi: 10.1016/j.sandf.2012.02.010

Shanmuganathan RT, Oades JM (1984) Influence of anions on dispersion and physical properties of the A horizon on a red brown earth. Geoderma 29:257–277

Simpson Leslie (2013) Workshop to launch the Global Soil Partnership: towards prevention and restoration of degraded soils in Central America and The Caribbean 30 September – 3 October 2013. FAO Havana Cuba, www.fao.org/fileadmin/user_upload/GSP/docs/1_Agenda_Workshop_Cuba.pd. Accessed 29 June 2015

Sumner ME (1993) Sodic soils: new perspectives. Aust J Soil Res 31:583–750

Susanto (1992) Effect of gypsum, agricultural lime, and organic matter on the infiltration characteristics of a red-brown earth. Unpublished Master of Agriculture thesis, The University of Adelaide, Australia

Tisdall JM, Oades JM (1982) Organic matter and water-stable aggregates in soils. J Soil Sci 33:141–163

Warkentin BP (1982) Clay soil structure related to soil management. Tropical Agri Trinidad 59:167–177

Zhang Y, Qu Y, Wu S (2001) Engineering geological properties and comprehensive utilization of the solid waste (red mud) in aluminium industry. Environ Geol 41:249–256

Structural Amelioration of Indurated Gypsum-Treated Bauxite Red Mud Tailings Using Bacterially Mediated Organic Treatments

Mark Anglin Harris

Abstract Calcium sulphate dihydrate ($CaSO_4 \cdot 2H_2O$) (gypsum) is often used as an amendment to improve soil structure. As this treatment eventually increases soil strength, thereby exacerbating conditions for root growth, reclamation of red mud tailings using this treatment can become counterproductive. Several years (7) after lithification by added gypsum, the effect of crushing, and added phyto-organics, on some physical properties of red mud wastes (RMW) was examined under laboratory conditions. It was hypothesized that mechanically breaking up the hardened masses would increase their permeability. However, on the contrary, crushing these rock-hard masses into 2-mm particles did not increase permeability. A phyto-organic treatment on the other hand greatly improved permeability of the crushed entities such that infiltration rates increased from 0 to 10 cm/h. After 24 weeks of incubation, the structure of the newly formed phyto-organically treated gypsum-hardened RMW aggregates was also substantially improved over that of the controls, as a 3-fold increase in resistance to disintegration during wet sieving was exhibited by the phyto-organically treated samples. The increase in water infiltration by combining crushing with phyto-organic infusions was ascribed to a decrease in particle dispersion, creating more stable clay clusters, which in turn stabilized micro channels and pores. This treatment increased the plant-growth-medium potential of the age-hardened gypsum-treated red muds, thereby justifying prior atmospheric carbonation and lithification of the red muds. Conversely, it was concluded that the singular crushing of CGR in the waste dumps to create a growth medium is potentially wasteful and ineffective. However, as CGR was seemingly of greater durability than soft limestone marl, it promises effectiveness as a stable road-bed material.

Keywords Bauxite wastes · Guyana · Gypsum-organic amendments · Jamaica WI · Organic remediation · Virgin islands

---

Original article: **Structural improvement of age-hardened gypsum-treated bauxite red mud waste using readily decomposable phyto-organics**. *Environmental Geology*, Volume 56, No. 8, February 2009.

M.A. Harris, *Geobiotechnological Solutions to Anthropogenic Disturbances*, Environmental Earth Sciences, DOI 10.1007/978-3-319-30465-6_10

Highlights

- Decomposable phyto-organic amendments increase infiltration in bauxite mine waste.
- Crushing of lithified carbonated gypsum-treated bauxite waste to create a growth medium is ineffective and potentially wasteful.

1 Introduction

1.1 Red Mud

Globally, the waste product from aluminium refining amounts to approximately 75 million tonnes and 120 million tons every year with less than half being re-used (Rout et al. 2013). Storage of this unutilized red mud occupies vast tracts of usable land, thereby polluting land, air and water. It is of concern that thousands of years are required to form the natural soil, yet construction of high embankments, under-passes, flyovers, and tailing dams use and/or occupy vast acreages of top soil (Rout et al. 2013).

1.2 Caribbean Bauxite

Being largely derived from aluminium tri-hydrate $Al(OH)_3$ (Sutar and Mishra 2014) or gibbsite (Fig. 1), alumina is extracted from Caribbean bauxite by the well-known Bayer process, the main characteristic of which is the digestion of the bauxite in caustic liquor.

Fig. 1 Gibbsite, $Al(OH)_3$, an ore of aluminium, formed on the surface of Miocene Limestone (Jamaica). Each small circle in the background is 1 mm in diameter. *Photo* Dwight Rose, Northern Caribbean University

However, being insoluble in caustic liquor, the impurities produce the insoluble red mud residue. The red mud liquor being strongly alkaline with a high pH, requires neutralization to a pH below 9 with an optimum value of 8.5–8.9 before becoming environmentally benign (Hanahan et al. 2004). Neutralization of red mud to pH around 8.0 is optimal because the chemically adsorbed Na is released, alkaline buffer minerals are neutralized and toxic metals are insoluble at this pH (Glenister and Thornber 1995).

Caribbean bauxite ore comprises 15 % of the world's karst reserves (Lyew-Ayee and Lenik 2008), and bauxite mine acreage is expanding there. For every ton of alumina processed, 1 ton of red mud waste is released. The red mud waste stored in ponds is therefore increasing, and more waste ponds are being constructed. Pumped from the factory as a slurry, the liquid phase contains significant concentrations of sodium hydroxide, sodium aluminate and sodium carbonate, and is alkaline in nature with an initial pH often as high as 13. However, several years of subsequent carbonation may substantially reduce such high levels of alkalinity (Table 1).

For a small island which mines 150,000 ton of bauxite annually, the expeditious return of such ponds to a de-toxified state which enhances re-vegetation has long been a required objective. Thus in karst bauxite areas of Jamaica, some red mud ponds at waste sites have been treated with gypsum. Similar treatments have been carried out in Western Australia. Application of gypsum to sodium-affected substrata can improve the physico-chemical status of bauxite tailings by increasing micro-aggregation (Harris and Rengasamy 2004), thereby improving infiltration, and, potentially, agricultural usage. But gypsum-treated red mud waste exposed to the weather leads eventually to atmospherically carbonated layers of high strength and hydraulic imperviousness (Harris 2008). Plant growth is inhibited by these physical obstacles to root establishment and extension.

On non-gypsum-treated red mud waste, one successful approach, e.g. in Western Australia, is covering it with 20 cm of topsoil followed by re-vegetation (Graham and Fawkes 1992). It is reasonable to conclude that this process would be even more successful on the more chemically benign (lowered pH) gypsum-treated red muds. Unfortunately, in Jamaica, available topsoil sources of the required magnitude required for increasingly large red mud waste ponds are unavailable, being largely under plantation crops such as sugar cane and bananas, or used for intensive market gardens.

Table 1 Some properties of bauxite red mud prior to and after gypsum treated layer and subjacent layer at Kirkvine Pond 6, Jamaica

Properties	Aged red mud	Gypsum-treated (0–15 cm depth)	Gypsum + phyto-organics (0–15 cm depth)
pH (saturated extract)	9.8	8.3	7.4
EC (saturated extract/mS cm)	2500	2000	5060
CEC (meq/100 g)	0.5	5.3	13.6
Ca (ppm)	15,000	3700	4400
K (ppm)	22,000	860	1308
Mg (ppm)	0	1	9

1.3 Aim

The application of gypsum to a Jamaican RMW several years before the present study seemingly allowed sufficient time for complete flushing by natural rainfall, and the effect of gypsum on sodic soil material has been well established in the literature. Under moist conditions Harris (2001) and Harris and Rengasamy (2004), respectively found improvements of physical properties of mine spoils (<2-mm diameter sized) after the addition of organics, and gypsum with organics. For this study, it was therefore hypothesized that (1) reducing particle sizes by crushing the gypsum-hardened masses should allow for greater permeability, thereby improving intrinsic physical properties such as water movements and (2), such an effect should be maintained or even improved if readily decomposable phyto-organic amendments were also applied. The aim of this study is therefore to determine the effectiveness of such treatments.

2 Materials and Methods

2.1 Study Area

2.1.1 Climate

Jamaica is a small island of less than 10,000 km^2, and has a tropical climate with two pronounced wet seasons. Data supplied by the Meteorological Office in Kingston indicate that average annual rainfall for the period 2000–2014 was 2300 mm in the northeastern section, and amounts ranging from 800 to 1500 mm in the rest of the country, of which 75 % fell between May and October. Maxima temperatures vary from 30 to 35 °C and night minima temperatures vary from 13 to 25 °C. In the northeast, relative humidity usually exceeds 80 percent, and in the south, less than 80 %.

Mean annual precipitation in the study area ranges from 1700 to 2200 mm and mean annual potential evapotranspiration ranges from 1200 to 1400 mm (Greenberg et al. 1996). As precipitation exceeds evapotranspiration, the soil moisture regime is udic. Like the remainder of the island, rainfall distribution is bimodal, with the driest period from December through February and another dry period during June and July. Mean annual air temperature is about 22 °C, with a difference of less than 5 °C between average daily winter and summer temperatures (Greenberg et al. 1996). The soil temperature regime is therefore iso hyperthermic (Ahmad 1995; Ahmad et al. 1999).

2.2 Review of Prior Gypsum Application at Study Site (Kirkvine, Jamaica)

A partially dried 10-year-old red mud site at the Kirkvine waste dump near Mandeville, Jamaica (Fig. 1) was treated with gypsum in 1996 by O'Callaghan et al. (1998). It was chosen for the present study because their efforts to grow plants there even after they followed the gypsum treatments with sewage applications had mixed success (Fig. 2).

The above workers used gypsum, though lime was more readily available and less costly in the karst bauxite areas. The choice of gypsum over lime was due to the comparatively low activity of $CaCO_3$ as alkalinity increased. The solubility of gypsum is thus less sensitive to pH than that of lime. At a pH of 10 and higher, lime, in contrast to gypsum, is insoluble (Wong and Ho 1992), and bauxite red mud wastes (RMW) have an initial pH of at least 11 (Bardossy 1982). Yet, ironically, a major long-term effect of the added gypsum in this case was to have exacerbated the physical condition of the red mud waste by having consolidated it into even more hardened masses. O'Callaghan et al. (1998) applied gypsum (on the surface) at 40 t ha^{-1} after untreated solidified red mud waste (RMW) was broken up by a tractor into 2–5-cm diameter particles. Though they had applied poultry sewage following gypsum application, decomposition of the poultry sewage was slow, and most of the colonizing plant species suffered stunted growth after six months. Results were thus unsatisfactory. Here and elsewhere, workers found that other mainly pre-decomposed manures such as sewage sludges and composts had been unsuccessful in red muds: Wong and Ho (1992), and O'Callaghan et al. (1998) found that sewage sludge decomposition was slow enough as to have required supplementary fertilization. Such un-activated sewage sludges are largely comprised of the waste products and recalcitrants of bacterial degradation. Decomposition of such materials had been already far advanced when applied to RMW, such already "worked over" substrates leaving only small proportions of nutrients to be degraded by bacteria. Microbial activity is therefore slowed down in such substrates, thereby requiring high rates for even modest structural improvements of inorganic growth media. On the other hand, when readily

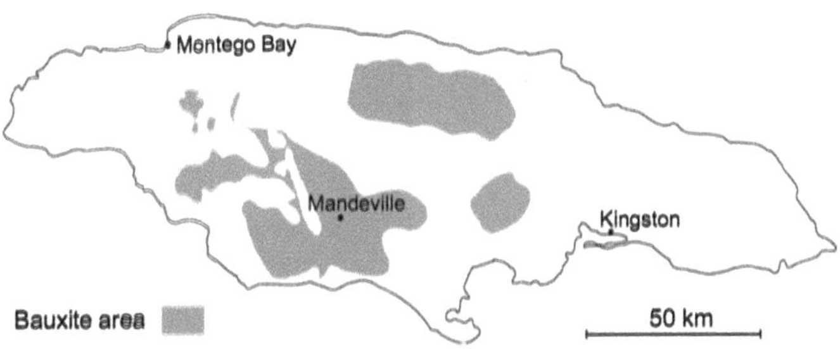

Fig. 2 Sketch diagram includes Kirkvine location near Mandeville, Jamaica, with surrounding bauxite areas. *Source* Jamaica Bauxite Institute

decomposable organic material was used as an activator (or "spike") for sewage sludge, the increase in microbial activity was high enough to have favourably modified the physical properties of even the sludge itself (Harris 2001). He therefore noted that decomposable organic treatments caused flocculation and aggregation of the sludge containing decomposed organic particles.

A field study of Kirkvine Pond 6 in 1996 (O'Callaghan et al. 1998) showed the initial pH of the mud to be as low as 9.8, indicating that most of the caustic had been already atmospherically carbonated prior to gypsum treatment in 1996 (O'Callaghan et al. 1998). In red mud sites elsewhere, Zhang et al. (2001) also found that long conserved samples of red mud underwent re-crystallization of $CaCO_3$, causing diagenetic hardening. Nevertheless, the RMW of this study, prior to gypsum addition, was much softer and subsequently of lower strength prior to gypsum addition. This is explained later. O'Callaghan et al. (1998) also found that to reduce pH to 8.0, the most effective and economical gypsum rate was 20–40 t ha^{-1}, and in 1996, gypsum was spread at a rates ranging from 10 to 60 t ha^{-1} on the 2-year-old red mud pond. The purpose was to reduce the pH and to ameliorate the soil characteristics by exchanging Ca with the Na in the red mud. Poultry litter and inorganic fertilizer were also added. In reference to pre-gypsum RMW, O'Callaghan et al. (1998) found that ploughing would have made little difference in reducing particle size except to "turn over the lumps of hard mud". However the strength of this "hard mud" had increased greatly within 7 years of their gypsum application (Fig. 3), and in 2003, soil strength tests in Pond 6 indicated a dry, brick-hard surface environment in the above gypsum-treated locations from the surface to a depth of 15 cm, but below 15 cm depth the material was less strong, as gypsum treatment had had little impact below 15 cm (Fig. 3). For successful re-vegetation, a method of ameliorating such gypsum-hardened surface entities was therefore required. Due the scarcity of topsoil in Jamaica (as it is in other Caribbean areas), mud lakes in Jamaica must employ strategies which do not require the application of topsoil. Thus for this study, a crushing process was selected.

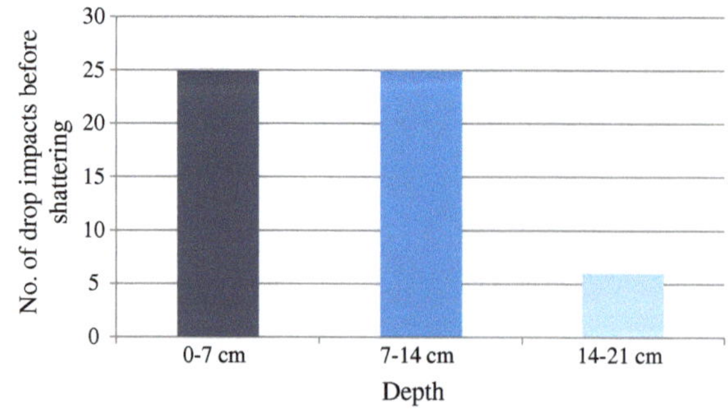

Fig. 3 Strength of red mud waste seven years after surface applied gypsum treatment. No. of drop impacts before shattering

2.2.1 Rationale for Gypsum

Though more quickly acting in soil than gypsum (Shainberg and others 1989), in hot humid areas under tropical rains, such as Jamaica, $CaCl_2$ dissolves so quickly that it is usually leached out of the soil layers in a relatively short period, and $CaCl_2$ is costly. Because the sparingly soluble gypsum remains in soil for a longer time, and opportunities are thereby provided for solvation to occur over many more effective exchange steps, it is both efficient and economical (Oster 1982). Further, the Cl^- ion can be toxic to plants.

Though lime is more readily available and less costly in the karst bauxite areas, the choice of gypsum over lime is due to the comparatively low activity of $CaCO_3$ as alkalinity increases. The solubility of gypsum is less sensitive to pH than that of lime. At a pH of 10 and higher, lime, in contrast to gypsum, is insoluble (Corbett 1969), and bauxite red mud wastes all have a pH of at least 11 (Bardossy 1984). When gypsum is added in the soil, the adsorbed sodium on the clay colloids is replaced by calcium. This is because during gypsum applications, sodium sulphate, a neutral salt is formed, rather than sodium carbonate, as would have been the case with lime applications (Corbett 1969), according to the following equation:

$$2\text{Na-clay} + CaSO_4 \rightarrow \text{Ca-clay} + Na_2SO_4$$

The sodium sulphate maintains the pH close to 7 until the replacement reaction has proceeded enough to allow time for the calcium to dominate the colloid, with a marked soil structural improvement (Seatz and Peterson 1968). The excess free sodium salts (such as Na_2SO_4) then have to be leached by repeated irrigations.

2.3 Experiment #1

Properties of gypsum-treated red mud Seven years after an application of 40 t ha^{-1} gypsum, samples of stony agglomerates ranging from 2–5-cm diameter were taken from Pond 6, at Kirkvine, on the Manchester Plateau, Jamaica. Table 1 shows some main features of red mud so obtained.

2.3.1 Strength Tests

To measure the shear strength of the "soil" a penetrometer was applied. However, the cone tip of a penetrometer failed to penetrate individual gypsum-treated agglomerates on the field surface. A "drop" method of testing soil strength was therefore adopted such that twenty replicates each from 0–7, 7–14 to 14–21 cm-depths were released from a predetermined height (3 m) onto a concrete floor. The number of drops required to shatter at least half of the agglomerate was recorded.

2.3.2 Agglomerate Treatments

In the present study, samples of Pond 6 gypsum-treated material were crushed and passed through the apertures of a 2-mm diameter sieve. Containers were polyacrylate bottles with a locking air valve affixed in the opening, for measuring (from the outside) internal gas pressure. The pH and EC readings (units) were obtained from 1:5 soil/water ratio extracts. Quantities of sodium and potassium were assessed by flame photometry, magnesium by AAS, and calcium by titration. Samples were watered to 70 % field capacity and incubated for 24 weeks in cylindrical polyacrylate containers with or without the incorporation of young, fresh, grounded (<1 mm), readily decomposable shoots of a wild legume (*Cordia sebestena*). This plant was chosen for its local luxuriant growth and high N content, the latter property hastening its decomposition. To reduce vertical settling and to facilitate a constant equilibration throughout the sample, each container was stored with its long axis in the horizontal direction.

2.3.3 Gas Pressure Tests

To compare rates of microbial respiration during incubation, the CO_2 pressure in the flexible-sided poly-acrylate sample containers was measured with a portable digital air pressure gauge (Accutire MS 4000). The gauge was fitted and held on the valve until the pressure was recorded. Readings were taken before and after incubation.

2.4 *Experiment #2: Re-treatment and Topsoil-Free Re-vegetation*

By 2005 the physical structure of the gypsum-treated area had attained high soil strength to a depth of 15 cm, thereby restricting root extension with attendant stunted vegetation and bare patches. In contrast to the 1998 treatment, for the 2005 treatment, readily decomposable (not decomposed) organic matter was concurrently added with the gypsum. This concurrent 2005 treatment was applied because Harris and Rengasamy (2004) found that gypsum, though increasing hydraulic conductivity of sodic subsoils, by itself produced only micro-aggregation of clay particles, in contrast to results seen when gypsum added concurrently with decomposable organic matter, macro-aggregated clay particles in addition (Harris and Rengasamy 2004). Based on the lack of free Ca^{2+} ions lost by years of carbonation, Harris (2005; personal communication) (1) Calculated a new gypsum requirement at the equivalent of 40 t ha^{-1}, (2) recommended a decomposable organic additive concurrently be applied (chicken manure at 0.6 kg m^2) and (3) stipulated the concurrent application of irrigation. This regime, except for irrigation was applied in 2005. Fortuitously however, in that year, rainfall totals exceeded expected levels.

3 Results and Discussion: Experiment #1

3.1 Water Movements and Infiltration

The terms permeability and infiltration are not synonymous. Infiltration describes the entry of water into soil, whereas permeability describes the ease with which water or other materials move through soil (O'Geen et al. 2006). Thus, whilst infiltration occurred on the gypsum-treated red muds (GRMW) retrieved from 30 cm depth, permeability was not exhibited. It was concluded that at that the ploughing depth of the applied gypsum had not penetrated the soil fabric at that depth.

During infiltration tests, the crushed gypsum-treated red mud samples (CGR) failed to facilitate drainage of water through its pores, as pores immediately sealed up after saturation point was reached. Therefore, against expectations, crushing and incubation of impermeable gypsum-hardened red mud waste did not increase its hydraulic conductivity, which remained at 0 cm/h (Fig. 4).

Yet, many workers (Susanto 1992; Rengasamy and Olsson 1993; Sumner 1993; Sekhon and Bajwa 1993; Harris and Rengasamy 2004) observed dramatic increases in hydraulic conductivity after applying gypsum to alkaline soils at similar rates. The large increase of 70 % in electrical conductivity for the phyto-organically treated samples was contemporaneous with its dramatic increases in cation concentration of K: 6-fold, Mg: 9-fold, Na: 1.7-fold, Ca: 1.18-fold (Fig. 5). It is noted here also that the total increase in poly-cations with K (non-dispersive) ions was four-fold that of Na (Fig. 5). Added poly-cations slow down the dislocation of clay particles which block micro-pores (Quirk and Schofield 1955). Table 2 shows that destruction of at least half of each agglomerate (level 3 dispersion) causing cloudiness had occurred for the CGR samples, while the particle dispersion of the phyto-organically treated CGR samples (CPGR) was substantially less. Spontaneous dispersion of particles could therefore have completely sealed pores in the crushed material, leading to an immediate and total restriction of water movement throughout the lower levels of the sample. Particles resulting from dispersion were invisible to the naked eye as a "cloud", and only approximately half of the agglomerate remained after cloudiness was exhibited.

It is thus reasonable to state that prior to dispersion, the silt and clay-sized particles which became dispersed had occupied approximately 50 % of the agglomerate. Goldberg and Glaubig (1987) found that for soils, even as little as 1 % of the total clay

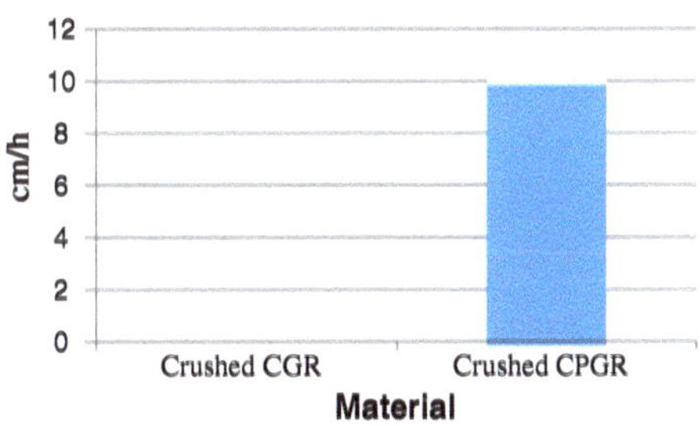

Fig. 4 Hydraulic conductivity of incubated gypsum-hardened bauxite red mud waste

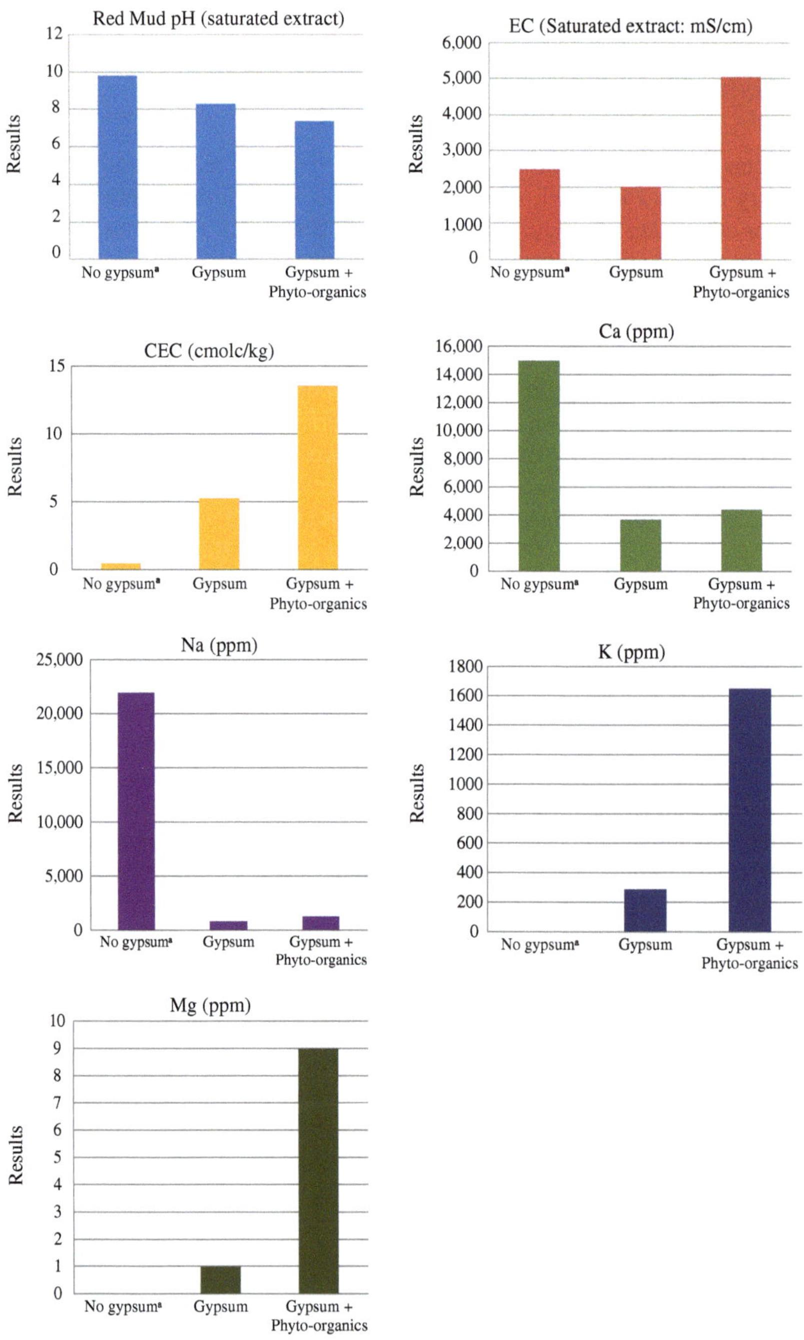

Fig. 5 Some properties of bauxite red mud waste prior to (No gypsum[a]) and after gypsum treatment. After O'Callaghan et al. (1998)

Table 2 Modified Emerson dispersion of gypsum-hardened red mud agglomerates at 2 min

Treatment	Increasing dispersiveness (0, 1, 2, 3 or 4)
Crushed gypsum-hardened red mud (CGR)	2
Crushed CGR with phyto-organics (CPGR)	0

0 No dispersion—the aggregate is completely intact
1 Dispersion of some particles—cloudiness or grains close to the aggregates
2 Considerable dispersion—more than half of the aggregate is dispersed
3 Total dispersion of all particles—the aggregate ceases to exist

dispersed affects the water conductivity by blocking micro-channels in the soil mass. Thus, even the more open network achieved through crushing in this study could not facilitate the clearing of dislodged particles from the fine channels in the samples.

While no planned flushing regime had been applied to the CGR prior to or during the present study, the red mud waste storage sites in the field had had no outlets for the removal of leachates. Therefore, despite adequate post-gypsum rainfall having occurred during 7 years after treatment by O'Callaghan et al. (1998) in 1996, subsequent drainage and removal of then newly formed sodium sulphate had been restricted (the ground being of a flattish topography). This is the same material (CGR) subsequently used in the present study. After 7 years in the waste dump and recent crushing to <2-mm-diameter sizes, CGR (gypsum-treated red mud) still apparently contained substantial amounts of sodium which ideally should have been leached within weeks of the gypsum treatment. High levels of exchangeable sodium cause repulsion and movements between inorganic colloids in soils (Rengasamy et al. 1984) when resulting waste sodium sulphate is not concurrently (or soon after) flushed from material treated. On the other hand, after Harris and Rengasamy (2004) flushed and drained gypsum-treated sodic sub-soils repeatedly during ensuing weeks after dissolution of incorporated of gypsum, thereby leaching out newly formed sodium sulphate from substitution reactions, hydraulic conductivity of green-manured gypsum-treated sub-soils significantly increased. In situ gypsum treatment of bauxite waste without flushing is therefore neither a permanent nor effective corrective (and hence a wasteful process). This problem was exacerbated in this study, as any free calcium (for counteracting the effects of sodium) remaining in the exposed site in the years prior to this study would have been carbonated by atmospheric CO_2.

3.2 Gas Pressures

In this study, gas pressure tests on the flexible poly-acrylate containers before and at the end of incubation indicated a 10-fold increase for the CPGR samples at the end of incubation, compared with no detected increase for the CGR samples (Fig. 6). Increase of pCO_2 during such a 6-month incubation would have been expected to bring some Ca^{2+} into solution as pH was lowered (Fig. 6). Some of the newly released Ca^{2+} would have then replaced some Na^+ on the inorganic

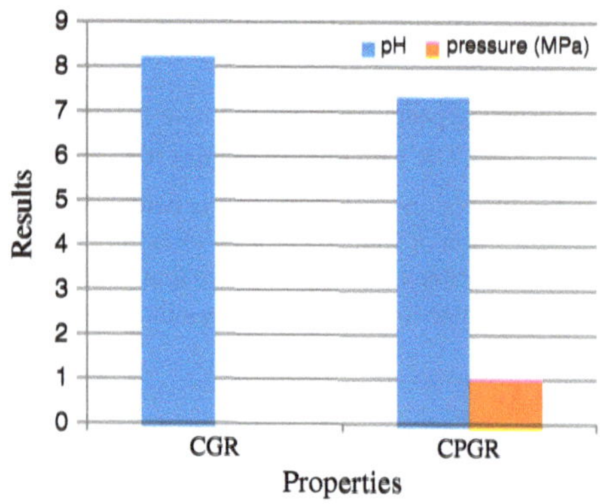

Fig. 6 Effects of incubation on pCO_2 of vessels for a gypsum-hardened bauxite waste

particles, thereby reducing particle repulsion. When readily decomposable phyto-organic waste material was incubated with CGR, the K_s increased significantly. The >10-fold gas pressure increase of the CPGR over CGR would have been due to microbial respiration in the containers. Phyto-organic material also decreased pH readings from, on average, 8.5–7.4 for CPGR (Fig. 5). By day 180, the indigenous $CaCO_3$ in the CPGR therefore could have become sufficiently solubilized under the continuous action of organic acids and/or H_2CO_3 caused by the increased respiration of microorganisms in response to the presence of a readily decomposable phyto-organic substrate. Tests on the samples showed increases in Ca ions up to >20 % of the original concentration. As this large proportional Ca increase is unlikely to have come from the phyto-organics alone when compared with similar studies (Harris and Rengasamy 2004; Sekhon and Bajwa 1993), the indication is that readily decomposable organics acted in the incubation also on the native calcium. The average size of dispersed materials in control samples was <50 μm (being cloudy), whereas, after phyto-organic treatment, the average particle size remained at 5 mm (being the size of the un-dispersed, stable agglomerates). Despite prior gypsum treatment, sufficient time, heat and rainfall for required chemical reactions, the major factor restricting infiltration in the mine tailings samples was the presence of dispersed inorganic particles. Thus gypsum treatment per se did not stop or reduce dispersion. Phyto-organics doubled the electrical conductivity of the samples by increasing the proportion of cations (Fig. 5) because gypsum is soluble in water to the extent of about one-fourth of 1 % (FAO 2014) and is, therefore, a direct source of soluble calcium. This reduction in the negative/positive charge ratio would have decreased mutual repulsion between negatively charged inorganic particles, thereby promoting interaction between clay particles to form stable clusters. However, this cation source would have decreased after 10 years in the ground, firstly, as some of the gypsum initially added (O'Callaghan et al. 1998) would have reacted with intrinsic sodium carbonate to convert into sodium sulphate (a neutral salt which does not contribute to high pH) as follows:

$$Na_2CO_3 + CaSO_4 \rightarrow CaCO_3 + Na_2SO_4 \text{ (leachable)}. \quad (1)$$

Secondly, Abrol and Dahiya (1974) showed that when gypsum was surface applied and leached, only a small fraction of the soluble carbonates reacted with applied calcium and that a major fraction of the soluble carbonates leached without reacting with applied gypsum. Under field conditions, one irrigation prior to application of an amendment would further ensure leaching of soluble carbonates, eliminating the need of additional quantities of gypsum for neutralizing the free sodium carbonate. If there is not enough natural drainage to the underground, or an artificial subsurface drainage system, leaching of the excess sodium by percolation of rain and/or irrigation water through the soil profile will be prevented. But, again, as stated before, for Kirkvine Pond 6, no devices had been installed in 1996 to facilitate the leaching of the sodium sulphate produced by the gypsum amendment applied by either O'Callaghan et al. (1998), or, 10 years later in 2005 by Lyew-Ayee et al. (2011). Therefore a build-up of harmful Na_2SO_4 would have been expected in the treated tailings since 1996. This would have helped to maintain a higher alkalinity in the subsoil layer compared with the upper and surface zone, maintained sodicity and impermeability in the subsoil, and, by extension, restricted plant growth to the mere 12 species found in the treated tailings in 2005 (Lyew-Ayee et al. 2011). Further, much of the vegetation encountered may be considered characteristic of the terra rossa soils of the bauxite plains of Manchester and St. Ann (Lyew-Ayee et al. 2011). Indeed, at a mere 20-cm soil depth, the gypsum treatment had exerted little or no effect in Kirkvine Pond 6 by 2005.

Moreover, gypsum often causes extreme soil compaction within a few years of it having been added to mine-tailings, even where the original pH had been acidic Harris (2001), or in alkaline red mud waste (Harris 2008). Further, natural carbonation increases as sequestered atmospheric CO_2 combines with the calcium, thereby producing large boulder-sized concretions and hard pans in the ground (Thompson et al. 1991). Such massive structures greatly diminish the total reactive surface area of the gypsum initially added as smaller particles, thereby decreasing drastically the available cations for the initiation of stable soil clusters.

3.3 Water-Stable Aggregation

Agglomerates created in this study from coalescence of <2-mm particles without phyto-organics were not resistant to physical pressure or electrolytic forces of high quality water, and broke down immediately on contact (at 0 s) under wet-sieving (Fig. 7). Harris and Rengasamy (2004) found that gypsum acting alone

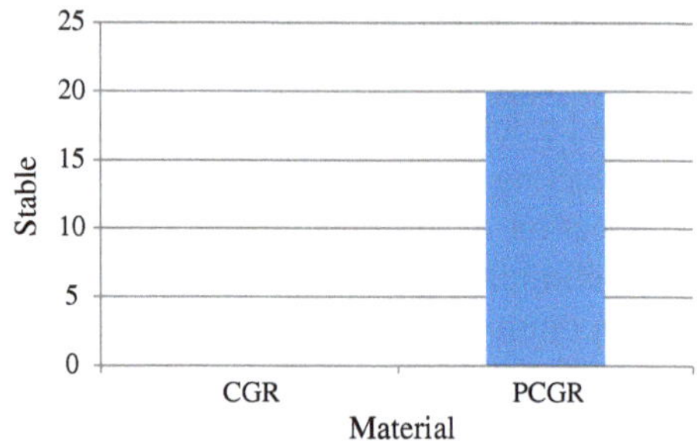

Fig. 7 Stability of crushed gypsum-hardened bauxite red mud during (CGR) during wet-sieving

did not create or stabilize large soil aggregates over 250 ml in diameter, but stabilized aggregates under 250 ml. Particle size also influenced the results. While Harris and Rengasamy (2004) had used <2-mm-diameter "normal" soil, particles, the <2-mm-size of the present study as stated above, was acquired by crushing larger lithified cobbles of 2–15-cm-diameter. According to O'Callaghan et al. (1998), such coarse agglomerates prevented excessive release of sodium embedded in them. Such particles, as remnants of dried, brittle red mud which underwent gypsum treatment, may never have been penetrated thoroughly by Ca ions from their gypsum field treatment. After crushing in the present study, some of the hitherto embedded sodium would have become exposed on the newly formed microsurfaces. As there had been no flushing of the crushed particles during incubation, the ESP of the particles would have been expected to be high throughout and after incubation in this study. Measurements of the CGR samples showed high levels of residual Na (Table 1), while, the phyto-organic additives increased these already high levels of Na by more than 60 % (Table 1). Yet, despite the extra sodium impregnation and an unchanged exchangeable sodium percentage (ESP) (Table 1), the phyto-organic material was able to create agglomerates, which were substantially more stable under wet sieving than those treated by crushing only. Harris and Rengasamy (2004) observed improvements in micro-aggregation of a phyto-organically treated sodic subsoil after gypsum treatment followed by repeated flushing of sodium sulphate waste. However, in that study where incubation lasted 12 weeks, phyto-organics did not significantly increase or cause stabilization of the soil particles in the macro-aggregate size range. In this study, the incubation period was longer—24 weeks, rather than 12, of the former study. The results of this present study therefore show that even the lack of a systematic flushing regime under field conditions after gypsum treatment on waste bauxite tailings may not prevent improved stabilization of the crushed waste after an extended incubation with phyto-organics. Moreover, by producing almost threefold increases in gypsum-treated red mud CEC (Table 1), the phyto-organic treatments of this study have substantially enhanced the potential for plant growth in such edaphically hostile environments.

3.4 Experiment #2: Organic Additives + Gypsum

The fresh gypsum and fresh chicken manure produced dramatic changes in vegetation within 24 months (Figs. 3, 4). High substrate strength facilitated by atmospheric carbonation had decreased initial gypsum treatment effects of O'Callaghan et al. (1998). Thus, with high sub-soil alkalinity at 15 cm-depth (Harris 2008) due to a lack of penetration of the 1996 gypsum treatment and poor sub-soil water movements (Harris 2008), unsuitable rooting conditions in 2005 prevailed in parts of Kirkvine Pond 6 (Fig. 8).

Yet, in 2007, i.e., two years later, (Harris and Evans 2007, personal communication) observed these previously bare patch areas and noted that rapid tree growth occurred, and by 2007 the ground was covered with trees, the whole area completely exhibiting thick lush vegetation (Fig. 9) where vegetation had been sparse or non-existent in 2005. A vegetation review (after 2005) on Pond 6 undertaken in 2011 encountered an additional 44 species, which is a composition comparable to a "dry limestone forest" in Jamaica (EAA 2011). Results of this so-called "topsoil free method" encouraged Rio Tinto Alcan to adopt a similar method on the remaining six ponds which still comprised bare red mud (EAA 2011).

This rapid change was attributed to invigoration by the re-addition of gypsum at a rate of 40 t ha^{-1}, a 50 % increase in the rate of chicken manure at 0.6 kg/m^2, and reseeding with Brachiaria, Bonavista Bean and Guinea Grass (EAA 2011). Yet, the change in vegetation had already occurred in its entirety in 2007, when the dramatic improvement in plant growth had been observed (Harris and Evans

Fig. 8 Large patches of un-vegetated ground in dewatered Kirkvine Pond 6, in 2005, 10 years after non-concurrent gypsum (40 t ha^{-1}) and chicken manure treatment applied without irrigation. *Source* Jamaica Bauxite Institute

Fig. 9 The same area as Fig. 3, in 2007. Heavily vegetated ground seen 2 years after concurrent gypsum at 40 t ha^{-1} and chicken manure treatment without irrigation but with heavy rainfall in early stages. *Source* Jamaica Bauxite Institute

2007) though it was reported to have initially observed in 2011 (EAA 2011). The quantitative and temporal boundaries of the new re-invigoration regime, was based on the gypsum calculations adopted by Harris (2005, personal communication) who determined the gypsum requirement of 40 t ha^{-1} which was actually applied (Harris 2012), and which differed from the previous 1996 treatment as follows:

(1) For the 1996 treatment by O'Callaghan et al. (1998), organic matter was not applied until one year after gypsum had been added; by then atmospheric carbonation (loss of free Ca^{2+}) had substantially occurred in the upper surfaces adjacent to the potential organic additive. (2) The rate of organic additive was 50 % less than that of 2005.

The rapidity with which the trees dramatically gained height within just two years (2005–2007) indicates that root extension substantially occurred. But root extension could not have occurred in the absence of sufficient water in the root zone.

Further, based on poor rooting conditions at the >15 cm depth including the high clay content, poor water movements, and high soil strength (Harris 2008) and toxicity, tree roots could not have extended below 15 cm. Any root extension would have therefore been done laterally, providing that the 15 cm plough depth of 15 cm had remained unchanged during the 2005 treatments. However, apart from the application rate, no details of the 2005 gypsum incorporation procedures were disclosed in EEA (2011). It is therefore uncertain as to whether the gypsum was ploughed in for the 2005 reinvigoration, and, if so, to what depth. The 50 % increase in 2005 of the 1996 chicken waste manure rate from 4 to 6 t ha^{-1} would have increased the soil organic matter and hence the water holding capacity and macronutrients in the root zone over those of the 1996 treatments.

4 Conclusions

4.1 Experiment #1

Gypsum treatments create a problem of lithification due to atmospheric carbonation in many red mud waste sites. Crushing such material fails to improve its water movements or produce edaphic aggregates of sufficient water-stability. This was not the case when phyto-organics were infused into the same age-hardened gypsum-treated RMW. The significant increase in aggregate water-stability was attributed mainly to the release of Ca^{2+} cations into the crushed agglomerates. The cations came mainly from the phyto-organics, but some may have come from the dissolution of components in the GRM, including from solubilizing of the native $CaCO_3$ by organic acids. Therefore, plans and programs to crush gypsum-hardened red muds by heavy machinery in the field to improve infiltration, without the addition of readily decomposable organics, would seem to be not only potentially costly but ineffective. However, reclamation can be accelerated by crushing, followed by phyto-organic treatment. It is also postulated, from the results of this study, that, initially it would be more effective to flush away the excess sodium soon after gypsum treatment. Such action, if followed by ploughing in of readily decomposable phyto-organics, may obviate the need for expensive crushing in the ensuing years.

4.2 Experiment #2

In 1996, 40 t ha^{-1} emerged as the optimum gypsum application rate for red mud waste at a particular water content. Repeating this treatment of calcium sulphate after 10 years substantially increased the growth rate and biodiversity of vegetation in an aged atmospherically carbonated gypsum-treated red mud. Reinvigoration occurred. This suggests that edaphic conditions in the growth medium prior to 2005 had been unsatisfactory for several years. Even though 40 t ha^{-1} of calcium sulphate was found to be the optimal rate, the initial effect had lessened over several years. Depletion of Ca^{2+} (1) due to atmospheric carbonation and (2) having been used up in the quenching process, and the presence of inadequate levels of organic matter as the low CEC levels (Table 1) indicate, are likely factors which had stunted vegetation growth.

5 Geobiotechnological Applications

- The expeditious closure of low-water content red mud dumps is a priority for environmental integrity. But atmospheric lithification after shallow-ploughed gypsum applications provide a surface seal which restricts evaporative losses of water from high viscosity underlying red mud layers. This prolongs the period for land restoration. Further, ploughing to a depth of only 15 cm restricts water movements and root extension. The treatment regime causing the reinvigoration promises to decrease the time required for closure of red mud tailings ponds. To increase the rate of tree growth and climax vegetation in such environments, it is recommended that root growth occurs also below 15 cm. The recommendations therefore are:

- Increase plough depth from 15 cm to 30 cm. This will require doubling the gypsum rate.
- Bio-solids addition at a rate of 4 t ha^{-1} is insufficient. Increase the rate of bio-solids from 4 t ha^{-1} to between 6 and 50 t ha^{-1} (again, doubling the rate to coincide with the doubling of plough depth).
- Gupta et al. (1984) studied the effect of organic materials on the dispersion behaviour of soils and inferred that at high ESP, the role of organic matter in improving soil physical properties was somewhat questionable. However when applied in conjunction with inorganic amendments or when applied in soils of mild sodicity, organic materials have always proved beneficial and therefore their use in the reclamation of sodic soils occupies an important place. Justification for organic matter with gypsum check 40 t ha could be wrong.
- Limited solubility of gypsum in water (0.25 % at 25 °C) is sometimes cited as a major drawback of this amendment when rapid reclamation is desired. Similarly, the quantities of water required for sodic soil reclamation are generally calculated on the basis of gypsum solubility in free water. For example, if a sodic soil requires a gypsum application of 12.5 t ha^{-1}, the quantity of water required to dissolve this quantity will be above 50 cm depth. That this is not likely to be the case in practice was shown by Abrol & Dahiya (1974) and Oster (1982). When gypsum is mixed in a highly sodic soil its solubility increases several fold because of the preference of exchange sites for divalent calcium ions compared to sodium ions. The higher the degree of sodium saturation the greater will be the dissolution of gypsum mixed in soil. Hira et al. (2004) observed that under 14 cm water were required to dissolve gypsum applied at 12.4 t ha^{-1} and leach the reaction products from the surface 15 cm soil in a highly sodic soil (ESP 94.0). This quantity is only about one-fourth the quantity of water calculated from considerations of solubility in water alone. This explains rapid vegetation growth in 2007 on pond 6.
- Whether or not water for irrigation is scarce, during the initial treatment apply the bio-solids sequentially (to reduce volatilization) and begin several months after the gypsum had been added at the first treatment. This allows time for rainfall to dissolve the gypsum and drop the pH to levels suitable for bacterial decomposition of added bio-solids.
- Apply the gypsum just prior to the rainy season.
- At the second (reinvigoration) phase apply the gypsum concurrently with the bio-solids, and just prior to the rainy season.
- Deep ripping >15 cm:

Some studies (Webster & Nyborg 1986) have shown gypsum application at high rates can decrease bulk density through increased soil aggregation. An Alberta study (Webster and Nyborg 1986) compared the effects of an 18 t ha^{-1} lime and 18 t ha^{-1} gypsum treatment on clod size and stable aggregate formation in a sodic soil. The gypsum treatment was superior to the lime treatment and the check in enhancing these characteristics, but only when tillage deeper than 15 cm was used. Tilling deeper than 18 inches improved soil physical characteristics where natural gypsum deposits underneath the sodic layer were mixed with the soil above.

References

Abrol IP, Dahiya IS (1974) Flow-associated precipitation reactions in saline-sodic soils and their significance. Geoderma 11:305–312

Ahmad R (1995) Landslides in Jamaica: extent, significance and geologic zonation. In: Barker D, Mcgregor DFM (eds) Environment and development in the Caribbean. Geographical perspectives. The Press, University of the West Indies, Barbados, pp 147–169

Ahmad R, Jerome D, James PM (1999) Landslide loss reduction: a guide for the Kingston metropilitan area, Jamaica. Unit for Disaster Studies, Department of Geography and Geology, The University of the West Indies, Jamaica, UDS Publication no. 6

Air Assessments (2004) Pinjarra Alumina Refinery (2007) efficiency upgrade dust management plan for the residue disposal area Alcoa world alumina Australia https://www.alcoa.com/.../Dust_Management_Plan_FINAL_Feb_07.pdf. Accessed 5 June 2015

Alcoa (2004) Air Assessments, 2004

Bardossy G (1982) Karst bauxites. Elsevier Scientific Publishing Co, Amsterdam

Corbett J (1969) the living soil: the processes of soil formation. Martindale Press for the Geography Teachers' Association of N.S.W., West Cono, N.S.W.

Evans (2011) EAA—Ref: bauxite residue mgmt.: best practice, Apr 2013. European Aluminium Association. www.world-aluminium.org

FAO (2014) Sodic soils and their management (www.fao.org/docrep/x5871e/x5871e05.htm). Accessed 2 June 2015

Glenister DJ, Thornber MR (1995) Alkalinity of red mud and its applications for management of acid wastes. Chemica (85):100–113 (41)

Goldberg S, Glaubig RA (1987) Effect of saturating cation, pH, and aluminium and iron oxides on the flocculation of kaolinite and montmorillonite clay. Clay Miner 35:220–227

Graham GA, Fawkes R (1992) Red muds disposal management at QAL. In: Proceedings of An international bauxite tailings workshop, Pereth, Western Australia. Australian Bauxite and Alumina Producers

Greenberg WA, Wilding LP, Daniel MC, Sylvan C (1996) Infiltration and erosion potential on reclaimed bauxite mine lands in Jamaica. http://natres.psu.ac.th/Link/SoilCongress/bdd/symp38/1277-t.pdf

Gupta RK, Bhumbla DK, Abrol ID (1984) Sodium-calcium exchange equilibria in soils as affected by calcium carbonate and organic matter. Soil Sci 138: 109–114

Hanahan C, McConchie D, Pohl J, Creelman R, Clark M (2004) Stocksiek C chemistry of seawater neutralization of bauxite refining residues (red mud). Environ Eng Sci 21:125–138

Harris MA (2001) The effect of sewage sludge and green manure onthe erodibility of two mine tailings materials. Environ Geol 40(11/12):1352–1367

Harris MA (2008) Structural improvement of age-hardened gypsum-treated red mud wastes using readily decomposable phyto-organics. Environ Earth Sci 56(8):1517–1522. DOI: 10.1007/s00254-008-1249-5

Harris M (2012) NCU scientist points to recognised solution for alumina dumps. Daily Observer February 11 2013

Harris MA, Rengasamy P (2004) Sodium affected subsoils, gypsum, and green manure: interactions and implications for amelioration of toxic red mud wastes. Environ Geol 45(8):1118–1130

Harris MA, Evans J (2007) Personal observation of Kirkvine Pond 6. RPS Consultants observation trip to Kirkvine with Northern Caribbean University (NCU)

Harris M, Evans J, Panton C, Wright V (2005) On the rejuvination of the o'callagahn topsoil-free growth experiment at Kirkvine Pond 6, Jamaica. RPS Consultants Conference, Terra Nova Hotel, Kingston Jamaica 2005

Hillel D (1982) Applications of soil physics. Academic Press, New York

Hira GS, Jalota SK, Arora VK (2004) Efficient Management of Water Resources for Sustainable Cropping in Punjab. Department of Soils, Punjab Agricultural University, Ludhiana

Lyew-Ayee PA, Stephan TL (2008). In: Archaeology and geoinformatics: case studies from the Caribbean edited by Basil A. Reid. University of Alabama Press. eISBN: 978-0-8173-8053-3

Lyew-Ayee PA, Persaud SD, Evans KA (2011) ICSOBA international seminar on bauxite residue

Mage SM, Porder S (2012) Parent material and topography determine soil phosphorus status in the Luquillo Mountains of Puerto Rico. Ecosystems. Springer Science & Business Media, New York. doi:10.1007/s10021.012.9612.5

Meteorological Office of Kingston Jamaica (2006). The hurricanes of 2005. www.metservice.gov.jm. Accessed April 3, 2016

O'Callaghan WB, McDonald SC, Richards DM, Reid RE (1998) Development of a topsoil-free vegetative cover on a former red mud disposal site. Alcan Jamaica Rehabilitaion Project Paper

Oster JD, Frenkel H (1980) The chemistry of the reclamation of sodic soils with gypsum and lime. Soil Science Society of America Journal. 44(1): 41–45. doi: 10.2136/sssaj1980.03615995004400010010x

Oster JD (1982) Gypsum use in irrigated agriculture—a review. Fert Res 3:73–83

Quirk JP, Schofield RK (1955) The effect of electrolyte concentration on soil permeability. J Soil Sci 6:163–178

Rengasamy P, Olsson K (1993) Irrigation and sodicity. Aust J Soil Res 31:821–837

Rengasamy P, Greene SRB, Ford GW, Mehanni AH (1984) Identification of dispersive behavior and the management of red-brown soils. Aust J Soil Res 22:413–431

Rout SK, Sahoo T, Das SK (2013) Design of tailing dam using red mud Science.gov (United States)

Sekhon BS, Bajwa BS (1993) Effect of organic matter and gypsum in controlling soil sodicity in rice and wheat-maize systems irrigated with sodic waters. Agric Water Manage 24:15–25

Shainberg I, Sumner ME, Miller WP, Farina MPW, Pavan MA, Fey MV (1989) Use of gypsum on soils: a review. Adv Soil Sci 9:1–111. Doi:10.1007/978-1-4612-3532-3_1

Sumner ME (1993) Sodic soils: new perspectives. Aust J Soil Res 31:583–750

Susanto M (1992) Effect of gypsum, agricultural lime, and organic matter on the infiltration characteristics of a red-brown earth. Dissertation, University of Adelaide, Australia

Sutar H, Mishra SC (2014) Progress of red mud utilization: an overview. Am Chem Sci J 4(3):255–279

Sutar H, Mishra SC, Sahoo SK, Chakraverty AP, Maharana HS (2014) Progress of red mud utilization: an overview. Am Chem Sci J 4(3):255–279

Thompson TL, Hissner WLP (1991) Micromorphology of calcium carbonate in bauxite processing waste. Geoderma 48(1):31–42

Wagner GR (1997) Asbestosis and silicosis. Lancet 349(9061):1311–1315

Webster GR, Nyborg M (1986) Effects of tillage and amendments on yields and selected soil properties of two solonetzic soils. Can J Soil Sci 66:455–470 Ref: www.world-aluminium.orgBauxite residue mgmt: European Aliminium association: Best Practice (April 2013)

WHO (2013) Airborne dust—WHO—World Health Organization. www.who.int/occupational_health/publications/en/oehairbornedust3.pdf

Wong JW, Ho GE (1992) Viable technique for direct re-vegetation of fine bauxite refining residue. In: Proceedings of international bauxite workshop, Perth, Australia, November

Zhang Y, Qu Y, Wu S (2001) Engineering geological properties and comprehensive utilization of the solid waste (red mud) in aluminium industry. Environ Geol 41(3/4):249–256

Management of a Potential Mine Capping for Reclamation of Open-Cut Mines: Responses to Decomposable Organic Treatments

Mark Anglin Harris and Pichu Rengasamy

Abstract Extraction of mineral ores, and some engineering projects commonly entail removal of large tracts of valuable top-soil often called "overburden." A potential mine overburden capping material from an alfisol with different management histories was treated to assess its physical responses to phyto-organic amendments during storage. Although there were small increases in hydraulic conductivity, phyto-organic amendments did not greatly improve the physical properties of the alfisol which had previously been cultivated for several years, as changes in stability and water retention were generally not statistically significant. However, very substantial increases of stability (about 25 %) and water retention (about 40 %) occurred in the same soil which had previously been under a long-term permanent pasture regime. As top-soils can quickly deteriorate when removed and stored, prior soil management should be taken into account where top-soil is to be removed and used as a capping after mineral ore extraction.

Keywords Infiltration · Open-cut mines · Overburden · Mine capping

Highlights

- Mine capping improved with pre-mining treatment of pasture grass
- Substantial increases of stability (about 25 %) and water retention (about 40 %) occurred in the same soil which had previously been under a long-term permanent pasture regime
- Prior soil management should be taken into account where top-soil is to be removed and used as a capping after mineral ore extraction.

Original article: **Treatment of a potential soil capping for open-cut mines (part I): Influence of prior management on responses to a phyto-organic amendment**. *Environmental Geology 46, no. 1, 15–21 (7).*

M.A. Harris, *Geobiotechnological Solutions to Anthropogenic Disturbances*, Environmental Earth Sciences, DOI 10.1007/978-3-319-30465-6_11

1 Introduction

1.1 Edaphic Problems of Mine Waste Rock

Mining and related activities cause drastic perturbation of terrestrial ecosystem (Das and Maiti 2005), leading to severe soil degradation (Davies et al. 1995). Many mine wastes are structure-less entities (Fig. 1), prone to crusting and low in organic matter and macronutrients (Waygood and Ferreira 2009), with contaminants including salts, metals, metalloids, and radionuclides (Sheoran et al. 2008). Such harsh conditions often prevent colonization by plants (Lottermoster 2003). Mine tailings are a significant health risk to nearby populations and they require novel remediation approaches that are economical and of low input. In arid regions, mine tailings and their associated contaminants are prone to wind-borne dispersion and water erosion. Capping mine tailings sites with soil, gravel, or even cement is an accepted although often impermanent way to reduce wind and water erosion. Revegetation is considered a cost-effective and more permanent alternative to these capping strategies. The challenge is that mine tailings have no nutrients or soil structure.

This is exacerbated by conditions of drought and salinity in arid and semi-arid regions. The end results of mineral ore extraction are commonly mine tailings and waste rock dumps lacking organic matter, and with weak structures prone to surface sealing during rainfall events. In a study assessing the shelf life of mine-soils in India, Ghose (2005) observed a 47 % decrease in SOC in the first year, followed

Fig. 1 Mine waste rocks from a pyrites mine. Note the lack of top-soil and undeveloped soil structure. Adapted from: South Australia Government Documents at: www.minerals.dmitre.sa.gov.au

by gradual decrease until a steady-state level was achieved at 80 % lower SOC than the initial concentration after 6 years. Therefore, the potential to increase the C capital of mine-soils is significant (Ussiri and Lal 2005). Additionally, the mineral separation process is only partially efficient, and after the milling processes, some of the metal-containing minerals are left behind as small tailings particles. As a result of these combined factors tailings do not readily support plant growth and can remain barren for decades or longer.

Research data from other mine-sites demonstrate that in some areas, waste rock weathers rapidly to form suitable materials for revegetation (Sheoran et al. 2008). Some of the important metallic micronutrients that are essential for plant growth are Fe, Mn, Cu, and Zn. These micronutrients are available in the soil due to continuous weathering of minerals mixed with primary minerals. Nevertheless, these metals are more soluble in acidic solution (Sheoran et al. 2010), and therefore can dissolve to form toxic concentrations that may actually hinder plant growth (Barcelo and Poshenrieder 2003; Das and Maiti 2005). Moreover, though many native species are adapted to low nutrient availability, in agricultural land use acidity merely equates to dysfunctional systems, lacking water and high nitrate levels if the rainfall is substantial (Sheoran et al. 2010).

Environmental losses

The effects of mine wastes can be multiple, such as soil erosion, air and water pollution, toxicity, geo-environmental disasters, loss of biodiversity, and ultimately loss of economic wealth (Wong 2003; Sheoran et al. 2008). The overburden dumps include adverse factors such as elevated bioavailability of metals; elevated sand content; lack of moisture; increased compaction; and relatively low organic matter content (Sheoran et al. 2012). Long term mine-spoil reclamation requires the establishment of stable nutrient cycles from plant growth and microbial processes (Singh et al. 2002; Lone et al. 2008; Kavamura and Esposito 2010).

An increase in the concerns for environment has made concurrent post-mining reclamation of the degraded land as an integral feature of the whole mining spectrum (Ghose 1989). For example, in order to avoid soil erosion and to ensure that vegetation is returned to the land, many companies have water drainage practices during mining operations, separate removal of top-soil (50 % reuse it directly after the mining operation) and over 60 % have their own nursery plant facilities (Singh 1994).

The top-soil gets seriously damaged during mineral extraction (Singh et al. 2002). The consequences of physical disturbance to the top-soil during stripping, stockpiling, and reinstatement cause unusually large N transformations and movements, eventually with substantial loss. Management of top-soil is therefore important for reducing N losses and increasing soil nutrients and microbes (Singh et al. 2002). There is severe loss of SOC due to enhanced mineralization, soil erosion, leaching, reduced/lack of inputs of SOM (Fig. 2), mixing and compaction, which decrease physical protection against decomposition (Ussiri and Lal 2005). Losses of SOC of as much as 80 % of the original pool have been observed from scraped top-soils (Ghose 2005).

Fig. 2 Rapid loss of SOM immediately begins on soil sub-aerial exposure due to clearing for mining operations

1.2 Purpose of Mine-Capping

Though the reason for capping is to either (a) increase infiltration to facilitate revegetation, or (b) decrease infiltration to decrease outflow of toxic mine materials, in either case, vegetation directly facilitates the objective (Waygood and Ferreira 2009). This is because the greater proportion of vegetation extraction of moisture by roots reduces the outflow of mine wastes in case (a) above (Waygood and Ferreira 2009). With botanists indicating that most grasses and trees have 80 % of roots within the upper 300 mm of soil with 90 % of biomass being within 600 mm of surface (Waygood and Ferreira 2009), the potency of a well-vegetated top-soil for mine-capping is clear. Nevertheless, this does not preclude vegetation from extracting water from deeper soils where practical.

1.3 The Ideal Mine Capping

For waste rock, infiltration is often so rapid (Fig. 1) as to cause high water stress for any vegetation it supports. The ideal capping should therefore possess adequate water retention and infiltration, and the ability to resist breakdown and erosion from water and wind. Revegetation constitutes the most widely accepted and useful way to reduce erosion and protect soils against degradation during reclamation (Sheoran et al. 2010). However, Mentis (2006) has shown that soil capping systems placed without consideration of the need for additional organic carbon, appropriate additions of nutrients, proper pasture management and detailed planning are often not successful in the longer term. In terms of final land use, he cites several problems on capping systems at various opencast mines associated with disturbed soils, depletion of organic carbon and nutrients and low soil pH. These factors have resulted in limited progression of vegetation to climax species, and a final land use that is less sustainable than before.

Ideally, therefore, to compensate for potential deterioration during storage, the aim should be to sufficiently improve the surface soil prior to its removal of overburden and top-soil. This is critical because top-soil is at times unintentionally mixed with subsoil during the open-cast mining process, thereby depleting the top-soil of nutrients (Harris and Omoregie 2008). On the other hand, mixing with subsoil is at times deliberate. For example, in the Witbank area of South Africa, 600 mm of soil cover would generally be considered adequate, with the stripped material being placed over spoils after they have been levelled, typically as one mixed layer of soils and sub-soils (Waygood and Ferreira 2009). The area is then grassed with a mix of species. The mixing of the soil layers should reduce hydraulic conductivity and lower the influx of water into the subjacent mine-spoil material.

Secondly, there are limits to the efficacious thickness of stored top-soil. At a depth of about 1 m in the stockpile, the number of anaerobic bacteria increases, whereas those of aerobic bacteria decrease (Harris et al. 1989). This inhibits nitrification due to poor aeration within the stockpile, leading to an accumulation of ammonia in the anaerobic zones. If a high level of ammonia is present in a reinstated soil, the amount of nitrate generated is likely to be much greater than the normal. Consequently there is high potential for N loss to the environment via leaching or/and denitrification (Johnson and Williamson 1994). Nitrate leached to water courses is not only a threat to aquatic environment and drinking water supplies (Addiscott et al. 1991) but if nitrogen is lost from soil in the form of gaseous nitrogen or nitrous oxides, this will contribute to degradation of the ozone layer (Isermann 1994; Davies et al. 1995).

1.4 Other Drawbacks of Top-Soil Storage

In addition to a loss in the breakdown of organic matter, stockpiling causes many other deleterious changes (Fig. 3) including a marked drop in the earthworm population (Johnson et al. 1991) which affects soil nutrients, bulk density and water holding capacity (Rana et al. 2007). It has been reported that stockpiling techniques, and the wholesale removal of the top-soil layers, reduce the chances for succession for much of the pre-existing vegetation (Jacobson 1999). Plant fragments from pre-existing vegetation are lost or greatly reduced. The seed bank is

Fig. 3 Untreated top-soil added to this open-cast bauxite mine had been stored (unmaintained) prior to spreading to increase infiltration and facilitate revegetation. Here the process is compromised by wind and water erosion after spreading, and hence is unsuccessful. Adapted from: Google Maps

also reduced, and what does remain must compete for the reduced nutrients with microbes. These microbes become highly competitive as the base of stockpiles become anaerobic (Jacobson 1999).

Fresquez and Aldon of the USDA Forest Service (1984) noted that top-soil stored for years, and, especially the mining overburden material, has little biological resemblance to the undisturbed surface soil and that the resulting reductions to the fungal genera and microorganisms result in an unstable and unbalanced soil ecosystem. Prolonged storage was also a part of the research conducted by Harris and Birch (1989). They concluded that prolonged storage intensifies the loss of the bacterial element of the soil.

In some operations such as bauxite mining, and in keeping with increasingly demanding environmental standards, the removed top-soil is commonly stored, to be used later as a capping for the exposed earth materials. The period between the initial removal of top-soil and final laying of the same over the reclaimed is too often over-extended. Hence, properties of stockpiled soil continually deteriorate and ultimately become biologically non-productive if it is not preserved properly (Ghose 2005). He advised against top-soil storage, especially in the long term, for a time length by which the mine-spoil cannot maintain its sustainability for suitable plant growth without biological reclamation, and, also, maintenance of growth of aerobic bacteria. Often, after a storage period of just six months, top-soil to be used as capping becomes sterile, after heavy losses in organic matter and microorganisms (Hannan 1978; Rana et al. 2007), thereby becoming less effective as a growth medium. Finally, bulk density of productive natural soils generally ranges from 1.1 to 1.5 g cm^3. High bulk density limits rooting depth in mine soils, and at lower levels of thickly stored top-soils (Williamson and Johnson 1991).

1.5 Effects on Mycorrhizae

Arbuscular mycorrhiza fungi are ubiquitous soil microbes occurring in almost all habitats and climates. The hypha network established by mycorrhizal fungi breaks when soils are initially moved and stockpiled (Gould et al. 1996). It is well documented that mycorrhizal associations are essential for survival and growth of plants and plant uptake of nutrients such as phosphorus and nitrogen, especially in P deficient derelict soils (Khan 2005).

According to Six et al. (2004), microbial secretions and fungal hyphae have generally been found to be extremely important components of macro-aggregate development. A study conducted in Derbyshire, England, of the relationship between aggregate stability and microbial biomass in three restored sites, showed a linear relationship between the health of the microbial community and the quality of soil structure (Edgerton et al. 1995). Mycorrhizae hyphae form an extensive network in soil, the hyphae being covered with extracellular polysaccharides that form soil aggregates. These aggregates are held intact by the roots so that they do not collapse in water. This system forms pore spaces and drainage channels. Based on observations in agricultural practices and previous reclamation research, most macro-aggregation is destroyed during top-soil stripping and salvage prior to mining as roots and fungal hyphae holding macro-aggregates together are disrupted (Wick et al. 2008). Soil disturbance associated with surface mining has a negative effect on both saprotrophic and arbuscular mycorrhizal fungal (AMF) populations (Mummey et al. 2002).

Secondly, mycorrhizal symbiosis is documented as protection against pathogenic fungi (Tate and Klem 1985). Moreover, the microbial community is responsible for the development of a soil structure conducive to the various biogeochemical cycles (Tate and Klem 1985). Mycorrhizal fungi are therefore a very important part of the microbial community. These fungi are often reduced or destroyed by stockpiling. They also found that deep stockpiles create both high and low moisture problems, which limits soil microbial respiration. To maintain a healthy microbial community, soil moisture must have some constancy in order for fungal propagules to survive (Tate ans Klem 1985). Thus Rana et al. (2007) found that recovery of the soil microbial community is crucial to successful land reclamation and ecosystem restoration.

1.6 Management History

As long storage is inimical to the integrity of removed top-soils, the nature and prior land use of the top-soil considered for mine waste capping may significantly influence its effectiveness. In models, soil organic carbon (SOC) is often grouped into three conceptual pools to characterize long- and short-term changes in soil C storage and decomposition rates (Schwendenmann and Pendall 2008). Active and slow pools are regarded as being sensitive to land management (McLauchlan and Hobbie 2004) and depth (Fierer et al. 2003).

The passive, resistant, or stable pool is considered to be weakly sorbed C (Schwendenmann and Pendall 2006), composed of aliphatic compounds, often mineral-stabilized, with mean residence time (MRT) in the range of 10^3 years (Paul et al. 2001). As soil management largely determines active and slow pools of organic matter levels in soils, and exposure through storage decreases soil organic matter levels in the short-term, it is important that prior to ore extraction, the management history of the top-soil be ascertained, and its effective treatments upon removal be considered.

Adequate top-soil management is thus considered the most important factor in successful rehabilitation of mining projects if the objective is to restore the native ecosystem of the project area. The top-soil contains the majority of the seeds and other plant propagules (such as rhizomes, lignotubers, roots etc.), soil micro-organisms, organic matter and much of the more labile (more readily cycled) plant nutrients. The top-soil from all areas being cleared would ideally be retained for subsequent rehabilitation. However, removal and storage reduces the integrity of top-soil. Therefore, as an abundance of soil organic matter is found under pasture lands (Fig. 4) in humid zones (Schwendenmann and Pendall 2006), the treatment

Fig. 4 Pasture lands contain both active and slow pools of organic matter as short and long-term nutrients for microbial life. After removal as overburden, this helps to maintain soil vigour during storage

and character of the top-soil prior to its removal largely determines its effectiveness as a mine-capping.

1.7 Purpose of Top-Soil In Situ Amendments

Top-soil is a scarce commodity, and it is never stored in the majority of potential sources; instead it is borrowed from nearby areas for the reclamation of the degraded mined-out areas (Sheoran et al. 2010). Also, in a tropical climate where 90 % of rainfall is precipitated within three months of the rainy season, storing of the top-soil and preservation of soil quality remains problematic (Sheoran et al. 2010). The large and relatively rapid decline in SOC pool in mine-soils indicates that there is a potential to enhance the rates of C sequestration in these soils with management practices that reverse the effects of degradation on SOC pools (Ussiri and Lal 2005).

1.8 Problem Area #1: Guyana (Bauxite Mining)

According to the EPA Guyana (2014) report, the East Kurubuka bauxite layer lies approximately 57.8 m below the surface and the West is approximately 38.1 m below the surface (EPA Guyana 2014). During the preparation of a site for open-cast mining, the surface soil is removed to get access to the bauxite. Therefore the local bauxite industry has excavated much more overburden earth to make contact with their bauxite (Fig. 5) than that of Jamaica.

Moreover, whereas the Jamaican ore is physically a soil, the Guyana bauxite exists as a rock (Fig. 6). This presents greater challenges for revegetation due to the lack of permeability, high strength and total absence of organic matter.

1.8.1 Gold Mining Effects in Guyana

An inspection of various mining activities in Guyana—manganese (no longer exploited), bauxite, and gold—shows that invariably mining destroys valuable stands of forest and wildlife habitat. Lang (2014) observes that medium and small-scale gold mining as currently practised and regulated inflict severe environmental, health, and social damage on the areas and people near such mining operations. Areas around mines resembled a moonscape of barren, mounded sand and mud (Fig. 7).

Since small-scale miners typically wash the top-soil away in order to get to the gold-bearing clayey soil underneath, the sites of former mines were very infertile and incapable of supporting regenerated rainforest. Failure to replant immediately after clear-felling of trees has caused the topical clay soils to dehydrate, sometimes irreversibly, and to become very hard and impermeable to water, thereby lowering

Fig. 5 A Guyana bauxite mine. Bauxite is located much more deeply; >3-fold the depth of Jamaica mines. Note the deep lakes formed in the mined out locations, thereby presenting almost insurmountable obstacles to land rehabilitation. *Source* Demarara Bauxite Company (DEMBA)

Fig. 6 Guyana bauxite is often in the form of a pink rock, contrasting with the soft, earthy, easily-worked Jamaican ore. Mine capping therefore is more costly and logistically and edaphically more difficult in Guyana. *Source* Demarara Bauxite Company (DEMBA)

Fig. 7 A gold mine excavation in Guyana. The uneven, pock-marked surface with pools at varying levels presents difficulties for subsequent mine-capping operations. *Source* Barrick Gold 2012

the soil potential to support growth of trees (Singh 1994). On the other hand, high rainfall rapidly erodes the top-soil and creates difficulties for tree establishment. In Guyana, vast tracts of scrubland are burned for cattle ranges. Very soon the soils lose their fertility and new ranges are created by burning more land.

Soils are stripped of their vegetation, scarring the landscape (Fig. 8). According to Lang (2014), large unsightly excavations and stockpiles of overburden are dumped on the top-soil. The loose soils cause dust pollution and affect all living organisms in the area. The heaps of overburden are prone to leaching and erosion, polluting the waters. There are mines where the soil thickness is extremely limited due to poor utilization of soils, or thin soils in the pre-mining environment.

1.9 Problem Area #2: Bauxite Pits in Jamaica

Bauxite-excavated pits in Jamaica exhibit low SOC, soil crusting, and low fertility (Harris and Omoregie 2008). Further, it is known that aluminium is most active in fixing phosphate at a pH of 5.0–5.5. Iron is especially active below pH 4.0 where phosphate is very strongly fixed. This apparently excludes tropical soils where pH exceeds these values, but pH values in soils can vary widely within short distances (Kundu and Ghose 1997), especially with sudden elevations changes as occurs often in Jamaica. Further, calcium is primarily responsible for phosphate fixation in alkaline soils where fixation peaks around pH 8.0. However, of the three major fixation processes—precipitation by iron, aluminium, or calcium—phosphorus is

Fig. 8 Guyana: scarred landscape after bauxite mining has removed soil and overburden. *Source* http://epaguyana.academia.edu

least unavailable when it is fixed by calcium (Ludwick 1998), and the reclaimed bauxite overburdens of Jamaica having karst terra rossa origins, are high in calcium. The structure of a mined-out pit depends largely on the shape and structure of the ore-body, such that the pit can vary from being a shallow saucer-shaped basin to a wide, deep hollow with vertical walls (Sheoran et al. 2010). The shallow basins are graded to a gentle slope ending at the approximate middle of the mined-out area or terraced to that point. Deeper deposits are graded to the toe of the vertical walls and serve as collection ponds for water (Sheoran et al. 2010).

1.10 Improving Stored Top-Soils

Several methods of treatment are known to improve top-soils. One cheap source of organic amendments is green waste or "green manure". In laboratory trials, Harris and Megharaj (2001) recorded substantial improvements in the physical properties of green manure-treated sludge used as a growth medium in mine-waste dump materials. Courtney et al. (2013) found that within one year, bio-solids-amended and vegetated sites exhibited improved physico-chemical properties and increased aggregate stability. They state that further improvement of these properties in subsequent years and evidence of non-labile carbon and nitrogen accumulation indicate that development of root systems and soil communities drive pedogenesis in restored industrial sites. In this study, the influence of prior management history, period of storage, and decomposing green manure amendments, on the physical properties of potential top-soil capping material are investigated.

1.11 Research Area

The Brukunga Mine of South Australia is a de-commissioned iron pyrites mine where re-vegetation attempts encountered stunted growth and death due to low pH, organic matter, and bacterial levels (Harris 2001; Harris and Megharaj 2001). Elevations range from 150 to 300 m above sea level, and pyrites is geologically the dominant mineral. The climate of the research area is moderately warm (25–35 °C summer, 12–18 °C winter, average), and the average total annual rainfall is 650 mm. The study area is in the Bremer River drainage basin in the Mount Lofty Ranges Watershed which drains to Lake Alexandrina (Fig. 9). The stream supplies irrigation water for various horticultural activities, and facilitates grazing and dairying. The total area of the watershed is approximately 900 km^2.

The Brukunga Mine runs for approximately 2000 m parallel to the western bank of Dawesly Creek. Waste rock is heaped along this area (Fig. 10).

On the eastern side, a large tailings dam has been capped to reduce acidic mine leachate (pH 2.5) generation. Leachate is caught in a number of dams and pumped back to an acid neutralization plant above the tailings dam. Though microbial digested sewage sludge has improved vegetation on some plots, the danger of heavy metal pollution to ground and surface waters, and community resistance to the obnoxious nature of such material (with dwelling houses and the Town Centre less than a kilometer away) discourages its large-scale use at this site. The discharge of effluent from the community septic tank of the Brukunga township is an additional source of pollution to Dawesly Creek. Therefore, a cheap, effective "clean material" would be a more acceptable capping.

Though the study area is located in South Australia, this experiment is applicable to several sites in the Caribbean. This is because sulfidic mine waste dumps in the Caribbean exhibit similar pH values whilst at least doubling the rainfall totals (FAO 2006) of the research area in Australia, and AMD is exacerbated by increased precipitation (Aubertin and Bussiere 2001). The climate of the Dominican Republic is tropical with a temperature range between 18 and 32 °C with generally high humidity. The heaviest rainfall period is between May and November with an accumulation of, more or less, 3 m of water (Nelson 2000).

1.12 Pyritic Mines: A Caribbean Perspective

At Pueblo Viejo Gold Mine in the Dominican Republic (Fig. 11), Trade Wind effects produce higher rainfall intensities than for South Australia tailings (Figs. 1, 9 and 12).

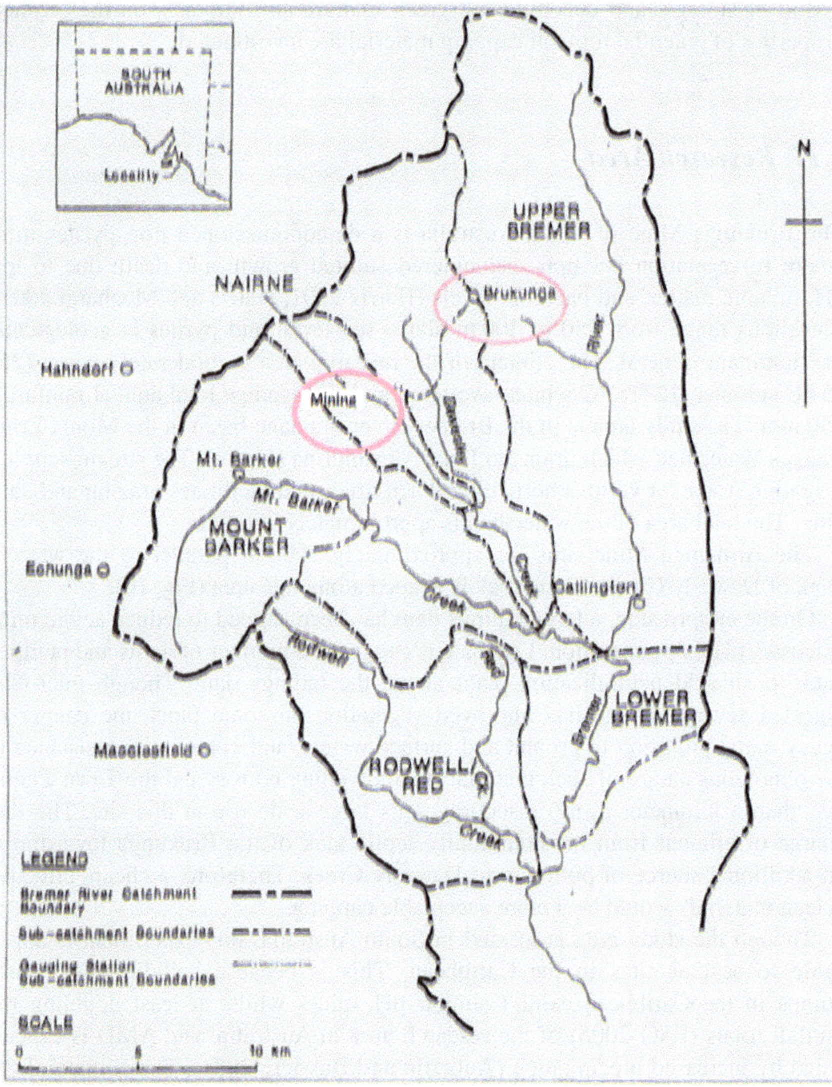

Fig. 9 Location of the Brukunga Mines of South Australia

The "high sulfur" central Cuban provinces of Las Tunas and Ciego De Avila and the inactive Zn–Pb mine of Santa Lucia in western Cuba (Romero et al. 2010) also contain areas of acidic mine drainage. Nevertheless, Alfaro et al. (2015) note that studies assessing the environmental quality of soils in Cuba are very scarce, especially involving various metals on a nationwide scale. Most studies, they report, on Cuban soils, focus on a small number of metals under specific situations (Romero et al. 2010). Nevertheless, as is the case for Pueblo Viejo, capping such mines is imperative to reduce the rate of environmental degradation (Figs. 13 and 14).

Fig. 10 Brukunga Mine, South Australia. Waste rock heaped along a creek bank remains unvegetated. Rainfall maximum here is unimodal (winter only). Comparing this result with that of the more intense bimodal rainfall regime of a similar pyrites mine in the mountainous Trade Wind region of Dominican Republic could reveal greater environmental impacts. *Source* South Australia Government Documents at: www.minerals.dmitre.sa.gov.au

Fig. 11 Pueblo Viejo gold mine, Dominican Republic. Note staining at right by leachate from atmospheric exposure of pyritic waste rocks. Heavy, frequent cumulonimbus clouds seen at top of photo signify a more humid climate compared to that of South Australia (see Fig. 9), indicating more moisture for sulfuric acid generation in the ground. *Source* Barrick Gold Corporation

Fig. 12 Climatic conditions for the Dominican Republic exacerbate the dissolution of exposed pyrites waste rock

Fig. 13 Effects of phyto-organics, and management history on hydraulic conductivity of an Alfisol incubated at 80 % field capacity (water content)

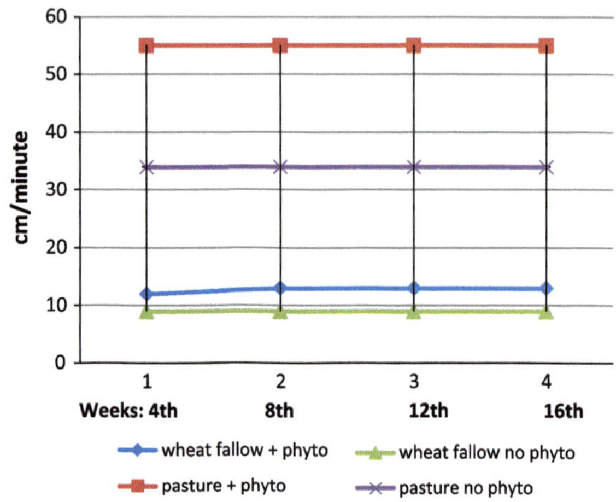

Fig. 14 Effects of storage period, phyto-organics, and management history, on stability of soil aggregates

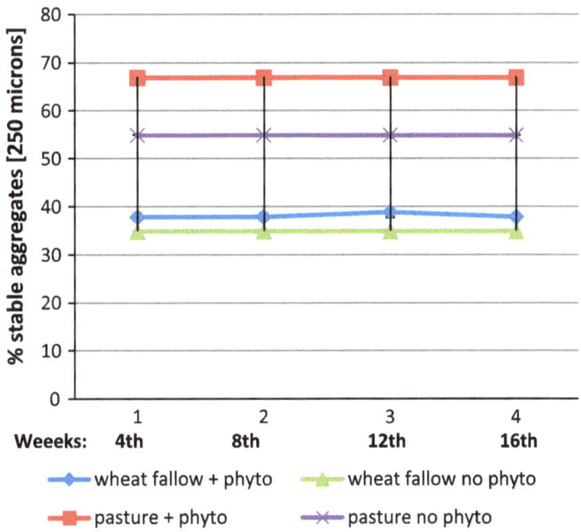

2 Materials and Methods

2.1 Soil Preparation

Near the foothills of the Adelaide Hills, 30 km south-east of the Brukunga Mine is the Long-term Land-use Rotation Experiment of Adelaide University, South Australia, on an Alfisol (Table 1). Samples of the Alfisol were collected with a scraper to a depth of 15 cm on a random basis, in turn, from the wheat/fallow, continuous wheat, and permanent pasture plots. Each of the above management histories had been in progress for the previous 30 years. After air-drying, the soil clods were broken and passed through a 2-mm sieve and thereafter stored in air- tight

Table 1 Properties of the South Australian Alfisol (0–15 cm)

Sub-soil property	Value
Soil pH	6.8
Organic carbon (%)	1.5
Sand	60[a]
Silt	25[a]
Illite	40[b]
Kaolinite	40[b]
Iron oxide	20[b]

[a]Particle size distribution (%)
[b]Clay mineralogy (% of clay fraction)

containers for 6 weeks. The leguminous plant, *Vicia sativa*, having the capacity to grow well on impoverished soils (Pieters 1927), was grown in a glasshouse in beds containing soil from the wheat/fallow, continuous wheat, and permanent pasture management plots. All seedlings were inoculated during germination, grown for 5 weeks and harvested prior to flowering. Twenty plants were grown in each bed. After the harvesting of all shoots on the same day, the plant parts were cut into small pieces (3-cm lengths) to increase the area of contact with soil particles. As it was anticipated that the incorporation process of the plant parts would entail a time period of at least one week, immediately after harvest the needed aliquots (of 80 g for each 1 kg of soil) were placed into plastic bags and tightly sealed against desiccation until the time of incorporation a few hours later. The following treatment particulars were adopted:

1. Plant parts incubated: young shoots only, as Amato et al. (1983) found that in soils, shoots decomposed faster than root or stem material.
2. Soils selected on the basis of their management history: wheat fallow, continuous wheat, and permanent pasture.
3. Water regimes used in soils: (a) 80 % of field capacity, as Linn and Doran (1984) found that moisture levels below 60 % of, at, or in excess of field capacity of soils, significantly retarded decomposition of buried plant materials, (b) alternate wet/dry water cycles (i.e. weekly alternations of watering to 80 % field capacity followed with one week of drying, thereby more realistically simulating water regimes under field conditions. Incorporation was achieved by spreading thin layers of soil between alternate thin bands of fresh plant parts. Equal weights of young stems and leaves were applied. After incorporation, the surface of the soil was firmly pressed down, in an effort to simulate the action of a roller, under field conditions, to facilitate better contact between soil particles and the potentially decomposing substrate.

The study maintained a temperature of 25 °C in a controlled growth chamber consisting of a sealed walk-in cabinet. Three replicates of each treatment were randomly arranged on tables for the duration of 4, 8, and the 12-week incubation storage, which simulated soils under storage conditions. A timed, calibrated dripper system was used once per week. To reduce the impact from the dripper, and reduce evaporation, the soil was covered with an inert plastic mulch separated from the soil surface by a nylon cloth. As an added precaution against water loss, and to ensure complete equilibration, the pots were tightly covered with plastic sheeting over the next two days. Three techniques were used to assess the soil structural stability and hence, changes to its capping suitability resulting from the various treatments: water stable macro-aggregation, water retention, and hydraulic conductivity.

2.2 Macro-Aggregation

For water stable macro-aggregation, the method of Kemper and Rosenau (1986) was adapted as follows: All samples were exposed after incubation storage until air-dry. The soils were gently crushed with a wooden mallet and passed through a 5-mm sieve. Three subsamples of 25 g were taken from each replicate. The samples were then placed on a nest of sieves of the following mesh diameters: 1000-, 500-, 250- and 125 μm. Samples were slowly lowered into a cylinder of distilled water until the water made contact with the bottom of the soil layer and then immersed in the cylinder and oscillated vertically at a stroke length of 2 cm with a frequency of 30 strokes per minute for 5 min. The material was carefully washed from each sieve into containers, oven dried at 105 °C for 48 h and each re-weighed. Water-stable macro-aggregation (WSMA) was calculated as the amount of material left on the 250 μm sieve plus the amount left on all sieves above this size; this was then corrected for the weight of sand >250 μm and expressed as a percentage of the original weight of aggregates.

2.3 Hydraulic Conductivity

Rate of water movement in soils was to be measured because a low infiltration rate results in a correspondingly high rate of runoff, with high volumes of runoff causing severe erosion in a short space of time, especially on the sloping reshaped mounds characteristic of open pit extraction sites. According to Hannan (1978), replaced top-soil in such situations is the first material to be lost. For the determination of the rate of water movement through the soil, the saturated hydraulic conductivity (K_{sat}) was determined for each treatment utilizing a constant hydraulic head in the following manner: 50 g of loose oven dried aggregate were placed in a "Perspex" cylinder. The soil surface was protected from the direct impact of the hydraulic head by using a filter paper. A constant head of 15 cm of water was maintained on each replicate, over a 2-h period during which four measurements were taken every 30 min of the amount of water conducted (q, cm^3) in that time period (t, h) period. The K_{sat} (cm h^{-1}) of three replicates for each treatment was calculated as:

$$K_{sat} = q(L)/H(At) \qquad (1)$$

where q was the volume of water (cm^3) collected per unit time, t(h), A(cm^2) was the cross-sectional area of the cylinder, L was the length of the sample, and H (hydraulic head) was (L + 15 cm).

2.4 Water Retention

The importance of moisture retention in stored top-soils has already been stated. Air-dried aggregates were equilibrated and measured, based on the method of Klute (1986). The amount of water retained in soils as a result of each soil treatment was determined by the pressure plate apparatus and sintered funnel assemblies with a range of suctions applied to the soils at: 0.1, 1, 10, 100 kPa suction, respectively.

3 Results and Discussion

3.1 Saturated Hydraulic Conductivity

Green manure caused significant increases in K_{sat} of all stored top-soils, particularly for the permanent pasture (PP) soils, where the increases were more than three times those of either continuous wheat (CW) or wheat/fallow (WF) soils (Fig. 2). Even without green manure, the K_{sat} of the untreated PP samples was 3-fold that of the untreated WF and CW samples, indicating the inherent superiority in water transmissivity of the PP as mine-capping material, compared to those of the same soil under different management histories (Fig. 2). A high proportion of un-decomposed stems was observed for the treatments having a regime of 80 % field capacity. Thus, the succulent shoot material decomposed quickly in those samples, in the process releasing a greater quantity of binding agents, thereby causing greater porosity. This indicates that decomposition and formation of organic binding agents and thus some stabilization of structure had taken place in all soils in the presence of green manure and adequate moisture and air. Schwendenmann and Pendall (2006) point out that the 'active', 'labile' or 'fast' pool is assumed to be composed of microbial biomass and easily decomposable compounds (e.g., proteins and polysaccharides) from leaf litter and root-derived material with short MRT (from days to years), while the 'slow' or 'intermediate' pool is understood to consist of refractory components of litter, weakly held carbon, and MRT from 10 to more than 100 years. The increases in K_{sat} were maintained throughout the 12 weeks of incubation, with highest increases beginning at 4 weeks.

This trend seems to tie in with the observations of Mason (1977), who found that differences in N mineralization rates of incubated green plant material were relatively small up to the third week of incubation, after the largest increases in the first week.

It is of interest that K_{sat} increases under a wet/dry water regime were not found for any of the samples. This suggests a minimal amount of microbial degradation of the newly added decomposable organic matter in those samples during storage. Microbial life was not therefore as active under the wet-dry conditions of storage, after the end of four months of storage. This could have been an indication of low microbial populations. As such water regimes of alternate wetting and drying often prevail in field locations, these results seem to corroborate the findings of Hannan (1978), who reported depletions in microbial counts (possibly due to low water retention) leading to sterile conditions for soils stored in excess of six months.

3.2 Macro-Aggregate Stability

Large increases of approximately 25 % above the controls for the permanent pasture system occurred (Fig. 3). There were no substantial increases in WSMA among either of the previously cultivated systems, and the only significant change occurred when the WF and CW were subjected to green manure at a water regime of 80 % of soil field capacity (Fig. 3). The wet/dry regime was therefore ineffective. Utomo and Dexter (1981) found that providing there is an extra energy source for microbial activity, wetting and drying first increases the proportion of water stable aggregates >0.5 mm to a maximum value; however, when there is no extra energy source, wetting and drying steadily decreases water-stability. As the samples in this study were not inoculated, and were held at air-dry for 6 weeks in the first stage of storage with no external energy source, such a process could have severely depleted the already low levels of viable but dormant organisms in the CW and WF samples. Conversely, the microorganisms in the PP samples would have been more abundant, and therefore more of them would have been expected to have survived this air-dry stage.

An additional possible explanation for the meagre significance of the WSMA results for the CW and WF systems is that the method of determining WSMA might have been too severe. It is known that macro-aggregates are more susceptible than micro-aggregates to breakdown under prolonged physical stresses, particularly for the more recently formed aggregates (Tisdall and Oades 1982; Varadachari and Ghosh 1983). Thus, some aggregates formed as a result of the treatments in this study, could have been destroyed via the impact of the wet sieving process. This could have partially masked significant improvements, however slight, in WSMA for the CW and WF management samples. On the other hand, the process of measuring saturated hydraulic conductivity of samples involved a much lower expenditure of kinetic energy than that of wet sieving. Under this process, more effective preservation of weaker aggregates and thus a greater response for hydraulic conductivity from the cultivated soils would have resulted. Thus, the changes contributing to the increase in K_{sat} for those samples were likely to have been of increased micro-aggregation and porosity. K_{sat} therefore seems to be a more precise parameter and a more sensitive instrument compared to WSMA for determining minute changes in structure of such disturbed samples.

Due to the extensive root systems of pasture plants, the upper soil horizon in old pastures is almost all rhizosphere (Connell and Hadfield 1961). Among the three original soils used in this study, the microflora population would therefore have been highest in the permanent pasture soils. Conversely, the cultivated soils inherently would have contained low levels of organic matter, owing to the dried out and resistant nature (more lignaceous) of the organic fraction in those soils; Golchin and others (1994), and Grierson et al. (1972), found that decreases in water-stability of aggregates have been associated with reductions in total amount of carbon in the soil in semi-arid areas of South Australia. Swift (1992) found that pasture soils contained dormant microorganisms in far greater numbers than those of cultivated soils. In such a favourable environment, the addition of green manure, by providing an abundant nutrient source for microorganisms, would have triggered a "priming" effect for dormant microorganisms in the pasture soils of this study. This in turn could have released a greater amount of binding agents, compared to the CW and WF soils. A higher level of biological activity therefore, compared to cultivated soils, improved the soil structure. However, whereas the increases recorded by Swift (1992) were short-lived and reduced after 4 weeks, in the present study, increases in WSMA and K_{sat} for the PP soils were maintained for up to 12 weeks during storage. This is likely to be a result of the complexity and far greater size of the substrate molecules in this study, compared to those of glucose, and hence a longer lasting effect by green manure which is not as quickly consumed.

3.3 Water Retention

Green manure did not significantly increase water retention in any of the samples. The only statistically significant trend seen in this study is that of the higher rate of water retention as it relates to prior soil management, and that without the need for any green manure. This is not surprising for the cultivated soils, as little decomposition was evident in their samples. For the PP samples, this lack of response, in spite of substantial decomposition is not surprising, as the green manure changed its macro- and micro-aggregation that largely influences water-holding characteristics of a soil. Nevertheless, an increase in density of organic matter would have increased its water-holding capacity, but the green manure organic matter added did not necessarily cause an increase in density of the pre-existing organic fraction in the PP. Only if the density of the organic matter in the treated samples increased, would the water retention of the PP samples also increase. It is possible that the organic matter added to the PP samples by the green manure, being recently degraded, was not as dense, and therefore no more water retentive than the typical rhizosphere-rich inherent organic matter typical of soils with a management history of permanent pasture, as found by Connell and Hadfield (1961).

4 Conclusions

Alfisols previously under a management system of permanent pasture are inherently superior, and show much more significant improvements in physical characteristics when organically amended, compared to cultivated Alfisols. During short-term removal and storage for up to three months, permanent pasture soils show no signs of sterility, whereas stored, previously cultivated soils are largely unresponsive to green manure treatments. Therefore, soils left in the field under grazing for prolonged periods prior to removal for capping may be more suitable and responsive as a growth medium than soils previously cultivated.

Based on large value differentials between K_{sat} and macro-aggregate stability after green manure treatment for cultivated soils in this study, recently formed aggregates of cultivated soils are weaker than those recently formed in permanent pasture soils. Therefore, to preserve physical integrity, added care may be required when removing and transporting capping material from soils with a prior history of cultivation, after application of organic amendments. If un-pastured overburden is to be removed and placed under storage, its effectiveness is increased if first placed under a management system of permanent pasture.

5 Geobiotechnological Applications

5.1 Choose Tyre-Mounted Machines

A tyre mounted mining machine (scrapers) rather than crawler mounted (dozers) to dig stored soil and minimize compaction (Sheoran et al. 2010). They observe that transporting soil from the stockpile to the reclamation site on a conveyor belt with trundling action improves soil structure by breaking up massive aggregates. As smaller aggregates continue to tumble, they tend to acquire an agglomerative skin of fine particles, which promotes loose soil structure (Sheoran et al. 2010).

5.2 Choose Grasses

Grasses, particularly C4 ones, can offer superior tolerance to drought, low soil nutrients and other climatic stresses, and hence would satisfy the prior management requirements of mine-capping. Roots of grasses are fibrous that can slow erosion and their soil forming tendencies eventually produce a layer of organic soil, stabilize soil, conserve soil moisture (Sheoran et al. 2010).

Fig. 15 Top-soil should be stored low (<2 m in height) and planted with fibrous rooted (grassed) vegetation as above

5.3 Use Waste Rock

Research data now available from other mine-sites demonstrate that in some areas, waste rock weathers rapidly to form suitable materials for revegetation. Consideration may also need to be given to the re-spreading techniques used, with a view to minimizing soil compaction that can inhibit later revegetation.

5.4 How to Stockpile Soil

In cases where stripped soil cannot be re-spread immediately, it should be stockpiled. Stockpiles should be established as close as practicable to areas to be rehabilitated (SVDPI 2004). They should be low (Fig. 15) (generally less than 2 m in height), gently battered and located away from drainage lines. Importing of soil is not recommended, especially in areas of native vegetation, because of the risk of the imported soil introducing weeds or plant diseases.

Top-soil stripped ahead of mining is applied to the reshaped surface in an even layer generally not less than 100 mm. Top-soil is placed using rear dump haul trucks and spread with dozers or graders. Once spread, the top-soil surface is disc or chisel cultivated to create a textured surface which assists in trapping surface runoff, provides seed entrapments and creates microclimates favourable for seed germination (SVDPI 2004). Where bio-solids are used, cultivation also integrates the top-dressing material. Increasingly, the aim of mine residue rehabilitation is moving towards ecosystem reconstruction rather than vegetation establishment.

5.5 Role of Capping Systems

Top-soil is used to cover poor substrate and to provide improved growth conditions for plants. Stockpiling of top-soil in mounds during mineral extraction has been shown to affect the biological, chemical and physical properties of soil (Johnson et al. 1991; Davies et al. 1995). Long term studies have indicated that capping systems deteriorate with time and often do not perform as expected. As a result, costs for managing water makes from mined out areas are likely to be higher than may otherwise have been expected. Future maintenance requirements of capping systems have generally not been seriously addressed (SVDP 2004).

5.6 The Function of Increased Aggregate Stability

Soil structure plays a dominant role in the physical protection of SOM by controlling microbial access to substrates, microbial turnover processes, and food web interactions (van Veen and Kuikman 1990). Relatively labile materials become physically protected from decomposition by incorporation into soil aggregates (Gregorich et al. 1991; Golchin et al. 1994) or by deposition into microbially inaccessible micro-pores.

Decreased soil stability can lead to increase in bulk density because the matrix does not resist slaking, dispersion by water and the forces imparted by wheels, hooves and rainfall. This, in turn, leads to decreased aeration and water infiltration rate and the development of anaerobic conditions. N losses by denitrification may follow under such an environment.

5.7 Ideal Compaction Levels

The surface four feet of mine soil material should be easily-weathered overburden, meaning that most rocks and boulders break apart and decompose quickly to fine soil materials. The soil texture of the fine-earth fraction should be loamy to sandy, and the mine soil should be low in total salts; mine soils should be moderately acid (pH 5.5–6.5) when native hardwoods are being planted. Most importantly, the mine soil must also be left uncompacted to a depth of four feet (Daniels 1999).

Loosely constructed, or "fritted", subsoil is very important to plant root systems. The extent of the root system determines a plant's ability to maximize its surface area and access a greater volume of water and soil nutrients. Plants grown in fritted subsoil have root patterns with extensive vertical and lateral penetration (Sheoran et al. 2010). Soil storage is for very short periods, periodically opening up and aerating the soil while stockpiled or permanently aerating, allowing drainage with a network of pipes and use of nitrification inhibitors after restoration are the operations that may in part ameliorate the problem (Davies et al. 1995).

5.8 What Is the Objective?

The top-soil is severely damaged if it is not mined out separately in the beginning with a view to replace it on the filled void surface area for reclamation in order to protect the primary root medium from contamination and erosion and hence its productivity (Kundu and Ghose 1997). Sendlein et al. (1983), however, indicate that systematic handling and storage practice can protect the physical and chemical characteristics of top-soil while in storage and also after it has been redistributed into the regarded area.

The three major macronutrients, namely nitrogen, phosphorus and potassium are generally found to be deficient in overburden dumps (Sheoran et al. 2008). Adequate top-soil management was considered the most important factor in successful rehabilitation of the project since the objective is to restore the native ecosystem of the project area. The top-soil from all areas being cleared would be retained for subsequent rehabilitation.

5.9 Soil Organic Carbon (SOC)

Organic matter is the major source of nutrients such as nitrogen, and available P and K in unfertilized soils (Donahue et al. 1990). A level of organic carbon greater than 0.75 % indicates good fertility (Ghosh et al. 1983). The level of organic carbon in overburden was found to be 0.35–0.85 %. Organic carbon is positively correlated with available N and K and negatively correlated with Fe, Mn, Cu, and Zn (Maiti and Ghose 2005). The potential of SOC sequestration in mine-soils depends on biomass productivity, root development in the subsoil and changes in mine-soil properties resulting from overburden weathering (Haering et al. 1993).

5.10 Increase Microbe Population

Microbial activity decreases with depth and time as top-soil continues to be stored during mining operations (Harris et al. 1989). Microbial activity, measured in ATP (adenosine tri phosphatase) concentrations, plummets to very low levels within a few months. Response to glucose is slower by microbes at all depths, suggesting that metabolic rates decrease with time (Visser et al. 1984). Once the soil is removed from the stockpile and reinstated, aerobic microbial population rapidly reestablishes, usually higher than the normal level (Williamson and Johnson 1991) and nitrification restarts at higher than the normal rates.

Microbial activity declines when soil layers are disrupted and is slow to resume independently (Williamson and Johnson 1991). Soil microbes include several bacterial species active in decomposition of plant material as well as fungal species whose symbiotic relationship with many plants facilitates uptake of nitrogen and phosphorus in exchange of carbon. They produce polysaccharides that improve soil aggregation and positively affect plant growth (Williamson and Johnson 1991). Sites with an active soil microbe community exhibit stable soil aggregation, whereas sites with decreased microbial activity have compacted soil and poor aggregation (Edgerton et al. 1995).

Soil enzymes activities have been used as sensitive indicators for reflecting the degree of quality reached by a soil in the reclamation process. Ceccanti et al. (1994) note that a direct measurement of the microbial population is the dehydrogenase activity. Dehydrogenase is an oxidoreductase, which is only present in viable cells. This enzyme has been considered as a sensitive indicator of soil quality in degraded soils and it has been proposed as a valid biomarker to indicate the changes in soil management under different agronomic practices and climates. Measurement of soil hydrolases provides an early indication of changes in soil

Fig. 16 In uneven topography, nutrients are readily leached from stored top-soils. After open-cast mining, newly exposed mine-soils require substantial inputs of fertilizers to maintain a new plant community

fertility since they are related to the mineralization of such important nutrient elements as N, P and organic carbon (Ceccanti et al. 1994).

When soil layers are removed and stockpiled, the bacteria inhabiting the original upper layers end up on the bottom of the pile under compacted soil (Williamson and Johnson 1991). A flush of activity occurs in the new upper layer during the first year as bacteria are exposed to atmospheric oxygen. After two years of storage there is little change in the bacterial numbers at the surface, but less than one half the initial populations persist at depths below 50 cm (Williamson and Johnson 1991).

There is a little decrease in viable mycorrhizal inoculum potential during the first two years of storage (Miller et al. 1985). Viability of mycorrhizas in stored soils decreases considerably and possibly to the levels 1/10 those of the undisturbed soil (Rives et al. 1980). Miller et al. (1985) observes this as an indication that soil water potential is a significant factor affecting mycorrhizal viability. When soil water potential is less than −2 MPa (drier soil), mycorrhizal propagules can survive for greater lengths of storage time; when soil water potential is greater than −2 MPa, length of storage time becomes more important. In drier climates, deep stockpiles may not threaten mycorrhizal propagule survival. In wetter climates, shallow stockpiles are more important to maximize surface-to-volume ratios with regard to moisture evaporation.

5.11 Role of Fertilizers

All newly created mine soils, and many older ones, will require significant fertilizer applications for the establishment and maintenance of any plant community (Fig. 16).

Acknowledgements The partial funding of this research by the Northern Caribbean University, Mandeville, Jamaica, is gratefully acknowledged. The authors also thank Dr. Rowl Twidale of the Department of Soil and Land Systems, School of Earth and Environmental Sciences, The University of Adelaide, South Australia, for useful comments on this manuscript.

References

Addiscott TM, Whitmore AP, Powlson DS (1991) Farming fertilizers and the nitrate problem. CAB International, Wallingford, UK 176 pp

Alfaro MR, Montero A, Ugarte OM, do Nascimento CWA, de Aguiar Accioly AM, Biondi CM, da Silva YJAB (2015) Background concentrations and reference values for heavy metals in soils of Cuba. Environ Monit Assess (2015) 187:4198 doi:10.1007/s10661-014-4198-3

Amato M, Jackson RB, Butler JHA, Ladd JN (1983) Decomposition of plant material in Australian soils. Aust J Soil Res 21:563–570

Aubertin M, Bussiere B (2001) Geotechnical news 2001 http://www.enviro-geremi.polymtl.ca/pdf/articles/GN-ChairRev.pdf). Accessed 6/7/2015

Barcelo J, Poschenrieder C (2003) Phytoremediation: principles and perspectives. Contrib Sci 2(3):333–344
Ceccanti B, Pezarossa B, Gallardo-Lancho FJ, Masciandaro G (1994) Bio tests as a marker of soil utilization and fertility. Geomicrobiol J 11:309–316
Connell RP, Hadfield JW (1961) Agriculture. Whitcombe & Tombs Ltd., Auckland, pp 125–187
Courtney R, Harrington T, Byrne KA (2013) Indicators of soil formation in restored bauxite residues. Ecol Eng 58:63–68
Das M, Maiti SK (2005) Metal mine waste and phytoremediation. Asian J Water Environ Pollut 4(1):169–176
Davies R, Hodgkinson R, Younger A, Chapman R (1995) Nitrogen loss from a soil restored after surface mining. J Environ Qual 24:1215–1222
Donahue RL, Miller RW, Shickluna JC (1990) Soils: an introduction to soils and plant growth, 5th edn. Prentice-Hall, 234 p
Edgerton DL, Harris JA, Birch P, Bullock P (1995) Linear relationship between aggregate stability and microbial biomass in three restored soils. Soil Biol Biochem 27:1499–1501
EPA Guyana (2014) (http://www.epaguyana.org/index.php/2014-02-10-20258?task=document.viewdoc&id=26). Accessed 15/12/14
FAO Fertilizer use by crop in Cuba (2006) www.fao.org/docrep/006/y4801e/y4801e05.ht. Accessed 7/6/15
Fierer N, Allen AS, Schimel JP, Holden PA (2003) Controls on microbial CO2 production: a comparison of surface and subsurface horizons. Glob Change Biol 9:1322–1332
Ghose MK (1989) Land reclamation and protection of environment from the effect of coal mining operation. Mine Technol 10(5):35–39
Ghose MK (2005) Soil conservation for rehabilitation and revegetation of mine-degraded land. TIDEE—TERI Inf Dig Energy Environ 4(2):137–150
Ghosh AB, Bajaj JC, Hassan R, Singh D (1983) Soil and water testing methods: a laboratory manual. IARI, New Delhi, pp 31–36
Golchin A, Oades JM, Skjemstad JO, Clarke P (1994) Soil structure and carbon cycling. Aust J Soil Res 32:1043–1068
Gould AB, Hendrix JW, Ferriss RS (1996) Relationship of mycorrhizal activity to time following reclamation of surface mine land in western Kentucky. I Propagule and spore population densities. Can J Bot 74:247–261
Gregorich EG, Voroney RP, Kachanoski RG (1991) Turnover of carbon through microbial biomass in soils with different textures. Soil Biol Biochem 23:799–805
Grierson IT, Kijne JW, Greenland DJ (1972) Changes in some physical properties of red-brown earths of different texture associated with increasing content of organic matter. Exp Rec 6:16–22
Haering KC, Daniels WL, Roberts JA (1993) Land reclamation. Changes in mine soil properties resulting from overburden weathering. Virginia council on the environment, Richmond. USDI Ofc of Surface Mining, Reclamation and Enforcement under Coop Agreement HQ51-GR87-10022. J Environ Qual 22:194–200
Hannan JC (1978) Coal: rehabilitation of coal mines in the Hunter Valley of New South Wales. In: Rummery RA, Howes KMW (eds) Management of lands affected by mining. Division of Land Resources Management, Commonwealth Scientific and Industrial Research Organisation, Australia, pp 111–119
Harris MA (2001) The effect of sewage sludge and green manure on the erodibility of two mine tailings materials. Environ Geol 40(11/12):1367–1376
Harris JA, Birch P (1989) Soil microbial activity in opencast coal mine restorations. Soil Use Manage 5(4):155–160
Harris MA, Megharaj M (2001) The effect of sludge and green manure on hydraulic conductivity and aggregation in pyritic mine tailings materials. Environ Geol 41(3/4):285–296
Harris MA, Omoregie SN (2008) Post-mining deterioration of bauxite overburdens in Jamaica: storage methods or subsoil dilution? Environ Geol 54(1):111–115

Harris JP, Birch P, Short KC (1989) Changes in the microbial community and physio-chemical characteristics of top-soils stockpiled during opencast mining. Soil Use Manag 5:161–168

Isermann K (1994) Agriculture's share in emission of trace gases affecting the climate and some cause oriented proposals for sufficiently reducing this share. Environ Pollut 83:95–111

Jacobson (1999) Personal communication, Ministry of the Environment, Jamaica Bauxite Institute, 1992. The bauxite/alumina industry and the environment. Hyde, Held & Blackburn Ltd., Kingston, Jamaica

Johnson DB, Williamson JC (1994) Conservation of mineral nitrogen in restored soils at opencast mines sites: I. Result from field studies of nitrogen transformations following restoration. Eur J Soil Sci 45:311–317

Johnson DB, Williamson JC, Bailey AJ (1991) Microbiology of soils at opencast sites. I. Short and Long- term transformation in stockpiled soils. J Soil Sci 42:1–8

Kavamura VN, Esposito E (2010) Biotechnological strategies applied to the decontamination of soil polluted with heavy metals. Biotechnol Adv 28:61–69

Kemper WD, Rosenau RC (1986) Aggregate stability and size distribution. In: Klute A(ed) Methods of soil analysis, 2nd edn. American Society of Agronomy, Soil Science Society of America, Madison, pp 425–442

Khan AG (2005) Role of soil microbes in the rhizospheres of plants growing on trace lement contaminated soils in phytoremediation. J Trace Elem Med Biol 18(4):355–364

Klute A (1986) Methods of soil analysis, 2nd edn. American Society of Agronomy, Soil Science Society of America, pp 425–442

Kundu NK, Ghose MK (1997) Soil profile characteristic in rajmahal coalfield area. Indian J Soil Water Conserv 25(1):28–32

Lang C (2014) Gold mining in Guyana: is this what the low carbon development strategy looks like? http://www.redd-monitor.org/2014/06/06/gold-mining-in-guyana-is-this-what-the-low-carbon-development-strategy-looks-like/). Accessed 23/12/14

Linn D, Doran JW (1984) Effect of water-filled pore space on carbon dioxide and nitrous oxide production in tilled and non-tilled soils. Soil Sci Soc Am J 48:1267–1272

Lone MI, He ZL, Stoffella PJ, Yang X (2008) Phytoremediation of heavy metal polluted soils and water: progress and perspectives. J Zhejiang Univ Sci B 9(3):210–220

Lottermoster BG (2003) isbn: 3-540-00-526-9. Springer, Heidelberg, pp 3–11

Ludwick AE (1998) Don't forget soil-test calibration. News and Views. A regional newsletter published by the Potash & Phosphate Institute (PPI) and the Potash & Phosphate Institute of Canada (PPIC)

Maiti SK, Ghose MK (2005) Ecological restoration of acidic coal mine overburden dumps: an Indian case study. Land Contam Reclam 13(4):361–369

Mason CF (1977) Decomposition studies. Institute of Biology, no 14. Edwin Arnold Publishers, London

McLauchlan KK, Hobbie SE (2004) Comparison of labile soil organic matter fractionation techniques. Soil Sci Soc Am J 68:1616–1625

Mentis MT (2006) Restoring native grassland on land disturbed by coal mining on the Eastern Highveld of South Africa. S Af J Sci 102(5–6):193–197. ISSN 0038-2353

Miller RM, Carnes BA, Moorman TB (1985) Factors influencing survival of vesicular-arbuscular mycorrhiza propagules during top-soil storage. J Appl Ecol 22:259–266

Mummey DL, Stahl PD, Buyer JS (2002) Microbial biomarkers as an indicator of ecosystem recovery following surface mine reclamation. Appl Soil Ecol 21:251–259. doi:10.1016/S0929-1393(02)00090-2

Nelson CE (2000) Gold mineralisation and volcanic dome emplacement, Pueblo Viejo district, Dominican Republic. Miner Deposita 35:511–525

Paul EA, Morris SJ, Bohm S (2001) The determination of soil C pool sizes and turnover rates: biophysical fractionation and tracers. In: Lal R, Kimble JM, Follett RF, Stewart BA (eds) Assessment methods for soil carbon. Lewis, Boca Raton, pp 193–206

Pieters A (1927) Green manuring. Principles and practice. Wiley, NY, pp 70–95

Rana S, Stahl PD, Ingram LJ, Wick AF (2007) Soil microbial community structure in reclaimed soil under different vegetation in Wyoming mine-lands. Proceedings of the 2007 National Meeting of the American Society for Mining and Reclamation, Gillette, WY, June 2–7, 2007
Rives CS, Bajwa MI, Liberta AE (1980) Effects of top-soil storage during surface mining on the viability of VA mycorrhiza. Soil Sci 129:253–257
Romero FM, Prol-Ledesma RM, Canet C, Alvarez LN, Perez-Vazquez R (2010) Acid drainage at the inactive Santa Lucia mine, western Cuba: natural attenuation of arsenic, barium and lead, and geochemical behavior of rare earth elements. Appl Geochem 25(5):716–727
Schwendenmann L, Pendall EG (2006) Effects of forest conversion into grassland on soil aggregate structure and carbon storage in Panama: evidence from soil carbon fractionation and stable isotopes. Plant Soil 228:217–232
Schwendenmann L, Pendall EG (2008) Response of soil organic matter dynamics to conversion-from tropical forest to grassland as determined by long-term incubation. Biol Fertil Soils 44:1053–1062. doi:10.1007/s00374-008-0294-2
Sendlein VA, Lyle YH, Carison LC (1983) Surface mining reclamation handbook. Elsevier Science Publishing Co Inc, New York
Sheoran AS, Sheoran V, Poonia P (2008) Rehabilitation of mine degraded land by metallophytes. Min Eng J 10(3):11–16
Sheoran V, Sheoran AS, Poonia P (2010) Soil reclamation of abandoned mine land by revegetation: a review. Int J Soil Sediment water 3(2). ISSN: 1940-3259
Sheoran V, Sheoran AS, Poonia P (2012) Phytoremediation of metal contaminated mining sites. Int J Earth Sci Engg 5(3):428–436
Singh JG, BGEN (1994) Forestry: third international conference on environmental enforcement: the enforcement experience in Guyana on exploitation of Natural resources pp 205–210, Guyanawww.inece.org/3rdvol1/pdf/singh.pdf Accessed 8/5/15
Singh AN, Raghubanshi AS, Singh JS (2002) Plantations as a tool for mine-spoil restoration. Curr Sci 82(12):1436–1441
Six J, Bossuyt H, Degryze S, Denef K (2004) A history of research on the link between (micro) aggregates, soil biota, and soil organic matter dynamics. Soil Tillage Res 79(7):31
SVDPI (The State of Victoria, Department of Primary Industries) (2004) ISBN 0 7306 9495 X2004) http://www.energyandresources.vic.gov.au/earth-resources/licensing-and-approvals/minerals/guidelines-and-codes-of-practice/rehabilitation-and-environmental-aspects-of-mining-and-extractive-work-plans
Swift RS (1992) Effects of humic substances and polysaccharides on soil aggregation. Adv Soil Organic Matter Res 153–201
Tate RL, Klem DA (eds) (1985) Soil reclamation processes: microbiological analyses and applications. Marcel Dekker Inc, New York
Tisdall JM, Oades JM (1982) Organic matter and water-stable aggregates in soils. J Soil Sci 33:141–163
Ussiri DAN, Lal R (2005) Carbon sequestration in reclaimed mine-soils. ISSN: 0735-2689 print/1549-7836 online. doi:10.1080/07352680591002147. Crit Rev Plant Sci 24:151–165
Utomo WH, Dexter AR (1981) Changes in soil aggregate water-stability induced by wetting and drying cycles in non-saturated soil. J Soil Sci 33:623–637
van Veen JA, Kuikman PJ (1990) Soil structure aspects of decomposition of organic matter by microorganisms. Biogeochemistry 11:213–223
Visser S (1985) Management of microbial processes in surface mined land reclamation in Western Canada. In: Tate RL, Klein DA (eds) Soil reclamation processes, Marcel Dekker, New York, pp 203–341
Varadachari C, Ghosh K (1983) On humus formation. Department of Agricultural Chemistry and Soil Science, Calcutta University
Waygood CG, Ferreira S (2009) A review of the current strategies for capping mining spoils. In: Abstracts of the International Mine Water Conference Proceedings 19–23 October 2009 Pretoria, South Africa Produced by: Document Transformation Technologies cc Conference organised by: Cilla Taylor Conferences ISBN Number: 978-0-9802623-5-3

Wick AF, Stahl PD, Ingram LJ, Vicklund L (2008) Soil aggregate and aggregate associated carbon recovery in short-term stockpiles 08. National Meeting of the American Society of Mining and Reclamation, Richmond, VA, New Opportunities to Apply Our Science 14–19 June 2008

Williamson JC, Johnson DB (1991) Microbiology of soils at opencast sites: II. Population transformations occurring following land restoration and the influence of rye grass/fertilizer amendments. J Soil Sci 42:9–16

Wong MH (2003) Ecological restoration of mine degraded soils, with emphasis on metal contaminated soils. Chemosphere 50:775–780

Erodibility of Unconsolidated Mine Wastes Under Simulated Rainfall and Hydraulic Forces After Organic Amendments

Mark Anglin Harris

Abstract Soil structure, in simple terms, is the physical relationship between the solid, liquid, and gaseous phases of soil. The arrangement of soil particles into peds or aggregates determines the size and shape of soil voids, or pores, and this greatly influences the movement of water and gases in soil. As a result, soil structure can have a substantial impact on plant growth and determines erosion rates. An experiment was conducted to determine the effectiveness of green manure and sewage sludge to reduce the erosion of (1) a mine tailing and (2) its leachate residue following lime treatments. Green manure (GM) was the only treatment that stabilized each material to some extent. In response to green manure plus sludge (SSGM) or sludge (SS), the mine tailing had the smallest increases in stability, where neither treatment significantly changed its poor structure. For the tailings, the effectiveness of the treatments was in the order (descending) of GM > SS = SSGM. For leachate residue material, the effectiveness of the treatments was in the order (descending) of GM = SSGM > SS. When SS was treated (as if it were a soil), the GM, and SSGM treatment combination yielded the most stables aggregates, because the number of simulated raindrops needed for aggregate breakdown was always at least 25 % higher than that of all other samples except the leachate residue control. The resistance of the leachate residue to raindrop impact was attributed to thixotropic effects caused by its gypsum fraction. For all materials, stability against the counted number of water drops (CND) was not significantly increased by sludge. However, after incubation with GM, sludge resisted 50 % more water drops than GM-treated mine tailings or tailings leachate residue. In the present study GM-impregnated sludge resists erosion to a far greater degree than the sludge to which GM had not been previously applied. This has implications for the treatment of sludge prior to its application to mine spoils.

Keywords Green manure · Mine erosion · Mine tailings · Slope stability

---

Original article: **The effect of sewage sludge and green manure on the erodibility of two mine tailings materials**. *Environmental Geology* Volume 40, Issue 11, pp 1367–1376.

M.A. Harris, *Geobiotechnological Solutions to Anthropogenic Disturbances*, Environmental Earth Sciences, DOI 10.1007/978-3-319-30465-6_12

1 Introduction

1.1 Caribbean and Humid Tropics

In a tropical cyclone-susceptible region such as the Caribbean, large volumes of soil can be eroded in severe storms even where there is no land disturbance (Kaspar and Singer 2011). Further, changes in land cover and use over the past 50 years due largely to ore mining in Cuba, Jamaica, Dominican Republic, St. Croix, and Guyana and Haiti have exacerbated the potential soil erosion rate. The extent of gullying or erosion may be used to indicate how effective a structure encapsulates waste material such as low grade mineralized rock or tailings. If erosion results in gullying which could expose tailings and other waste material, enabling contamination of the environment, then the design of the landform obviously needs to be re-assessed (Australian Government 2011). Yet, notwithstanding the benefits of "best practice" landscape macro-morphology, it is resilience at the microstructure levels including micro- and macro-aggregation, which determines resistance to gullying (Mamo and Bubenzer 2001).

1.2 Definition and Description of Soil Erosion

Soil erodibility is a measure of soil susceptibility to detachment and transport by the agents of erosion (Cerda 1998). In general, soil erosion is a three-step process. It begins with particle detachment, which is followed by particle transport and finally by deposition of transported particles in a new location. The first two steps are influenced to a large extent by the nature and properties of the soil (O'Geen et al. 2006). The inherent properties of a soil play a major part in the ability of water to detach and transport its soil particles. These characteristics contribute to an index of soil erodibility (O'Geen et al. 2006). Soil erosion is a result of several factors, including rainfall intensity, steepness of slope, length of slope, vegetative cover, and management practices (O'Geen et al. 2006).

Losses of P from agricultural systems to surface waters are largely dependent on the amount of surface runoff and sediment transport that occurs. Thus the already low phosphorus in many Caribbean soils is exacerbated by exposure to high intensity rainfall particularly during frequent hurricane storms. Phosphorus is transported in runoff as soluble P and particulate P (Sharpley 2000). Particulate P consists of P bound to soil sediment and P contained in organic matter. Sharpley (2000) summarized research on the effect of cover crops on P losses and found that reductions in total P losses, which consist mostly of particulate P, ranged from 54 to 94 % (Kaspar and Singer 2011).

1.3 Characteristics of Erodibility

Fundamental to any assessment of the stability of post-mining rehabilitated landform is the ability to quantify soil erosion rates. Four major soil properties that govern erodibility—texture (particle size distribution), structure, organic matter content, and permeability—have been identified through nationwide studies performed by USDA–ARS using rainfall simulation tests (USDA–NRCS 2005). USDA–NRCS uses these soil properties to derive the K-factor, used in all soil surveys published in California (O'Geen et al. 2006). The measurement of these properties is used to predict the soil's potential for erosion by water. Thus the K-factor is synonymous with soil erodibility.

O'Geen et al. (2006) cited soil texture as an important property contributing to soil's erodibility. Thus soils with a high content of silt, very fine sand (0.05–0.10 mm in diameter), or expanding clay minerals tend to have high erodibility. Erodibility is low for clay-rich soils with a low shrink-swell capacity because these clay particles mass together into larger aggregates that resist detachment and transport. These include the kaolinite-rich bauxite soils located in parts of the Caribbean such as Western Jamaica (karst bauxites), St. Croix in British Virgin Islands, and Haiti. Sandy soils with large amounts of fine, medium, or coarse sand particles (0.10–2.0 mm in diameter) also have low erodibility. Sand particles lack the ability to aggregate together, but because most sandy soils are highly permeable, water runoff is low, hence erosion is often slight (O'Geen et al. 2006).

In addition, the large grain size of sandy soils means that it takes more energy to transport its particles than those of finer-textured soils (Flanagan 2002). Medium-textured soils (loamy soils) tend to be most erodible because they have high amounts of silt and very fine sands. These soils tend to have moderate to low permeability and low resistance to particle detachment. Hence, if disaggregated, small particles (silts and clays) are easily transported (O'Geen et al. 2006).

Undecomposed organic residues present at the soil surface protect the soil against raindrop impact. Highly decomposed organic material in the soil, called humus, acts as a glue to bind soil particles together into aggregates. Soils that are higher in SOM are more resistant to erosion (Flanagan 2002).

Permeability is a measure of the rate at which water percolates through a soil. It is a function of texture, structure, and soil bulk density. Water rapidly enters highly permeable soils, reducing runoff and, therefore, soil erosion (O'Geen et al. 2006). Thus the iron-rich inorganically cemented aggregates of mainly kaolinite clays of Caribbean bauxite soils create high levels of permeability and hence reduce erodibility. Yet, even for soils with such low erodibility, careless management can accelerate soil erosion.

The permeability of the subsoil largely determines infiltration rates. A subsurface horizon that is slowly permeable to water can cause a perched water table to develop during a large storm or irrigation event. When even a highly permeable soil is saturated because of a perched water table, infiltration slows down and surface runoff becomes a major path for hydrologic flow, increasing soil erosion

(O'Geen et al. 2006). For example, in a laboratory study, Harris (2008) detected hydraulic conductivity rates of 0.00 cm h^{-1} below the 15 cm plough-depth of a >10 year-old gypsum-treated red mud bauxite waste dump in Jamaica. He observed a substantial increase in hydraulic conductivity in the 0–15 cm gypsum-treated layer after treating it with decomposable phyto-organics. Yet, the impermeable conditions remaining at just 15 cm below the surface after such treatment in the field would have provided perched water table conditions during hurricane events, thereby potentially increasing erosion.

1.4 Mine Soils

Sheoran et al. (2010) observes the general unsuitability of mine soils with slopes greater than 15 % (Fig. 1) for intensive land uses such as vegetable or crop production, but they may be suitable for grazing and reforestation. Broad flat benches and fills with slopes less than 2 % often have seasonal wetness problems. Many benches with an overall gentle slope contain areas of extreme rockiness, pits, hummocks, and ditches (Sheoran et al. 2010). Average slope of most reclaimed modern mines is quite a bit steeper than the older benches, but the newer landforms are considerably smoother and more uniform in final grade. Bench areas directly above intact bedrock on older mined lands are usually fairly stable but may be subject to slumping (Sheoran et al. 2010).

Fig. 1 A steep minesoil having a slope exceeding 15 % from the horizontal. Note the thin soil, suggesting a condition before capping, or, the erosioin of applied capping. *Source* Jamaica Bauxite Institute

1.5 Raindrop Impacts, Bulk Density, and Runoff

Ideally, a soil can resist detachment under raindrop impact and retain its structural integrity during rainfall events (Cerda 1998). One of the objectives of mine-spoil rehabilitation is to minimize the potential for water and wind erosion (Marshall 1978). For a wide range of soil types Loch and Foley (1994) concluded that raindrop impact reduces infiltration of the surface layer even for soils of extremely high hydraulic conductivity. They observed that compaction of the surface layer by raindrops is the major mechanism that causes surface sealing (i.e. not necessarily aggregate breakdown), which increases waterlogging and reduces aeration of the soil.

Inorganic particles in mine tailings tend to pack to a high bulk density, resulting in low infiltration and low permeability, causing restricted plant root penetration (Mulligan and Dahl 1996). A low infiltration rate results in a high rate of runoff. On sloping tailings dumps, high volumes of runoff can cause severe erosion (Hannan 1978). Further, soil particles in the <0.5 mm diameter range are particularly prone to movement by wind (Mullins et al. 1990), and after the beneficiation process for mineral extraction, the particle size distribution in mine tailings is almost wholly restricted to the clay size (<0.002 mm diameter; Hore-Lacy 1978). Therefore stabilization at the microstructure level is imperative for mine-spoil resistance to erosion.

Larson and Pierce (1994) inventoried certain soil attributes based on their ability to accept, hold, and release nutrients and water, promote growth, provide habitat, and resist degradation. They divided these attributes between those important in the surface horizon and those critical in the (crop) limiting horizon. Surface horizon attributes were total and labile organic matter, nutrient supply, soil texture, surface horizon depth, structure, pH, and electrical conductivity (EC). Attributes important in the limiting horizon were texture, depth, structure, pH, and EC. However, organic additives can reduce erosion (Cerda 1998). For one virgin soil and one direct drilled soil, the presence of considerable partially decomposed organic material in the surface layer appeared to greatly reduce compaction of the surface layer, even when the soil was bare of cover (Loch et al. 1978).

Because inter-rill erosion results from the detachment of soil particles by raindrop impact, living or dead plant material that intercepts raindrops and dissipates impact energy will reduce inter-rill erosion (Ram et al. 1960). Inter-rill erosion is largely dependent on raindrop impact to detach soil particles, whilst rill erosion relies on the sheer force of running water (Flanagan 2002). Yet, annual crop plants, such as corn (*Zea mays L.*) and soybean, only provide significant canopy cover for four months or less each year. Additionally, crops such as soybean, cotton (*Gossypium hirsutum* L.), or corn harvested for silage often do not leave enough residues to fully protect the soil between harvest and development of the next crop canopy (Ram et al. 1960). The added organic amendments provide active organic compounds such as particulate organic matter (POM) that act as nucleation sites for the formation of macro-aggregates. Therefore for stored uncovered mine overburdens, aggregate stability of topsoil should be high enough to withstand the erosive energy of adverse weather conditions.

Rationale for green manure type

MacRae and Mehuys (1987) found that soil aggregate water stability may be increased by green manure, and Dabney et al. (2001) used winter cover crops to improve soil and water quality. But recently formed green manured soil aggregates however, are held together by bacterially synthesized gums that are themselves subject to bacterial attack (Allison 1968). The choice of green manure type was to be based on the following: one of the main factors affecting decomposition of cover crop residues and release of N is the C to N ratio of the residues (Hansen et al. 2000). Residues from grass or cereal cover crops have relatively high C to N ratios, decompose slowly, and in some cases immobilize NO_3 already present in the soil (Rosecrance et al. 2000). Legumes have lower C to N ratios than nonlegumes, decompose more quickly, and release or mineralize more N (Utomo et al. 1990; Ruffo and Bollero 2003). Using surface-applied ^{15}N-labeled cover crop residues, Kaspar and Singer (2011) showed that averaged over a 2-year period, the addition of high quantities of organic soil amendments such as animal manure, compost and green manure, as well as soil-building leguminous crops increases soil organic matter (SOM) levels and the formation and stability of aggregates (Aoyama et al. 2000; Janzen 2006; Tisdall and Oades 1982). Rapid changes in microbial biomass and aggregate size distribution in response to changes in organic matter management in grass pasture.

1.6 Aim

The aim of this study was to determine the effectiveness of the impacts of simulated raindrops on the erosion of mine-spoils treated with young green manure and sewage.

2 Materials and Method

2.1 Mine Tailings

Material from a dried out iron pyrite mine tailings heap of average particle size <5 μm in diameter that had remained un-vegetated for >30 years was randomly sampled at two locations with moderately sloping topography. Because of the restricted soil depth (<10 cm), the acquisition of conventional soil cores was not feasible. Soil samples were taken from the 0–5 cm surface layer, which is underlain by stones. This depth was representative of the whole soil. According to Fletcher (1978), the top 5 cm contains most of the organic matter in a soil. Some characteristics of the mine materials are shown in Table 1.

Table 1 Some characteristics of two pyrites tailings materials

Soil property (t ha^{-1})	Leachate residue	Mine tailings
Organic carbon % by weight	0.2	0.4
pH (1:5 sludge: water ratio)	7.5	3.5
EC (1:5) sludge: water extract (dS m^{-1})	0.17	1.6
Texture (%)		
Clay	12	16
Silt	16.4	15
Sand	69.5	67.1
CEC (meq)	12.7	12.9

2.2 Sludge

An aerobically digested air-dried sludge was crushed and passed through a 2-mm sieve. High rates of sludge application (>25 %) are currently applied in the field. Therefore, in this laboratory study, high sludge rates were likewise applied to the soils at 25 and 50 % of the weight of the soil. The powdered sludge was thoroughly hand-mixed with air-dry soil. Some chemical and physical properties of the sludge are shown in Table 2.

Table 2 Some physical and chemical characteristics of Murray Bridge sludge

Soil property (t ha^{-1})	Value
Organic carbon % by weight	1.0
pH (1:5 sludge: water ratio)	6.5
EC (1:5) sludge: water extract (dS m^{-1})	0.27
Total C (%)	1.24
Total N (%)	1.8
$CaCO_3$ (%)	2.5
Exchangeable cations (cmol (+)/kg)	
Ca	11.6
Mg	5.4
Na	0.43
K	0.95
Total	18.4
Clay (%)	40
Quartz sand (%)	39.2
Silt (%)	12.8
Gravel (%)	8

2.3 Green Manure

2.3.1 Rate of Application

Green manure was applied at a rate of 1.5 % w/w. This is because organic matter in soils should be at least 10 % of clay content (w/w; MacRae and Mehuys 1987). The two mine materials of the present study, contained 12 and 16 % clay respectively. Pinck and Allison (1961) found that any excess above 1.5 % of decomposable plant material in the soil did not cause an overall percentage increase in decomposition. The inhibitory effects on decomposition of large quantities (>1.53 % w/w) of green manure in the soil have been reported (Singh et al. 1992).

2.3.2 Preparation of Green Manure

Harvesting of the legume *Leucanea leucocephala* (Leucania) occurred just prior to flowering. Tissues of immature plants generally decompose more rapidly than those of mature plant materials. They were chopped in turn to <2 mm in a portable food processor operated at 1000 rpm. The decomposition rate increased with the finely divided material (Kroth and Page 1946; Alexander 1976; Singh et al. 1992). After measuring the decomposition of different parts of the medic plant, Amato (1983) found that the rate of decomposition in the field occurred in the following order: leaves > stems > roots > pods. The ground pods were decomposed just as easily as unground stems. However, MaCrae and Mehuys (1985) cautioned that coarse organic matter may prevent the formation of large aggregates by blocking the attractive forces between smaller aggregates. It has been concluded that incorporated coarse organic matter in soils can block water movement by isolating soil zones, thereby reducing infiltration (Jamison 1963).

Indices of erodibility

These procedures for determining erodibility included water-drop impact after Cerda (1998). During the experiment, drops of distilled water 4 mm in diameter were used to strike the aggregate. Drop size was measured by a graduated cylinder, which was placed under a counted number of drops. For a sphere,

$$V = 4/3\pi r^3 \qquad (1)$$

where r is radius, and:

$$r = \frac{\sqrt[3]{3V \cdot (1)}}{4\pi} \qquad (2)$$

The drop size was controlled by the diameter of the tap fitting. Dispersion (modified Emerson water dispersion test (MEWDT) was carried out after Emerson (1967). The measurement of spontaneous dispersion in the absence of any imposed external force will reflect the behaviour of surface soils during the rainfall events when the surface is effectively protected by plant material (Rengasamy et al. 1984). All the other tests in the present study simulate impacts on bare soil. This dispersion test was applied to ten replicate samples for each treatment. An aggregate was slowly and carefully immersed in 40 ml of de-ionized water. At intervals of 0, 2, 15, 360 and 720 min, the degree of aggregate dispersion was defined by the following four scales (Cerda 1998):

0 No dispersion—the aggregate is completely intact.
1 Dispersion of some particles—cloudiness or grains close to the aggregate.
3 Considerable dispersion—more than half of the aggregate is dispersed.
4 Total dispersion of all particle—the aggregate ceases to exist.

For simulated rainfall, A highly aggregated soil (0–10 cm) of the permanent pasture (red-brown earth) RBE plot from the Waite long-term trial, Waite Campus, University of Adelaide, was chosen as a reference. After simulated rainfall had been applied to the soils, samples were taken from the 0–5 mm depth and put into sealed plastic dishes. The proportion of water stable aggregates were then measured.

2.3.3 Turbidity and Water Stability After Rain

After a pre-determined time, an aliquot was taken for measurement of clay content using a Hach Ratio turbidimeter that had been calibrated for each soil. After the soils were subjected to rainfall, samples were taken from 0–5 mm depth, and put in sealed plastic bags. The proportion of water-stable macro-aggregates was measured by wet-sieving (while aggregates were still saturated).

3 Results

3.1 Water-Drop Impact Tests (CND)

Stability for aggregates under water-drop impact was done under air-dry conditions. The results of the counted number of drops (CND) tests show that treatments caused various levels of aggregate stability under water drops. GM is the only treatment that stabilized every material to some extent (Table 3). The stability of all the materials was increased by at least one treatment, but for green-manure-treated sludge or sludge, the mine tailings (MT) had the smallest increases in stability. None of them significantly changed the poor structure of mine tailings (Table 3). For the mine tailings material, the effectiveness of the treatments was in the order (descending) of GM > SS = SSGM = control. For leachate residue material (LR), the effectiveness of the treatments was in the order (descending) of GM = SSGM > SS. Its control actually resisted the CND to a far greater degree than any treatment. For SS (as a treated soil), the GM and green-manure-treated sludge treatments yielded the most stable aggregates. The resulting CND was always at least 25 % higher than that of any other sample except the leachate residue control.

Table 3 Mean number of drop impacts necessary to break down an aggregate distinguished by different treatments

Aggregates	Treatments				
	GM	SS	SSGM	Controls	LSD 0.05 (within treatments)
Silt	28^{c}	16b^{a}	27^{c}	150^{e}	10
Sand	30^{c}	5^{a}	8^{a}	6^{a}	6
CEC (meq)	44^{d}	n.a.	40^{d}	10^{a}	7

Column values with different letters are significantly different ($P < 0.05$) as determined by Tukey's HSD test. *LR* leachate residue; *MT* mine tailings; *SS* sludge; *GM* green manure; *LRGM* green-manured LR; *MTGM* green-manured MT; *SSGM* green-manured SS

For each material, stability against CND was not significantly increased by sludge. However, after incubation with GM, sludge withstood 50 % more drops than GM-treated mine tailings or leachate residue (Table 3). The high CND of 150 for the leachate residue control compared with an average of 24 for the other treatments suggests that those treatments were inimical to its structural stability. In contrast, the mine tailings control shows a CND of just 6, which indicates an inherently very weak structure that has been strengthened by the green manure treatment. The reference soil required a CND of 150. This is the ideal approximate CND result for a highly stable soil. Even though 2- to 5-fold stability increases were observed for mine-spoil materials, any stability increases against CND was far below the CND of the reference soil. Some treatments decreased the erodibility of the soils in this study. However, they are still very unstable compared with a soil exhibiting an ideal condition.

3.2 *Modified Emerson Water Dispersion Test (MEWDT)*

The results of the (MEWDT) test show that all the soils were improved by one or more of the treatments. Approximately half of the samples showed only a very small amount of stability, resulting in considerable dispersion after 2 min (i.e. >50 % loss of the aggregate; Table 4). On average, the index of dispersion increased from 1.6 at 0 min (immediate dispersion) to 2.3, 2.5, 2.9 and 3.6 after 2, 5, 360 and 720 min respectively. The most effective treatment ingredient against erodibility was the green manure, with the index of dispersion (average combined values) increasing at 0.2, 1.2, 1.8, 2.4 and 3.3 for 0, 2, 5, 360 and 720 min respectively (Table 4). The most ineffective treatment ingredient against erodibility was the sewage sludge, with an index of dispersion (average combined values) increasing at 2.3, 2.8, 3.1, 3.5 and 4.6 for 0, 2, 5, 360 and 720 min respectively. The green manure combined with sludge caused the least dispersion over time. The dispersion index increases were at 0.0, 0.6, 1.3, 2.0, 3.6 for 0, 2, 5, 360 and 720 min respectively. The dispersion indices for sludge, and green manure treatments were 3.26 and 1.5 respectively (Table 4). This shows that the sludge as a

Table 4 Mean values for the index of the modified Emerson water dispersion test (MEWDT)

Soil and treatment	Time				
	0 min	2 min	15 min	360 min	720 min
Reference soil	0	0	0	0	0
LR	1	3	3	4	5
LRGM	1	3	3	4	5
LRGMS	0	1	2	2	3
LRSS	4	4	4	4	4
MT	4	4	4	4	5
MTGM	1	2	2	2	3
MTGMS	0	4	1	2	3
MTSS	4	4	4	4	5
SS	4	4	4	4	5
SSGM	0	0	1	2	2

0 No dispersion; *1* dispersion of some particles; *2* aggregate partly dispersed; *3* considerable dispersion; *4* total dispersion
LSD within treatments not significant at the $P = 0.05$ level

soil amendment would have caused twice the amount of soil breakdown as a green manure amendment under a natural ground cover.

The decrease in erodibility caused by green manure for the mine tailings was dramatic. It decreased the MEWDT index of the MTGM by 100 % (Table 4). According to this test, untreated mine tailings (MT) were much more erodible than untreated mine leachate residue (LR; Table 4). In other words, the untreated leachate residue would be much more stable than the untreated mine tailings against breakdown under the cover of plant material, such as leaves, branches and mulches (Rengasamy et al. 1984).

The most stable soil was the reference soil, with aggregates that survived the immersion throughout the test period without any dispersion. It maintained this status (after the experiment) for more than 100 h (data not shown here). This reference soil was under grass sod for several years prior to its removal from the ground for this study, whereas all the other soil materials had been sub-aerially exposed.

3.3 Rainfall, Runoff, Sediment Removal and Turbidity

3.3.1 Rainfall Simulation

Low energy-high intensity (LEHI) rainfall produced results different to the water-drop test, where the drops were bigger and heavier, but of lower frequency. The treatments that withstood the CND test the best (such as green manure-treated sludge and green-manured mine tailings) released the smallest amount of sediment for the LEHI (Table 3). Those treatments that scored well on the LEHI test did not necessarily withstand the greater kinetic energy of the CND. The sludge-treated mine tailings (MTSS) could not resist the CND test and broke down after being impacted by only five water drops (Table 3). It had reduced sediment loss under low energy-high intensity rain (Table 3). Of all combinations, the green-manure-treated sludge that withstood the highest number of CND drops (44 and 40) held virtually all its sediments against removal by erosion under LEHI (Table 3).

3.3.2 Aggregate Stability After Rain (LEHI)

Visual observation of the aggregates showed high aggregation of the green-manure-treated mine soils and green-manure-treated sludge. The persistence of well-defined aggregates was observed, as opposed to the slaked, congealed mounds of all the controls except those of the leachate residue. The water-stable aggregation tests after rain for the sludge showed that green manure had increased aggregates of >2 mm significantly to about 60 % (Table 5). The application of green manure to the sewage sludge led to the creation and stability of aggregates of >2 mm against rainfall (Table 5).

These aggregates were not broken into smaller sizes even by a low energy (15 J m^{-2} mm^{-1})-high intensity (100 mm h^{-1}) rainfall event of 25-min duration. For the water drop test, the LR-control aggregates at air-dry withstood a CND more than five times that of the GM-treated LR, which is similar (150 water drops) to that of the reference soil. Owing to chemical reactions following $Ca(OH)_2$ treatment of the acid sulphate drainage leachate, the LR had become impregnated with $CaSO_4$. This could have helped to indurate the mineral particles in the LR during storage, but not in the MT samples. Therefore, the likely explanation would be the long storage period of the samples prior to the commencement of this study. The fact that green manure was not needed to stabilise the LR controls suggests that age-hardening occurred in these samples. Green manure in the MT also stabilised aggregates of >2 mm, but to a lesser degree (Table 5). All other controls were subjected to the same procedure, but in no instance did any aggregate of >2 mm survive the wet sieving in this study, including after rain. Sewage-sludge-treated tailings materials did not withstand the wet sieving following simulated rain.

Table 5 Effect of green manure plant species and soil type on water stability of aggregates after simulated rain (% remaining on 2 mm sieve)

Soil/ treatment	Leachate residue (LR)	Mine tailings (MT)	Sludge (SS)	LSD 0.05 (within treatment)
Legume	67	15	60	3
Non-legume	63.5	15	50.0	4.5
Sludge	0	0	0	NS[a]
Control	90	0	0	NS

[a]Not statistically significant

Except for the leachate residue, there are no significant differences between the stability of aggregates from mine materials treated with sewage sludge, on the one hand, and that of their controls. No sewage sludge samples contained any >2 mm aggregates after this treatment (Table 3).

3.3.3 Runoff

The effects of green manure and SS on water runoff, sediment erosion, and turbidity after simulated low energy-high intensity rain on soils are presented in Figs. 2, 3 and 4. There are significant differences between green-manure-treated soils (including sewage sludge) and controls from all treatments. Green manure decreased runoff, sedimentation and turbidity for all soils (Figs. 2, 3 and 4).

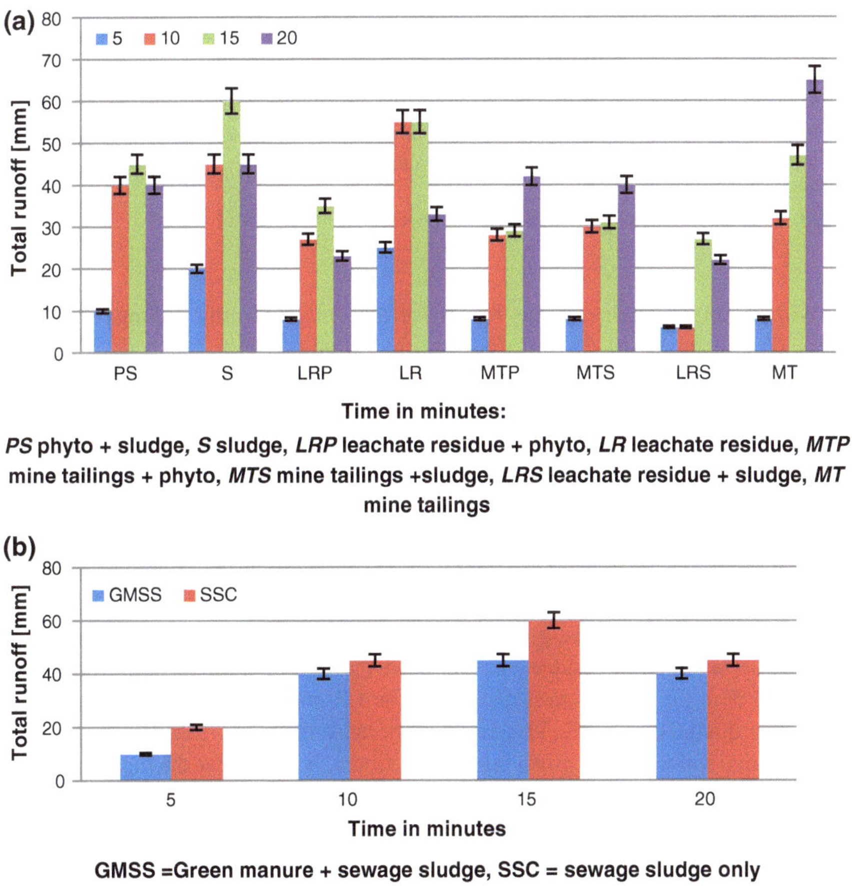

Fig. 2 **a** Runoff on mine tailings after treatment with phyto-organics and sludge. **b** The effects of green manure and sludge on the total runoff of mine substrates

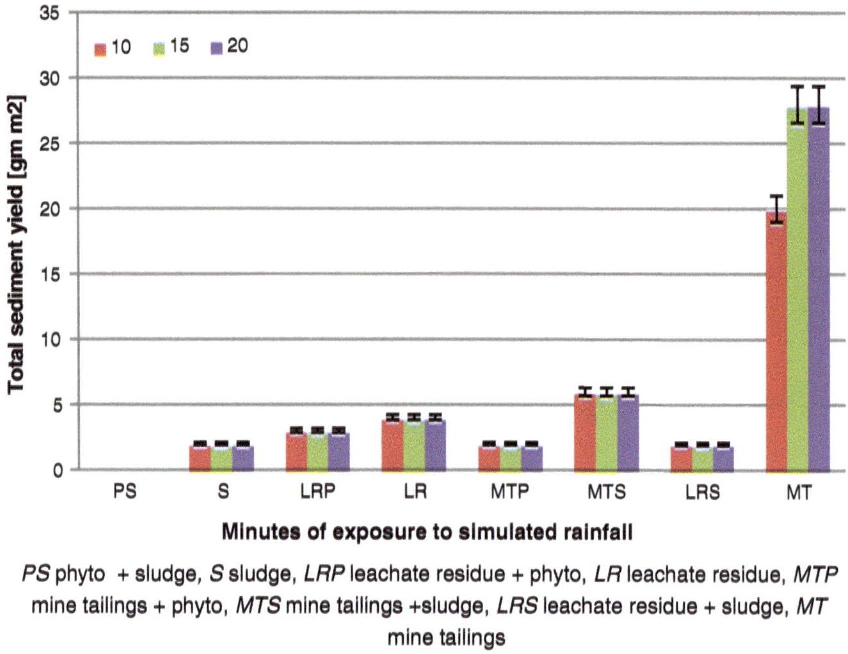

Fig. 3 Erosion of mine tailings after treatment with phyto-organics and sludge

Without any amendment, the runoff for the untreated mine tailings was highest at 66 mm in the 20–25 min, followed by 45 and 36 mm, respectively, for the untreated sludge for untreated leachate residue for the same time period (Fig. 2). For the untreated soil materials, the total runoff for the 25-min duration of rainfall in descending order occurred in the sequence of MT > SS > LR. Therefore the inherent infiltration capacity of the sludge was higher than that of the mine tailings. Of all the materials, mine tailings is the most prone to surface sealing and the most susceptible to erosion. This result is in accordance with an earlier study (Harris 2000), which showed the hydraulic conductivity of the untreated mine tailings to be 0 cm h^{-1}.

Relative to their controls, green manure decreased runoff by 42 % on mine tailings, 41 % on leachate residue and 16 % on the sludge (Fig. 2). Over the 25-min simulated rain, the effectiveness of GM on the LR and MT for runoff decrease was more than double on the SS. This could be because of the increased hydrophobicity of the sludge following the green manure treatment. It had by far the highest intrinsic amount of organic matter. The total runoff reduced by sludge in the leachate residue was 38 %. The runoff reduced by the sludge in the mine tailings was 15 % (Fig. 2). For runoff, each mine material had similar responses toward green manure, but different responses towards sludge. For the GM, runoff was greatly decreased for both GM and sludge, whereas the mine tailings was only improved substantially by the GM.

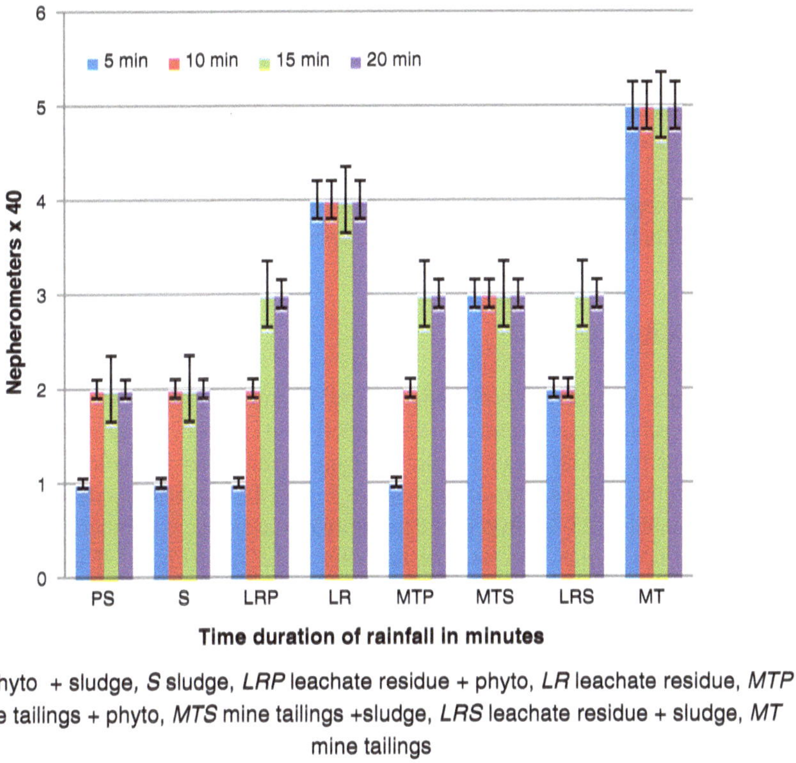

Fig. 4 Changes in turbidity of mine tailings after phyto-organic and sludge treatments

3.3.4 Sediment Erosion

The application of green manure in the mine tailings and leachate residue decreased sediment erosion significantly compared with their respective controls. Green manure decreased erosion by about 800 % on the mine tailings and 25 % on the leachate residue (Fig. 3). The effect of GM on sludge was most dramatic. It stopped the soil erosion completely (Fig. 3). Although only a relatively small amount of sediment was removed from the untreated sludge, no sediment was released from green-manure-treated sludge during a 25-min period under high intensity-low energy simulated rainfall (Fig. 3).

For the leachate residue material, the total sediment loss for green manured samples was similar to that of sludge-treated leachate residue samples. Sediment losses were approximately 60 % in each case. When compared with its controls, sediment loss from the mine tailings was reduced five times by the sludge but ten times by the green manure (Fig. 3). Although seemingly more impressive than the responses of the leachate residue, in absolute terms, there was much more sediment lost from all treated mine tailings samples compared with the treated leachate residue samples (Fig. 3).

The GM was more effective at reducing sediment loss than the SS, on both the leachate residue and mine tailings. Relative to their respective controls, GM was more effective on the mine tailings than on the leachate residue in reducing sediment erosion (Fig. 3). Of all the controls, the sludge lost by far the smallest amount of sediment under low energy high intensity rain. When incubated with green manure, sediment losses from the sludge were most drastically decreased, and became negligible. In terms of the reduction of sediment loss under rain, the GM-treated sludge samples proved vastly superior to all others in this study.

3.3.5 Turbidity After Rain

Green manure and sludge both reduced the amount of suspended particles after rain in the leachate residue and mine tailings (Fig. 4). However, the amounts of dislodged matter from GM-treated materials were much lower in the leachate residue than in the mine tailings. Part of the reasons is that the leachate residue has a lower native clay content (25 % lower). The turbidity of the sludge was inherently low. For SS, GM did not further decrease the turbidity under rain (Fig. 3). The GM-treated sludge samples acquired a greenish-brown coloration. This may have increased the turbidity readings such that for the GM-treated sludge, any reduction in turbidity by the GM was offset by the colour of the decaying organic matter from the GM.

Turbidity was highest for the untreated mine tailings control samples, being approximately 20 % higher than that in the leachate residue controls. In terms of sediment loss, the treatments were more effective for the mine tailings, even though the total turbidity from the treated leachate residue was not greater (Fig. 3). The sludge control and its treatments exhibited the least turbidity of all samples. This was not because of the GM (Fig. 3) and suggests that the sludge microstructure was the most stable. When averaged over the 25 min, the turbidity differences between the sludge and green manure treatments are not apparent for the leachate residue (Fig. 3). However, in the first 5 min of observation under rain, the turbidity of the SS-treated leachate residue and mine tailings samples was double that of the GM-treated leachate residue and mine tailings samples (Fig. 3).

4 Discussion

After the CND procedure, no sample treated with sewage sludge contained any >2 mm aggregates or parts thereof (Table 3). This shows that the aerobically aged sewage sludge incubated in this study is not an effective incorporated amendment against mine-soil aggregate breakdown by heavy rain. The sludge was aged subaerially for 2 years. Its organic matter particles would have been relatively small in size. It is in this period that, when soils are air-dried, they are most fragile and susceptible to high wetting rates that lead to erosion (Cernuda et al. 1954). Durgin and Chaney (1984) found that mutual repellence caused dislocation of particles if the organic matter particles in soil material were smaller than the clay particles. This seems to confirm the inability of old aerobic sludge to strongly bind the added soil particles, and the superiority of GM in this respect: the former largely consisting of old resistant organic matter, whereas the latter is a source of rapidly decomposable material for the ongoing and active microbial production of effective soil-binding cements. The high level of turbidity of the mine tailings after low energy rain confirms the ease with which its clay-sized fraction is dislocated. This is not aggregate breakdown, but clay dispersion, which causes surface sealing (Loch et al. 1978). Surface sealing has been linked with dramatic decreases in infiltration after rain (Mullins et al. 1990), which leads to increased runoff. For the mine tailings, the superiority of green manure to suppress the disintegration of aggregates when they are most fragile is indicated. Green manure was much more effective at reducing turbidity than sludge (Fig. 4). In general, GM has proven more effective than SS as an incorporated treatment for the erosion of the mine tailings materials.

The addition of sludge as dry matter at 50 % (w/w) meant that 0.75 % (w/w) of organic C was added to the mine-spoil material by sludge. Because the leachate residue and mine tailings had 0.1 and 0.3 % of native C, respectively, this meant that a 7.5- and 2.5-fold increase in organic C, respectively, occurred for these materials. Yet, the addition of sludge was not effective at improving the over-all erodibility of the mine soils. This disagrees with several studies that positively correlate the increase of organic C with the soil stability (Epstein et al. 1976; Oades 1988; Ekwue 1991; Evans 1996). The last above-mentioned study suggests that water erosion of soils in temperate climates is accelerated when soil organic matter falls below a 2 % threshold. In addition, MacRae and Mehuys (1985) found that soil organic matter must comprise at least 10 % of the clay content w/w for effective aggregation. Reducing water erosion is one of the main reasons for growing cover crops (Langdale et al. 1991), and, as stated above, soils are generally more susceptible to erosion when they are not covered with the canopies of living plants or their plant residues (Kaspar and Singer 2011). Because the clay contents of the mine tailings and leachate residue were 16 and 12 %, respectively, the above-mentioned percentage of organic matter was inadequate. In general, the addition of green manure improved the erodibility of all mine-spoil materials against water more effectively than the sewage sludge. This is in agreement with Cerda (1998), who explained this positive effect in terms of the cementing agents of decomposable organic matter in litter and exudates, citing the formation of macro-aggregates around plant roots. This also agrees with Kroth and Page (1946) who showed that the ongoing processes of in situ decomposition in soils that cement and stabilise soil particles cause aggregation. These findings indicate that the quantities of organic matter found in the sludge used for stabilisation in this study were insufficient in quantity and quality.

In the light of the above-mentioned statements, it may seem surprising that the organically depleted sludge samples exhibited the least turbidity (Fig. 4). Spontaneous breakdown leads to turbidity. However, the high quantities of soluble $CaCO_3$ used in the treatment of the sludge before this study, giving rise to Ca^{2+}, along with its small negatively charged organic fraction, would have caused linkages amongst particles that could resist turbidity. The mine tailings required just six water drops to disintegrate its untreated aggregates, and its dispersion in water was quicker. It is intrinsically more erodible than the leachate residue. Muneer and Oades (1989) confirmed that the presence of Ca^{2+} decreased the amount of dispersible clay. During treatment of the leachate, $Ca(OH)_2$ was the active reactant in the AMD neutralisation process. Nonetheless, the MEWDT test showed that even though the leachate-residue untreated aggregates were less dispersive than the mine-tailings-untreated aggregates, the GM-treated mine tailings throughout the test was always at least 50 % less dispersive than the GM-treated leachate residue.

The answer to this problem may lie in the different aggregate sizes. For treated and control samples, the size of the leachate residue units used for the MEWDT test was 6 mm in diameter, as opposed to 2.5–3 mm in diameter for the mine tailings. The intrinsic specific gravity of the leachate residue was 0.4 times that of the mine tailings. For each finite number of small aggregates in any mine tailings sample, there was the same number in a corresponding leachate residue sample. Each individual leachate residue aggregate was more than twice as large. Dexter (1988) found that the smaller the aggregate, the lower the ratio of joints to aggregate volume. Large aggregates are inherently more dispersive than small aggregates because of their greater number of joints per volume. The GM-treated leachate residue retained a high level of aggregate stability in the LEHI test. The Ca cementation in the smaller aggregates may have been stronger than the organic cementation for the larger aggregates. Had smaller leachate residue aggregates been used in this study, the leachate residue would have exhibited a lower level of dispersion than observed for the mine tailings samples. Warkentin (1982) pointed out that aggregation is flocculation plus; i.e. Ca^{2+} cannot prevent aggregate breakdown when there is no (organic) cementation of floccules. If they were true aggregates, the untreated (control) leachate residue entities without sufficient organic matter should not have resisted breakdown in the above tests. The LR control samples were likely comprised of thixotropically affected entities, referred to by Coughlan et al. (1973) as fortuitous agglomerates'. These fortuitous agglomerates survive wet-sieving but are not true aggregates. This may contribute partly to the high level of resistance to the wet-sieving amongst the GM-treated leachate residue aggregates following exposure to rainfall and strong resistance to the CND test.

5 Conclusion

Large improvements have been observed in the tests in the resistance to erosion. The cementation bonds associated with the green manure in both leachate residue and mine tailings, particularly in the leachate residue, were still weaker than those of the reference soil. Sewage sludge at a rate of 25 % (w/w dry matter), as used in the present study, did improve the erodibility of acid sulphate mine-spoil materials, but green manure at the rate of 2.5 % dry matter was more effective, especially under high energy water drops. It decreased the erodibility not only of the above-mentioned materials, but of the sewage sludge as well. In mine-spoil sites, leachate residue (LR) from acid neutralising plants is commonly stored in dams on flat ground, whereas the less cohesive mine tailings are often found on sloping ground. The effect of exposure on mine tailings is to increase the erosive forces, which increases the threat to the quality of soil, groundwater and streams in the catchment area of concern. Given the large index of erodibility for the untreated mine tailings, there is greater urgency regarding its rehabilitation compared with the leachate residue.

The main physical disadvantage of aerobically aged sludge as a top-dressing, especially in a sloping topography, is its susceptibility to removal by water and wind erosion prior to colonization by native flora. Sludge is normally added at high rates to soils, but the sludge itself may be weakly structured. The above-mentioned results show readily decomposable green manure to be a promising physical treatment for aerobically aged sludge prior to its addition to weakly structured mine spoils. In combination (incorporation) with mine tailings, the improvements by sludge were not as great as those caused by green manure. The GM-impregnated sludge resists erosion to a far greater degree than the sludge to which green manure had not been applied.

Therefore, where aged sludge is to be applied as a top-dressing on steep slopes, as it presently is in the case of Brukunga, the increased stability observed in this study should be exploited. If the sludge is incubated with green manure prior to spreading, a more long-lasting effect against detachment from raindrop impacts, down-slope slippage and erosion should be observed. There are also advantages for flat ground' application of green-manure-treated aged sludges. Because sludge has an inherently high WHC, waterlogged conditions would prevail in the wet season. This leads to restricted root extension and stunted plant growth. The prior improvement in granulation of the sludge would increase permeability, water movement and oxygen availability in the root zone. For flat areas, the pre-treatment of sludges with green manure also promises improvements for the rehabilitation of acid-sulphate mine-spoils.

6 Geobiotechnological Applications

6.1 Detecting Earth Movements

According to Department of the Environment Australia (2011), often the first sign that there may be a problem of soil instability is when interior doors become difficult to close and scrape the floor surface when being opened and shut. On floor surfaces covered with ceramic tiles or wood parquetry, tiles may become loose in particular areas for no apparent reason. If these features are seen in a room, check how flat the floor is by using a length of cord and a spirit level to make a horizontal reference line across the room. Note any areas where the floor surface appears to bulge upwards and the magnitude of the bulges. If floors are covered by rugs or loosely fitting carpets, it may be possible to pull the covering back to look at the underlying concrete floor. The presence of star-shaped radial cracks and/or white chalky deposits in areas with bulges may indicate damage is being caused by the oxidation of pyrite in soil beneath the concrete slab of the house. Fine 'hairline' cracks that are often seen on plasterwork inside houses are generally of no concern, and often result from minor shrinkage of plaster or settling of new brickwork. However, cracks more than one millimetre wide may be of concern, particularly if there is evidence that plaster is flaking from the edges of the cracks and is falling to the floor and if the wall surface across the crack is not level.

6.2 Slowing Down Erosion

Cover crops or surface residues (Brown and Norton 1994) can slow flow velocity at the surface by increasing hydraulic resistance. Lastly, because cover crop plants or residues are anchored to the surface by roots and because they hold other unanchored surface residues in place (Kaspar and Singer 2011), flowing water cannot easily move residues and expose the soil surface to shear forces of water.

Cover crops can reduce rill erosion by reducing the sheer force of flowing water or by increasing the resistance of soil particles to detachment. One way cover crops reduce the sheer force of runoff water is by reducing its volume through increased infiltration. This occurs because cover crops prevent surface sealing, increase storage capacity, and improve soil structure (Dabney 1998).

Martín-Moreno et al. (2013) outlined the importance of slope form and topography in erodibility: two topographic profiles, linear and concave, with two surface soil covers, subsoil and topsoil, were monitored for two hydrologic years. Sediment load, rill development and plant colonization from the four profiles were measured under field conditions. Their results show that, a thick and non-compacted topsoil cover on a linear slope yielded less sediment than carbonate colluvium or topsoil cover on a concave slope. The study also showed that vegetation establishment (important in erosion control) depends on topography. Plant cover was more widespread and more homogeneous on linear profiles with topsoil cover. On concave slopes, plant establishment was severely limited on the steepest upper part and favoured in the bottom. They note that management of topography and surface soil cover should be approached systematically, taking three outcomes into consideration: (i) topsoil can lead to a successful mining reclamation regardless of topography, (ii) created concave slopes can lead to a successful mining reclamation and (iii) topography determines the vegetation colonization pattern.

Acknowledgements The author is grateful to Associate Professor Vic Gostin (Department of Geology and Geophysics), Professor Martin Williams (Foundation Professor in the Department of Geographical and Environmental Studies), and Dr. P. Rengasamy (Department of Soil Science) of the University of Adelaide, for their constructive advice and assistance.

References

Alexander M (1976) Symbiotic nitrogen fixation. In: Alexander M (ed) Introduction to soil microbiology. Wiley, New York, pp 305–330

Allison FE (1968) Soil organic matter and its role in crop production. Elsevier, Amsterdam, pp 112–138

Amato M (1983) Decomposition of plant material in Australian soils. Aust J Soil Res 21:563–570

Aoyama M, Angers DA, N'Dayegamiye A, Bissonnette N (2000) Metabolism of 13Clabeled glucose in aggregates from soils with manure application. Soil Biol Biochem 32:295–300

Brown LC, Norton LD (1994) Surface residue effects on soil erosion from ridges of different soils and formation. Trans ASAE 37:1515–1524

Cerda A (1998) Soil aggregate stability under different Mediterranean vegetation types. Catena 32:73–86

Cernuda CF, Smith RM, Vicente-Chandler J (1954) Influence of initial soil moisture condition on resistance of macro-aggregates to slaking and water-drop impact. Soil Sci 77:19–27

Coughlan KJ, Fox WE, Hughes JD (1973) A study of the mechanisms of aggregation in a krasnozem soil. Aust J Soil Res 11:65–73

Dabney SM (1998) Cover crop impacts on watershed hydrology. J Soil Water Conserv 53:207–213

Dabney SM, Delgado JA, Reeves DW (2001) Using winter cover crops to improve soil and water quality. Commun Soil Sci Plant Anal 32:1221–1250

Dept. Environment Australia (2011) Post mining landform stability Australian gov dept of the environment. http://www.environment.gov.au/science/supervising-scientist/research/hgp/landform-stability. Accessed 3 Dec 2015

Dexter AR (1988) Advances in characterization of soil structure. Soil Till Res 11:199–238

Durgin PB, Chaney JG (1984) Dispersion of kaolinite by dissolved organic matter from Douglas fir roots. Can J Soil Sci 64:445–455

Ekwue EI (1991) The effects of soil organic matter content, rainfall and aggregate size on soil detachment. Cremlingen 4:197–207

Emerson WW (1967) A classification of soil aggregates based on their coherence in water. Aust J Soil Res 5:47–57

Epstein E, Taylor JM, Chaney RL (1976) Effects of sewage sludge and sludge compost applied to soil on some soil physical and chemical properties. J Environ Qual 5:422–426

Evans R (1996) Some soil factors influencing accelerated water erosion of arable land. Prog Phys Geogr 20(2):205–215

Flanagan DC (2002) Erosion. In: Lal R (ed) Encyclopedia of soil science. Marcel Dekker Inc., New York

Fletcher GK (1978) Mineral sands—mining effects and rehabilitation. In: Rummery RA, Howes KMW (eds) Management of lands affected by mining. Division of Land Resources Management, Commonwealth Scientific and Industrial Research Organization, Australia, pp 129–139

Hannan JC (1978) Coal: Rehabilitation of coal mines in the Hunter Valley of new South Wales. In: Rummery RA, Howes KMW (eds) Management of lands affected by mining. Division of Land Resources Management, CSIRO, Australia, pp 18–22

Hansen EM, Djurhuus J, Kristensen K (2000) Nitrate leaching as affected by introduction or discontinuation of cover crop use. J Environ Qual 29:1110–1116

Harris MA (2000) The effect of sewage sludge and green manure on the aggregation and hydraulic conductivity in pyritic mine tailings. Environ Geol 41:285–296

Harris MA (2008) Structural improvement of age-hardened gypsum-treated bauxite red mud waste using readily decomposable phyto-organics. Environ Geol 56:1517–1522. doi:10.1007/s00254-008-1249-5

Howes KMW (eds) (1979) Management of lands affected by mining. Division of Land Resources Management, Commonwealth Scientific and Industrial Research Organization, Australia, pp 18–32

Hore-Lacy I (1978) Sulfide ores: an overview of the effect of mining. In: Rummery RA, Howes KMW (eds) Management of lands affected by mining. Division of Land Resources Management, Commonwealth Scientific and Industrial Research Organization, Australia, pp 111–119

Jamison VC (1963) Water movement restriction by plant residues in a silt loam soil. Trans 7th Int Congr Soil Sci 1:464–472

Janzen HH (2006) The soil carbon dilemma: shall we hoard it or use it? Soil Biol Biochem 38:419–424

Kaspar TC, Singer JW (2011) The use of cover crops to manage soil (Tom.Kaspar@ars.usda.gov) (jeremy.singer@ars.usda.gov), USDA-ARS, National Laboratory for Agriculture and the Environment, doi:10.2136/2011.soilmanagement.c21

Kroth EM, Page JB (1946) Aggregate formation in soils with special reference to cementing substances. Soil Sci Soc Am Proc 11:27–34

Langdale GW, Blevins RL, Karlen D, Thomas DK, Tyler DD, Williams JR (1991) Cover crop effects on soil erosion by wind and water. In: Hargrove WL (ed) Cover crops for clean water, pp 15–22. Proceedings of the international conference, Jackson, TN, Soil and Water Conservation Society, Ankeny, IA 9–11 April 1991

Larson WE, Pierce FJ (1994) The dynamics of soil quality as a measure of sustainable management. In: Doran et al. (ed) Soil science society of America special publication 35, pp 37–52. Soil Science Society of America, Madison

Loch RJ, Foley JL (1994) Measurement of aggregate breakdown under rain: comparison with tests of water stability and relationships with field measurements of infiltration. Aust J Soil Res 32:701–720

Loch RJ, Cleary JL, Thomas EC, Glanville SF (1978) An evaluation of the use of size distributions of sediment in runoff as measure of aggregate breakdown in the surface of a cracking clay soil under rain. Earth Surf Proc Land 13:37–44

MacRae RJ, Mehuys GR (1985) The effect of green manuring on the physical properties of temperate-area soils. Adv Soil Sci 3:1–94

MacRae RJ, Mehuys GR (1987) Effects of green manuring in rotation with corn on the physical properties of two Quebec soils. Biol Agric Hortic 4:257–270

Marshall JK (1978) Workshop group summary. In: Rummery RA, Howes KMW (eds) Management of lands affected by mining. In: Proceedings of a workshop organised by the CSIRO Divisions of Land Resources Management and Mineralogy at Western Australian School of Mines, Kalgoorlie, pp 126–128

Martín-Moreno C, Duque J, Ibarra J, Castillo L (2013) Effects of topography and surface soil cover on erosion for mining reclamation: the experimental soil heap at El Machorro Mine, (Central Spain). Land Degrad Dev. doi:10.1002/ldr.2232

Mamo M, Bubenzer GD (2001) Detachment rate, soil erodibility, and soil strength as influenced by living plant roots. Part II: Field study. Trans ASAE 44:1175–1181

Mulligan DR, Dahl NW (1996) Native vegetation establishment—a bauxite mining perspective. In: Bellairs SM, Marris JM (eds) Proceedings of Workshop on Native Species Establishment on Mined Lands in Queensland, Brisbane, pp 14–24. Centre for Mined Land Rehabilitation, University of Queensland, Brisbane, 8–10 Nov 1995

Mullins CE, MacLeod DA, Northcote KH, Tisdall JM, Young IM (1990) Hard-setting soils: behaviour, occurrence, and management. Adv Soil Sci 11:37–108

Muneer M, Oades JM (1989) The role of Ca-organic interactions in soil aggregate stability. I. laboratory studies with 14C-glucose, CaCO3 and CaSO42H2O. Aust J Soil Res 27:389–399

Oades JM (1988) The retention of organic matter in soils. Biogeochemistry 5:35–70

O'Geen AT, Rachel E, David L (2006) Erodibility of agricultural soils, with examples in Lake and Mendocino Counties University of California Agriculture and Natural Resources, Publication 8194. at http://anrcatalog.ucdavis.edu

Pinck LA, Allison FE (1961) Adsorption and release of urease by and from clay minerals. Soil Sci 91:183–188

Ram DN, Vittum MT, Zwerman PJ (1960) An evaluation of certain winter cover crops for the control of splash erosion. Agron J 52:479–482

Rengasamy P, Greene RSB, Ford GW, Mehanni AH (1984) Identification of dispersive behaviour and the management of red-brown soils. Aust J Soil Res 22:413–431

Rosecrance RC, McCarty GW, Shelton DR, Teasdale JR (2000) Denitrification and N mineralization from hairy vetch (*Vicia villosa* Roth) and rye (*Secale cereale* L.) cover crop monocultures and bicultures. Plant Soil 227:283–290

Ruffo ML, Bollero GA (2003) Modeling rye and hairy vetch residue decomposition as a function of degree-days and decomposition-days. Agron J 95:900–907

Sharpley AN (2000) Agriculture & Phosphorus Management, Lewis Publishers/CRC Press, Boca Raton, Florida

Sheoran V, Sheoran AS, Poonia P (2010) Soil reclamation of abandoned mine land by revegetation: a review. Int J Soil Sediment water. 3(2):13. ISSN: 1940–3259

Singh Y, Singh B, Khind CS (1992) Nutrient transformations in soils amended with green manures. Adv Soil Sci 20:238–305

Tisdall JM, Oades JM (1982) Organic matter and water-stable aggregates in soils. J Soil Sci 33:141–163

U.S. Department of Agriculture, Natural Resources Conservation Service (2005) National soil survey handbook, title 430-VI

Utomo M, Frye WW, Blevins RL (1990) Sustaining soil nitrogen for corn using hairy vetch cover crop. Agron J 82:979–983

Warkentin BP (1982) Clay soil structure related to soil management. Trop Agric Trinidad 59:167–177

Modification of Hydraulic Conductivity and Aggregation in Pyritic Mine Tailings Materials with Sludge and Decomposable Phyto-Organics

Mark Anglin Harris and Mallavarapu Megharaj

Abstract The physical character of acid sulphate soils preclude agricultural usage for crop growth. An experiment was conducted to determine the effectiveness of green manure and sewage sludge to improve the aggregation and hydraulic conductivity of (1) a pyritic mine tailing and (2) its leachate residue. As sludge is often used as a growth medium in mine spoils, the ability of green manure to improve its physical characteristics was also investigated. Only green manure (GM) improved both characteristics of the materials to some extent. In response to green manure, plus sludge (SSGM) or sludge (SS), the mine tailing had the smaller increases in aggregation and infiltration. For the tailings, the effectiveness of the treatments was in the order of GM > SS = SSGM. For leachate residue, effectiveness of treatments was in the order of GM = SSGM > SS. Sludge treated with GM yielded the most stable aggregates, as the mass of aggregates remaining intact after wet-sieving was always at least 50 % higher than that of all other samples. After treatment with green manure, the hydraulic conductivity of the tailings leachate residue was three times as great as that of the tailings. Increases in hydraulic conductivity for the green manure-treated sludge were similar to those of the leachate residue. This study showed that GM-impregnated sludge resists wet-sieving to a far greater degree than the sludge to which GM had not been previously applied. This has implications for the treatment of sludge prior to its application to mine-spoils.

Keywords Erosion · Green manure · Infiltration · Mine tailings · Sludge

Highlights

- Sewage sludge + green manure may improve structure of acid sulphate mine tailings
- Green manure improves structure and aggregate stability of sewage sludge.

Original article: **The effects of sludge and green manure on hydraulic conductivity and aggregation in pyritic mine tailings materials**. *Environmental Geology*. 41: 285–296.

M.A. Harris, *Geobiotechnological Solutions to Anthropogenic Disturbances*, Environmental Earth Sciences, DOI 10.1007/978-3-319-30465-6_13

1 Introduction

The formation of pyrite is the basis for the genesis of potential acid-sulphate soils acid sulphate soils where pyrite comprises 2–10 % of the mass of these soils (van Breemen 1983; Queensland Govt. 2015). Mine tailings that have been through an extraction and beneficiation process are dominated by fine textured, non-flocculating, non-aggregating inorganic materials (Whyte and Sisam 1949). For such weakly structured materials in acid-sulphate mine-spoils, deteriorated mechanical behaviour, wet and dry, remain one of the main problems for the facilitation of revegetation (Bell and Jones 1987). Soil aggregation and aggregate stability are the chief physical properties, which affect its susceptibility against the impact of raindrops, runoff water and consequently soil erosion. Further, as chemical problems in the soil occur as a result of physical ones (Bell et al. 1989), the physical problems of soils must be corrected before the chemical ones. Though improved structure in soils increases aeration (Warkentin 1982) and reduces waterlogging (Oades 1993), thus reducing acidity, no amount (or combination) of plant nutrients will make up for poor physical properties (Hausenbuiller 1978).

1.1 Definition of Acid Sulphate Soils

Actual acid sulphate soils (ASS) are those soils which contain sufficient metallic sulphates, that have been exposed to oxygen in the air (Fig. 1), and the sulfuric acid produced is in excess of the soil buffering capacity (Queensland Govt. 2015).

Fig. 1 Actual acid sulphate soil (AASS). Acid mine drainage formed in metal mines release metals such as iron which form reddish oxides. *Source* Queensland Govt. Resource & Minerals (2002a, b)

The resulting strong acid may acidify soil, ground water and/or surface waters. Vegetation achieves optimal growth in soil at a neutral pH. When the soil pH drops below 5.5, reduced legume and forage growth occur due to metal toxicities such as aluminum or manganese, phosphorus fixation, and reduced population of N-fixing bacteria (Sheoran et al. 2010). ASS can occur naturally along the coast usually where land elevation is less than 5 m AHD. Such soils affect urban, transport, tourism, agricultural and industrial land uses (Fig. 2). In addition to the sulfuric acid, which can mobilize the naturally occurring metals in the soil, toxic quantities of aluminium and other heavy metals are produced, in forms that can be commonly released into waterways.

The acid corrodes concrete and steel infrastructure and, together with the metal contaminants, can kill or damage fish, other aquatic organisms, native vegetation and crops, degrade water quality, and contaminants can cause dermatitis, with dust from disturbed ASS potentially causing eye irritation.

When acid sulphate soils are undisturbed, they are dark blue-grey (sometimes black) and wet, with no structure (Figs. 2, 3), being often high in clay, but can be sands or sometimes even gravels (Queensland Govt. 2015). Estuarine environments, where sulphate-rich marine or oceanic water are diluted with organic-rich freshwater run-off from the land can therefore produce acid-sulphate conditions.

If water cannot drain from the soil however, it remains benign as in mangrove swamps (Fig. 2). By introducing hydrogen ions, acid mine drainage affects the acidity of a stream. As a side effect of their metabolism, sulphate reducing bacteria (SRB) make pyrite crystals (Queensland Govt. 2015). They can only survive when there is:

Fig. 2 Natural acid sulphate soil formation. This began with bacterial conversion of the sulphate into sulfide. The bacteria also need a source of energy provided by organic material such as decaying vegetation seen here. Continuous water-logging preserves anoxic conditions, thereby maintaining the blue-grey coloration of soil. Note structure-less (i.e., single grain structure) soil. *Source* Queensland Govt. Resource & Minerals (2002a, b)

(a)

(b)

Fig. 3 **a** The poor structure of acid sulphate soils in anoxic conditions. **b** Water movements in an acid sulphate soil define the oxidation state, i.e., eH—pH status of the soils such that close to the water surface pH is low and eH high, with the obverse increasingly occurring towards the *top*. *Source* Queensland Govt. Resource & Minerals (2002a, b)

- An anoxic environment—this means a soil that is saturated with water. The lack of oxygen protects both the bacteria and the pyrite they create.
- A supply of organic matter—usually in the form of decaying leaves as an energy source.
- The presence of sulfur (in the form of sulphate SO_4^{2-})—used by the bacteria to remove electrons to obtain energy. Thus sulphate-reducing bacteria use sulphate and produce hydrogen sulfide (H_2S) the way animals use oxygen and produce carbon dioxide. As sulphate is a normal component of seawater, acid sulphate soils commonly develop along coasts, and particularly where decaying vegetation exists (Queensland Govt. 2015).

Iron in its reduced form (Fe^{2+}) reacts with the hydrogen sulfide (H_2S) so produced, to form pyrite (FeS_2). As pyrite forms, bicarbonate (HCO_3^-) also forms from the same reaction (Queensland Govt. 2015). If it is not removed from the environment, the pyrite-forming reaction eventually slows and stops. Water movement (like tidal flushing of the system) will remove bicarbonate and leave the pyrite behind, allowing it to accumulate. Time is required for pyrite to accumulate. Unlike many other soil processes however, pyrite accumulation can happen very quickly. A few years can be enough time for a significant amount of pyrite to form.

1.2 The Pueblo Viejo: Dominican Republic

The Pueblo Viejo precious and base metal deposit consists of high sulfidation or acid sulphate epithermal gold, silver, copper, and zinc mineralization that formed during the Cretaceous Period island-arc volcanism. According to Smith and Stephenson (2011), the two main areas of alteration and mineralization are the Monte Negro and Moore deposits (Fig. 1). Pueblo Viejo is situated in the Los Ranchos Formation, a series of volcanic and volcani-clastic rocks that extend across the eastern half of the Dominican Republic, generally striking northwest and dipping southwest. The Pueblo Viejo Member of the Los Ranchos Formation comprises sedimentary deposits that range from coarse conglomerates to thinly bedded, carbonaceous sandstones, siltstones, and mudstones. To the south, the Pueblo Viejo Member is un-conformably overlain by the Hatillo Limestone Formation by means of a low angle, southwest dipping thrust fault (Smith and Stephenson 2011).

The Moore deposit is located at the eastern margin of the Pueblo Viejo Member, while the Monte Negro deposit is located at the north-western margin. Sulfide mineralization is primarily pyrite with lesser amounts of sphalerite and enargite. Pyrite mineralization occurs as disseminations, layers, replacements, and veins. Sphalerite and enargite mineralization is primarily in veins, but disseminated sphalerite has been noted in core (Smith and Stephenson 2011).

There were two stages of advanced argillic alteration, both associated with precious metal mineralization. A third stage of mineralization occurred when hydro-fracturing of the silica cap produced pyrite-sphalerite-enargite (Stage III) veins with silicified haloes. Stage III veins contain the highest precious and base metal values and are more widely distributed in the upper portions of the deposits. The most common vein minerals are pyrite, sphalerite, and quartz with lesser amounts of enargite, barite, and pyrophyllite (Smith and Stephenson 2011).

Gold is intimately associated with pyrite veins, disseminations, replacements, and layers within the zones of advanced argillic alteration. Gold values generally are the highest in zones of silicification or strong quartz-pyrophyllite alteration. These gold-bearing alteration zones are widely distributed in the upper parts of the deposits and tend to funnel into narrow feeder zones. The proportions of the different forms and carriers of gold vary throughout the Moore and Monte Negro deposits, but the majority of gold is found as sub-microscopic gold in microcrystalline, disseminated or porous pyrite (Smith and Stephenson 2011).

Pyrite is by far the most abundant sulfide scattered in the rock, commonly forming submillimeter-sized rounded to sub-angular, occasionally sub-euhedral grains along with minor chalcopyrite, enargite and sphalerite; pyrite grains are commonly fractured and variably corroded and contain fairly abundant chalcopyrite, enargite, sphalerite, covellite, rutile and gangue inclusions and infillings (Abat 2014). Sulfides are easily oxidized when exposed to the atmosphere or water, as occurs in the study area (Fig. 4).

1.3 Soil Quality

According to USDA NRCS (2003), soil quality is the ability of soil to perform certain functions, such as (1) effectively cycling nutrients, minimizing leaching

Fig. 4 Pueblo Viejo Mines, Dominican Republic showing weak soil structure caused by low levels of organic matter. Acidic conditions prevent the buildup of organic matter. Reddish oxidized zone in sulfidic waste rock occurred after atmospheric exposure and contact with moisture. Anderson and Cobb (2008) noted that as the surface water reacts with the exposed pyrite, the pH of the surface water decreases and further reacts with pyrite, causing the pyrite to oxidize rapidly. Lower pH can increase the speed of oxidation (Anderson and Cobb 2008). *Source* Barrick Gold

and runoff, while making them available to plants (2) maximizing water-holding capacity and minimizing runoff and erosion (3) adsorbing and filtering excess nutrients, sediments, and pollutants (4) providing a healthy root environment and habitat, and (5) providing a stable foundation for (man-made) structures. As acid sulphate soil conditions decrease the existence of the above requirements, amendments are necessary.

1.4 Sludge

Amending mine-soils with bio-solids increases SOC, CEC, soil nutrient levels and promotes soil ecosystem recovery after mining. Added OM increases aggregate stability and moisture holding capacity and decreases soil bulk density (Ussiri and Lal 2005).

Bio-solids are often an excellent amendment for mined soils where the original topsoil contains low OM, or where topsoil substitutes such as overburden material or subsoil are used (Ussiri and Lal 2005). Roberts et al. (1988) concluded that mine spoils treated with bio-solids at 56 Mg ha^{-1} sustained higher forage yield than mine-soils amended with lime and fertilized native topsoil.

Many mine-soils are excessively wet during much of the growing season because of poor internal drainage or due to compacted layers that perch water. Most high-value crop trees are intoleant of wet soils (Daniels 1999) Bio-solids, or sewage sludge, are the end products of wastewater treatment, which can be in a number of forms, including liquid, sludge cake, dewatered and dried, heat-dried, alkaline treated or composted (Walker 1994). Bio-solid products may be stabilized by lime or alkali, aerobically or anaerobically, composting, or heating to reduce pathogen level (Walker 1994).

Research has shown that sewage sludge as an amendment results in immediate improvement of soil physical conditions, increasing soil water content and retention, cation exchange capacity (Epstein 1975), and water percolation (Boyle 1990). Robertson et al. (1982) found that sewage sludge increased soil organic matter to a depth of 90 cm in a fine sandy loam. After its incorporation in a silt loam subsoil, Epstein et al. (1976) found that the proportion of stable aggregates for the sludge treatments was 34 %, compared with 17 % in the untreated soil control, the increases persisting for at least 6 months. Lunt (1959) increased moisture equivalent of soils by 13 % by incorporating composted sludge with oak chips. Hill and Montague (1977) concluded that in mine-soils, applied sludge may increase granulation, water-holding capacity and cation exchange capacity. In disused mine-soils, Seaker and Sopper (1984) used aerobically treated sewage sludge to accelerate plant establishment and growth, and achieved long-term productivity. Mine-soils amended with bio-solids at the Powell River Project in southwest Virginia produced higher forage quality and more persistent stands than mine-soils reclaimed with fertilizers (Daniels and Haering 1994). Yields from deep ripped and bio-solids amended plots exceeded control (undisturbed farmland) 1 year after soil reconstruction (Powell et al. 1986). Zhai and Barnhisel (1996) and Daniels and Haering (1994) observed increased corn yield when bio-solids were applied to the subsoil at a rate of 34 Mg ha^{-1} and incorporated with chisel plow before topsoil replacement.

However, the treatment of the sludge prior to its incorporation is of critical importance to the physical properties of the soil. Although it is firmly established that organic matter bonds stabilized aggregates against slaking and disaggregation, once these bonds break and disaggregation occurs, the organic matter acts as a de-flocculant (Emerson 1983). Thus, as decomposition proceeds in a sludge, individual organic particles decrease in size, and organic matter particles no larger than the individual clay edges cause clay particle dispersion. Treated sludges, themselves, largely the product of prior bacterial decomposition, are sometimes lacking in decomposable organics (as opposed to raw sludge), and thus may lack sufficient latent energy to sustain microbial production of sufficient soil binding gums and cements in the soil. This may also be partly the reason that significant changes to soil structure seem to occur only at high application rates of treated sludge. Kladivko and Nelson (1979), for example, found that sludge added at 56 t/ha, decreased bulk density significantly up to 5 years after its application over that in the untreated plots, but that lower rates (11.2–22.4 t/ha) had no effect on the soil physical properties after the same time period. Losses through volatilization are also lower at higher application rates because thicker layers mean smaller sub-aerially exposed surfaces. The total porosity of all treated plots was still significantly greater than that of the control in a sandy loam soil 1 year after incorporation of sludge at 150 t/ha (Pagliai et al. 1981). However, at such very high rates, treated sludges are potentially toxic when applied to soils, posing obvious risks to soil and groundwater environments (Alloway and Morgan 1986). For this reason, lower rates of sludge are environmentally desirable, though less effective. Therefore, if lower rates of sewage sludge were combined with equally or more effective, but less toxic organic amendments to achieve comparable ameliorative results, the toxic effects on the soil and groundwater environments would be lessened. Options include the addition of glucose, and other readily decomposable microbial energy sources that have been implicated in the indirect improvement of soil structure by stimulating increases in decomposition rates of microbially 'unpalatable' resistant organic residues in soil (Hayes 1980; Chaney and Swift 1984; Golchin et al. 1994). Chaney and Swift (1984) observed that incubation of a glucose solution with previously dried and crushed aggregates reformed stable aggregates through the production of microbial saccharides. Therefore, readily decomposable materials, combined with lowered rates of treated sludge may be just as effective at improving soil physical properties as high rates of treated but toxic sludge.

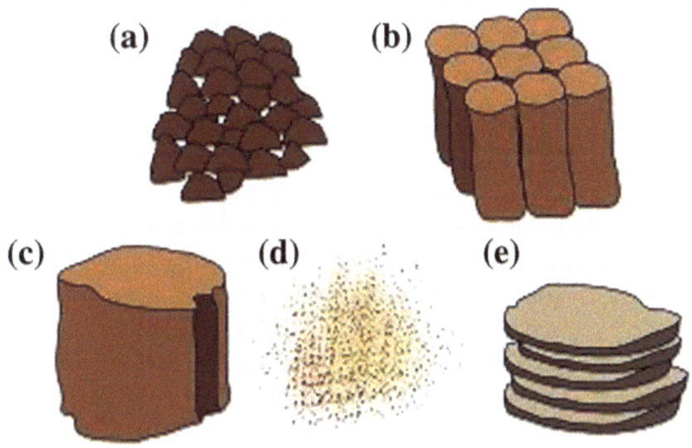

Fig. 5 The five forms of soil aggregation. **a** blocky, **b** columnar, **c** massive, **d** single grain, **e** platy. After FAO.org

1.5 Green Manure and Soil Structure

Additionally, compared with glucose, a cheaper and more ubiquitous source of readily decomposable organic matter is green manure, which can improve several physical properties of clayey soils. Clays have naturally reactive surfaces containing sites for the physico-chemical reactions that cause flocculation and cementation, but, according to Whyte and Sisam (1949), many mine-spoils are dominated by a single-grained structure, including sand (Fig. 5). However, Forster (1979) found that decaying organic matter colonized by bacteria on a sandy dune had a sticky surface, which included polysaccharides secreted by the microorganisms, thereby leading to the build-up and formation of discrete aggregates up to 6 mm in diameter.

As binding efficiency varies among microbial strains and environmental conditions (Maartens and Frankenberger 1992), the environmental conditions under which particular species thrive would seem to affect the physical properties of the soil. Bacteria can, therefore, play a major role in aggregating and stabilizing sand (Fig. 6) prior to colonization by higher plants. This has implications for sandy mine spoils (Fig. 6).

Manures have not been used extensively in reclamation because of the high moisture content and transportation costs involved in moving the bulk volume from their source to the mine sites (Ussiri and Lal 2005). However, manuring has the same advantages as bio-solids; however manures have higher N contents than bio-solids and pose no risks of high levels of micronutrients and heavy metals (i.e. Cd, Cr, Ni, Pb, and B) found in some bio-solids (Ussiri and Lal 2005).

The addition of high quantities of organic soil amendments such as animal manure, compost and green manure, as well as soil-building leguminous crops increases soil organic matter (SOM) levels and the formation and stability of aggregates (Aoyama et al. 2000; Janzen, 2006; Tisdall and Oades 1982). The addition of animal manure has been noted to influence aggregation, i.e. high manure application results in higher aggregate mean weight diameter (MWD) than no addition (Gulde et al. 2008; Min et al. 2003; Whalen et al. 2003). The added

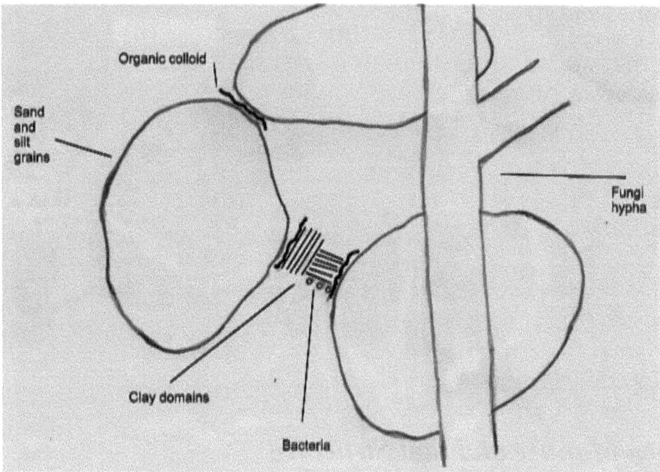

Fig. 6 Configuration of a typical soil aggregate showing organic binding materials. After FAO.org

organic amendments provide active organic compounds such as particulate organic matter (POM) that act as nucleation sites for the formation of macro-aggregates (Chivenge et al. 2011).

Bacterial populations can be re-established in mine-spoils by organic manures (Bitton and Boylan 1985). As mine-soils usually suffer from a paucity of microbial life (Seaker and Sopper 1988), the effectiveness of a sludge application can be gauged by the range and numbers of microorganisms in the topsoil, which determine the state of soil quality for plant growth. Seaker and Sopper (1984) found that microbial processes such as humification, soil aggregation and N cycling are essential in the establishment of productivity in mine-soils. The decomposition of plant materials in soil may also be enhanced by microorganisms introduced from a sludge inherently low in toxic heavy metals. Microbially derived polysaccharides can contribute up to 10 % and as much as 20 % of the organic matter in humus (Hayes and Swift 1978).

Based on measured microbial populations and activity, Bitton and Boylan (1985) found that ecosystem recovery rate on sludge-amended mine-sites appeared to be occurring at a more rapid rate than at inorganic fertilizer-amended mine-sites. On sludge-amended sites ranging in age from 1 to 5 years following sludge application rates of 120–134 t/ha (dry weight basis), Seaker and Sopper (1988) found that populations of aerobic heterotrophic bacteria, fungi, Nitrobacter, and respiration rate in the soil were significantly increased by the high rate of organic matter input. A possible explanation may be that organic matter in sewage sludge can stimulate urease activity (Moe 1967; Khan 1970; Zantua and Bremner 1976), which can help to activate and release any native soil N present in some mine tailings soils. This happens because urease is an inducible enzyme (Alexander 1976), whose activity is stimulated by even minute amounts of urea in the soil (Paulson and Kurtz 1969), such as that from readily decomposable organic matter or the urea present in sewage sludge. The urea activates genes encoding microbial enzymes required to catalyse the mineralization of otherwise inert N in such soils (Pinck and Allison 1961).

Thus, under certain environmental conditions, microbial activity in such soils is stimulated, leading to a flush of microbial activity and subsequent improvements in aggregation (Martin et al. 1955; Mehta et al. 1960). It is anticipated and hypothesized that green manure, when applied to fine-textured mine-spoil soils in combination with reduced quantities of treated sludge, will cause physical changes comparable with those documented elsewhere for larger quantities of sludge.

1.6 Aims

The aim of the present study was to identify the effects of sludge and green manure on some physical properties of an acid sulphate mine tailing, and its leachate residue formed after calcium hydroxide neutralization. The hypotheses for this study were:

Green manure will improve the granulation of a mine tailing.

Green manure will more effectively granulate mine tailings than aged sludge.

Green manure in combination with aged sludge will be at least as effective as the same quantity of green manure in the granulation of acid-sulphate mine tailings.

2 Materials and Methods

2.1 Treatment

Aerobically digested sewage sludge from the Murray Bridge treatment plant was applied to ®eld plots on an iron pyrites mine spoil site (Table 1).

Characteristics of the sludge are shown in Table 2. Additives in this sewage included 40 % clay and 2 % lime as $CaCO_3$. At three sites containing acid-sulphate mine tailings, a 4 cm-thick layer of sewage sludge was applied to the tailings dump surface.

Table 1 Some characteristics of some Brukunga sludge sites

Site	Age (months)	Amendment (at air dry)	Application date	Application rate ($Mg\ ha^{-1}$)	pH
2	2	sludge	Jan. 1998	110	3.4
2	18	sludge	Jan. 1997	110	3.1
3	24	sludge	Jan. 1997	110	3.0

Table 2 Some physical and chemical characteristics of Murray bridge sludge

Soil property (t ha^{-1})	Mean value (3 reps.)
Organic carbon % by weight	1.0
pH (1:5 sludge: water ratio)	6.5
EC (1:5) sludge: water extract (dS m^{-1})	0.27
Total C (%)	1.24
Total N (%)	1.8
$CaCO_3$ (%)	2.5
Exchangeable cations (cmol (+)/ kg)	–
Ca^{2+}	11.6
Mg^{2+}	5.4
Na^{+}	0.43
K^{+}	0.95
Total	18.4
Clay (%)	40
Quartz sand (%)	39.2
Silt (%)	12.8
Gravel (%)	8

Table 3 Some characteristics of to Brukunga mine tailings materials

Soil property	Leachate residue	Mine tailings
Organic carbon (% by weight)	0.2	0.4
pH (1:5 sludge: water ratio)	7.5	3.5
EC (1:5 sludge: water extract dS m^{-1})	0.17	1.6
CEC (cmol kg^{-1})	12.7	12.9
Texture %	–	–
Clay	12	16
Sand	16.4	15
Silt	69.5	67.1
Gravel	2.0	1.9

Field sampling

From locations previously treated with aerobically treated sludge, three replicate mine tailings samples for microbial counts were collected on all sites in January 1999, at 12–18 months after surface application of the sludge. These were mainly skeletal mineral particles of a sandy consistency, the main characteristics of which are shown in Table 3. Six equally spaced soil samples were collected at each site with a soil auger to a depth of 5 cm from a transect line through each site at random locations. Samples were placed in sealed plastic bags and refrigerated at 5 °C until analyzed.

2.2 Bacterial Assays

Each core was separately analysed. As an adjunct to direct counts of heterotrophic bacteria, an indirect method of assessing biological activity was applied. A piece of clean, white paper was perpendicularly buried amongst mine-spoil material in the middle of each sample, parallel to the sides of the container. To ensure close contact between paper and mineral particles, the particles were settled by dropping the container a predetermined number of times from a fixed height. With adequate moisture in the containers, each was incubated at 28 °C for 15 days. The paper remaining in each container was visually observed and weighed.

The populations of culturable heterotrophic bacteria were enumerated as colony-forming units (c.f.u.) from 10-fold serial dilutions of the soils plated out in triplicate on agar plates and colonies counted after 6 days of incubation at 28 °C (Megharaj et al. 1998). The bacterial population was estimated using 10 % full-strength Trypticase soy broth (BBL Co., USA) supplemented with 1.5 % agar. Counting of the colonies was done under a dissecting microscope. Numbers of organisms were calculated per gram of oven-dry tailings.

2.3 Field Collection of Tailings Leachate Residue

Two years after the residue from lime treated acid drainage had accumulated in a storage dam and progressively lost enough moisture on exposure to the air, air-dry samples of the solid residue (LR) were collected from the edges of a tailings dam. Characteristics of this dried material are listed in Table 3. It was not feasible to obtain samples from the centre of the dam because the material was still in its wet unconsolidated sludge form. The edge samples, however, were not representative of the average grain sizes of the whole dam because, during deposition, the silts and other fines settled in the centre of the dam with the coarser material surrounding them. Thus, samples were collected at random locations around the edges at three distances from the lake edge corresponding to a texture gradient. Three subsamples collected at each distance were composited to form a sample, and two composite samples were collected at each distance. Samples were securely fastened in plastic bags until required.

2.4 Mine Tailings

Characteristics of the mine tailings material are listed in Table 2. Areas of the iron pyrites mines tailings heap that had remained un-vegetated and were more than 30 years old were randomly sampled at two locations from an area of moderately sloping topography. Because of restricted soil depth (<10 cm), the acquisition of conventional soil cores was not feasible. Therefore, soil samples were taken from the 0–5 cm surface layer because of stoniness below this level in the mine-spoil. According to Fletcher (1978), it is the top 5 cm that contains most of the organic matter in a soil.

2.5 Sludge

An aerobically digested air-dried sludge (characterized in Table 2) was crushed and passed through a 2 mm sieve. High rates of sludge application (>25 %) are currently applied in the field. In this laboratory study high sludge rates were likewise applied to the soils: 25–50 % of the weight of the soil, to gauge the effects of the currently high field application rates on soil physical properties. The powdered sludge was thoroughly hand-mixed with air-dried soil. The mixture was watered initially with de-ionized water to 60 % water-holding capacity using a perforated U-shaped tube placed in each pot before the soil mixture was added. Each container was sealed, but re-opened every 3 days for 2 min to admit fresh air.

2.6 Green Manure

Green manure was prepared according to Harris and Ragusa (2000). The green manure rate used was 1.5 % w/w for this study. This is because organic matter in soils should be at least 10 % of clay content (w/w; MacRae and Mehuys 1985), and the two mine materials of the present study, contained 12–16 % of clay. It was suspected that a low pH, which could cause toxicity to microbial growth, existed in the mine tailings. Therefore, the fresh plant parts were partly fermented after grinding for 2 days prior to incorporation to increase the size of the viable microbial biomass inherent in the green manure.

2.7 Soil Physical Properties

Hydraulic conductivity (McCalla 1944), Water-stable aggregation (Kemper and Roseneau 1986), Spontaneous dispersion (Cerda 1998) were carried out after incubation at 80 % water-holding capacity (WHC) at ambient temperatures (20–30 °C) for a period of 60 days.

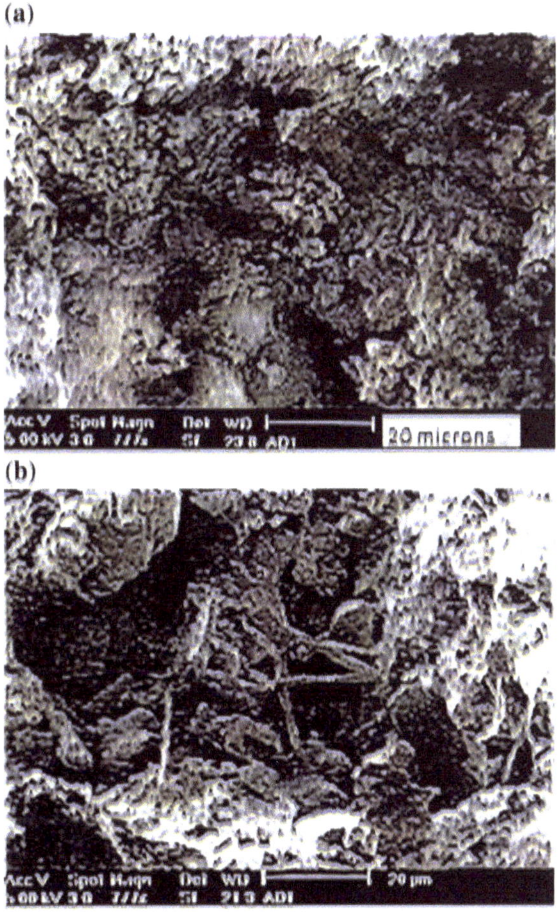

Fig. 7 Internal structure of aerobically treated sludge aggregate **a** before and **b** after the application of green manure

3 Results

3.1 Field Study: Aerobic Heterotrophic Bacteria

Decomposition of paper after burial for 30 days varied inversely with the age of the sludge-amended samples (Fig. 7). The papers acquired a darker grey with decreasing age of treatments. These changes are indicative of greater biological activity with the more recent the sludge treatment. Bacterial populations ranged from 8 to 35 × 10^6 g^{-1} (Table 4). Compared with the older sites, bacterial counts were substantially higher on site 1, reaching two to three times higher (Table 4; Fig. 8). The first-year peak and subsequent stabilization of bacterial populations is a typical response following organic matter additions to mine-spoils (Fresquez and Aldon 1986). When sludge is applied to land, there is an initial flush of microbial activity, followed by a decrease in microbial biomass as the more easily degraded

Table 4 Total counts of heterotrophic bacteria on a mine-spoil after sludge amendment (mean of three samples)

Site	Age (months)	Aerobic heterotrophic bacteria (g^{-1})	LSD 0.05 (g^{-1})
1	12	35×10^7	10×10^7
2	18	17×10^7	5×10^7
3	24	8×10^7	3×10^7

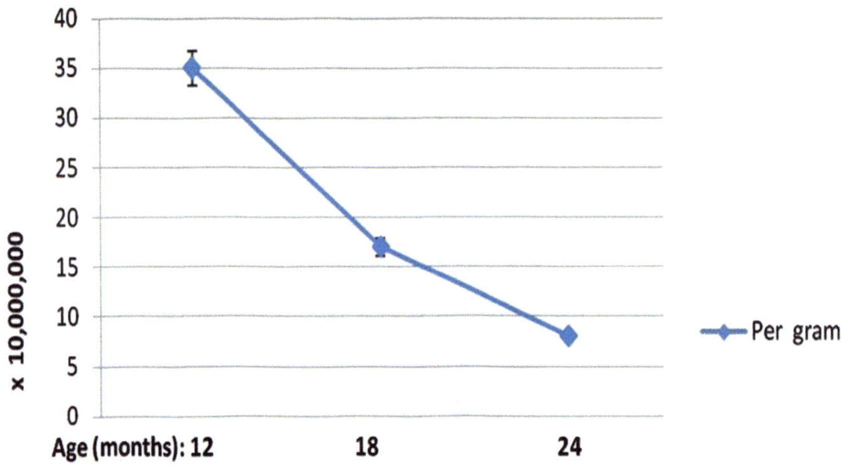

Fig. 8 Total counts of heterotrophic bacteria in a mine-spoil after amendments

materials and microbial metabolites are mineralized (Boyle 1990). Degradation of the sub-aerially exposed sewage sludge in the mine spoil would have been aerobic, and thus similar to composting. Applying infrared spectroscopy and nuclear magnetic resonance (NMR), Inbar et al. (1989) monitored the maturation of composted organic matter. They found, beginning with the raw manure, a decrease in the large carbohydrate fraction, while the proportion of the more resistant fractions such as the carboxyl, alkyl and aromatic C increased with time. This chronosequence gets progressively more unpalatable to microorganisms.

Considering the low initial pH of the mine tailings in this study ranging from 3.0 to 5.0 prior to lime additions, the microbial populations achieved 1 year after sludge application compare favourably with estimates of 1 to 34 $\times$ 10^6 g^{-1} reported for undisturbed soils (Wilson 1965; Alexander 1976; Miller and Cameron 1978; Visser 1985; Segal and Mancielli 1987).

Compared with the non-sludged sites, even the lowest estimates from the sludge-amended sites, which occurred on site 3, were approximately double that of the controls. This indicates that, without adequate organic matter input, microbial activity in unamended mine-spoils remains extremely low.

3.1.1 Sewage Sludge

For sewage sludge (SS), two rates were used: the equivalent of 50 and 100 t/ha. Differences between the effects of the two application rates were neither significant nor were there any significant increases in aggregation (>2 mm) for either the mine tailings (MT) or leachate residue after their sludge treatment (Table 5). It was seen previously (experiment 1) that even aged sludges contained much higher concentrations of soil microorganisms than highly acidic mine tailings. Yet this experiment shows that the sewage sludge, contrary to results of Epstein et al. (1976), made no difference in the aggregation of the mine materials (Table 5). Even though a better result was obtained when sludge was applied in combination with GM to the mine residues, it was still less than that which occurred when GM only was applied (Table 5). Therefore, as a treatment for the mine tailings or leachate residue, the green manure proved marginally better when added alone than when combined with the sludge (Table 5).

3.1.2 Hydraulic Conductivity and Dispersion

Whereas the saturated untreated mine tailings was totally non-transmissive to water under the constant hydraulic head of this experiment, the untreated saturated leachate residue conducted water through its fabric at 125 $cm^3\ h^{-1}$ (as was the case for the saturated untreated sludge). However, when incubated with green manure, the hydraulic conductivity of sewage sludge increased 11-fold (to 1400 $cm^3\ h^{-1}$), which approximated the rate of conductivity of the green manured leachate residue (1425 $cm^3\ h^{-1}$), and almost tripled that of the green manured mine tailings (500 $cm^3\ h^{-1}$; Table 6). Both green manures significantly increased hydraulic conductivity in either mine-spoil material, but more so for the leachate residue (Table 6). Sludge alone as a treatment did not increase hydraulic conductivity greatly for either mine material (Table 6; Fig. 9). Further, the green manure effect on leachate residue was at least a fourfold improvement over that of the sludge-treated leachate residue (Table 6). There was no apparent impact of

Table 5 Effect of green manure and material type on aggregation as a percentage of total weight after wet sieving (% retained on 2 mm sieve)

	Treatment #1: Leachate residue (LR)	Treatment #2: Mine tailings (MT)	Sludge (SS)	LSD 0.05
Control	0	0	0	NS
GM	18.5	11	50	8
SS (20 t ha^{-1})	0	0	–	NS
SS (40 t ha^{-1})	0	0	–	NS
SSGM	13.5	7.5	–	5

LR leachate residue; *MT* mine tailings; *SS* sewage sludge; *GM* green manure; *LRGM* green manured LR; *MTGM* green manured MT; *SSGM* green mnaured SS; *NS* not statistically different at the 0.05 level (within treatment)

Table 6 Effects of two manures on K_{sat} ($cm^3\ h^{-1}$) of two mine-spoil materials

Material	Sludge resisdue	Green manure	Sludge + green manure	Control	LSD 0.05 (within treatments)
MT	0^a	500^c	280^b	0^a	50
LR	200^b	1425^e	$770^{c,d}$	125^b	70
SS	150^b	1400^e	1400^e	150^b	80

MT mine tailings; SS sewage sludge; *LR* leachate residue, *MT* mine tailings. Values followed by different letters are significantly different at the ($P < 0.05$) level of significance as determined by Tukey's HSD

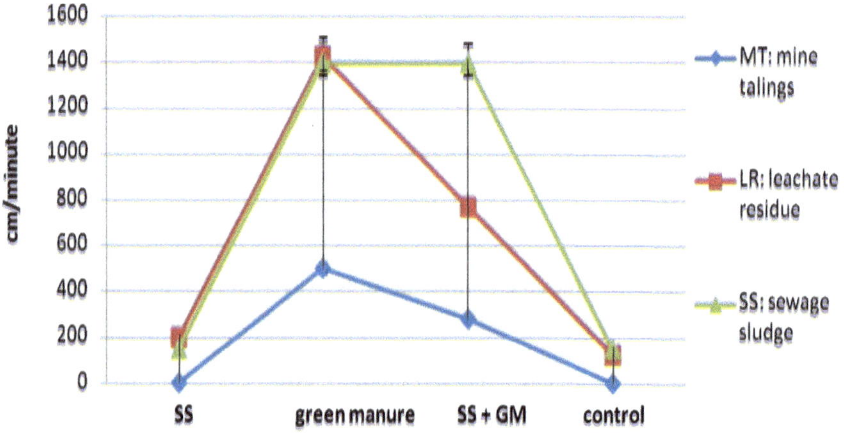

Fig. 9 Effects of two manures on hydraulic conductivity of two mine-spoil materials

sludge on the mine tailings because the originally complete pore blockage for the untreated mine tailings (0 cm h^{-1}) remained unchanged by the sludge treatment (Table 6). However, as was the case for aggregation, sludge, when combined with GM caused better results than those results caused by the sludge-only treatment (Table 6). Still, for the sludge-only treatment, it is seen that there was no resulting substantial improvement in the hydraulic conductivity of the leachate residue (or mine tailings) over and above that of the untreated (control) samples (Table 6).

Green manure significantly reduced spontaneously dispersed clay for all materials, but more so for the leachate residue as compared with the mine tailings (Table 7; Fig. 10). It caused a threefold reduction in dispersible clay for the sludge, but only a less-than twofold decrease was observed for mine materials. Therefore, it is not surprising that the greatest hydraulic conductivity (HC) increases were also those for the GM-treated sludge. The greatest impact of GM in this study was, therefore, on the sludge.

Table 7 Effect of green manure (20 t ha^{-1}) and material type on spontaneous dispersion of aggregates as a proportion of total material weight (clay dispersed: g kg^{-1})

Green manure (t ha^{-1})	Acid sulphate mine tailings (MT)	Leachate residue (LR)	Sludge (SS)	Control (reference soil)	LSD 0.05 (within treatments)
0	58.0^{e}	48.0^{d}	56.0^{e}	0^{a}	5.0
20	29.5^{c}	33.5^{c}	18.0^{b}	0^{a}	3.5

LR leachate residue; *MT* mine tailings; *SS* sewage sludge. *MT* mine tailings; *SS* sewage sludge; Values followed by different letters are significantly different at the ($P < 0.05$) level of significance as determined by Tukey's HSD

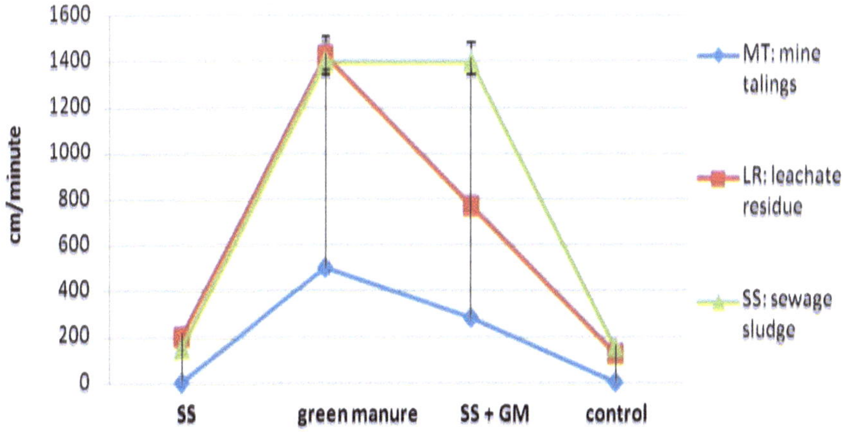

Fig. 10 Clay dispersion of treated aggregates as a proportion of total mass (g/kg)

4 Discussion

4.1 Mine Tailings and Leachate Residue

The greatest changes in the samples occurred not between treatments involving the two mine-spoil materials, but between the green manured sewage sludge and the green manured mine-spoil materials. Nonetheless, it was seen that the physical improvements measured (especially HC) of the leachate residue exceeded those of the mine tailings for all treatments applied.

4.2 Controls

All materials, including the sludge, contained high levels of spontaneously dispersible clay. The untreated mine tailings was significantly more dispersive than the untreated leachate residue. Because of prior neutralization treatment for the leachate residue, the proportion of free $CaCO_3$ and the more soluble $CaSO_4$ would have been substantially higher in the leachate residue compared with the mine tailings. In addition, there was already 25 % more clay in the mine tailings (Table 2). Several workers (Frenkel et al. 1978; Rengasamy 1983; Alperovitch et al. 1985) have found direct correlations between clay dispersion and the restriction of water movements in soils. Thus, McNeal et al. (1968) reported that the more clay mineral fraction present in the soil, the greater the dispersion, and hence the larger the reduction in hydraulic conductivity. Oster et al. (1980) found that although low levels of exchangeable sodium did not affect the structure of the clay, it did cause a large increase in clay mobility, and results of tests for spontaneous dispersiveness found the mine tailings to be more sodic than the leachate residue (Table 7). Thus, even though the sludge contained approximately three times the amount of clay as either the mine tailings or leachate residue, the amount of clay dispersed did not exceed that of the mine tailings, and was only marginally higher than that of the leachate residue. This means that, compared with the mine tailings or leachate residue, a much greater fraction of clay was not spontaneously dispersible in sludge. Native Ca^{2+} ions therefore suppressed the effects of sodicity for the leachate residue and sludge, but more so for the sludge. This is in agreement with Baldock et al. (1994), who found that soluble Ca^{2+} in the soil, which directly increased flocculation of clay particles, also reduced dispersion of the clay. At the <2.0 mm diameter for stabilized aggregates, the leachate residue contained more than double the stabilized aggregate mass of the mine tailings, at 40 and 15 % respectively (Fig. 2). Also, even though there was just a 4 % difference in the native clay contents of the mine tailings and leachate residue at 16 and 12 %, respectively (Table 3), there was a much greater difference in the hydraulic conductivity of both materials (Table 6). According to Frenkel et al. (1978), if the magnitude of swelling is not sufficient to seal pores, hydraulic conductivity is not affected and dispersion is minimal. However, the complete hydraulic blockage of the mine tailings controls (and substantial sealing for leachate residue controls) indicates that swelling and dispersion would potentially reduce hydraulic conductivity, but more so for the mine tailings control (16 % clay) than for the leachate residue control (12 % clay; Table 3). In the untreated or treated mine tailings and leachate residue, the following three conditions have, been seen, therefore, to have controlled HC:

(1) Where the stable <2 mm aggregate fraction was high, then pore blockage did not exist and HC was relatively high.
(2) Where both the stable <2 mm aggregate and stable >2 mm aggregate were low, HC was always very low.

(3) Where stable <2 mm and stable >2 mm aggregate fractions were high, HC was at least relatively high. Therefore, it is seems that although the occurrence of many stable aggregates of >2 mm always accompany a high HC (Tables 5 and 6), a total hydraulic blockage does not always accompany the complete breakdown of the 2 mm aggregate fraction. Thus, for this experiment, the sometimes poor correlation between low aggregate stability of the untreated leachate residue and its relatively high hydraulic conductivity is against the general impression (Epstein and Grant 1973) that aggregate breakdown causes surface sealing. Loch et al. (1978) concluded that it is the fine sized classes (<125 μm) that affect infiltration and surface-seal development, rather than the coarser sized classes, and that compaction of the soil layer, rather than increased aggregate breakdown, is a major cause of surface sealing by raindrop impacts. In this study, in spite of poor aggregate stability of >2 mm, the hydraulic conductivity of untreated leachate residue (and untreated sludge) was much higher than that of the untreated mine tailings (Table 6), which had much less Ca cations, and therefore a greater proportion of <125 μm particles.

4.3 Treated Samples

It is known that decomposition usually proceeds more readily in neutral than in acidic soil conditions (Strayer et al. 1981), and in the present experiment, the pH of the leachate residue was neutral, whereas that of the mine tailings was <3.5 pH units (Table 3). Stroo and Jencks (1985) found that the evolution of stable (and hence productive) mine-soils required active microbial populations and that decomposition of organic matter is highly restricted in low pH conditions because of inhibiting effects on microorganisms in the soil. Hayes (1980) found that the microbial populations involved in the transformation of organic matter in acidic soils are very different from those involved in soils that are neutral to alkaline. Nonetheless, it is clear that species of microorganisms that may have survived the highly acidic mine tailings environment would have been either insufficient in numbers or inefficient decomposers of the applied GM organic matter. On the premise of a lower biomass of active soil microorganisms, it would have been expected that decomposition products from green manure degradation would have been produced in the mine tailings to a lesser extent than occurred in the pH-neutral leachate residue. Warkentin (1982) also concluded that for macro-aggregation to occur there first must be micro-aggregates in the soil, because it is the linking of them by cements that creates macro-aggregates. Two factors could have helped to increase micro-aggregation (inherently, and during the incubation) in the leachate residue: (1) Ca^{2+} and (2) more bacteria than would have existed in the mine tailings because of its low pH.

Based on the above premise, the statistically significant existence of aggregation caused by the GM in the mine tailings material (Table 5) warrants an explanation. As seen in Table 7, there was some decomposition of the green manure during its incubation with the mine tailings. Ponnamperuma (1972) observed that all submerged soils will eventually acquire a neutral pH. In this study, the pH of the mine tailings in a 1:5 soil: water suspension was measured at pH 3.5 before incubation. Yet, by the end of incubation with GM, the pH had increased to 5.6, whereas the pH of the control soil (no green manure) had changed by a significantly smaller increment (Table 3). Therefore, decomposition of GM via microbial action must have occurred in the mine tailings of this study. This is in agreement with Yu (1985), who found that green manure increased the rate of pH increase in soils. Thus, he found that at just 3 days after incubation, the pH of a green manured submerged soil had moved from 5.3 to a value of 6.6 units. Thus, with an elevated pH for the GM-treated mine tailings, a less hostile environment for microbial life would have been created, thereby increasing decomposition of the organic matter.

4.4 Sludge and Green Manure

The green manured samples in this study were found to be less dispersive than the controls. The decomposing green manure led to the release of organic acids in the soil. Following the breakdown of rapidly decomposable organic materials in green manured soil, Sekhon and Bajwa (1993) detected increased concentrations of Ca^{2+} in the soil. In this study, the green manure decomposition would have similarly caused an increase in the concentration of Ca^{2+} in the soil. This would have increased the electrolyte concentration in the samples, thereby flocculating the clay particles and reducing dispersion. Prior to this study, the sludge was mixed with over 40 % inorganic constituents during its treatment before undergoing ageing (Table 2). Further, being an aerobically processed sewage sludge, its high proportion of older resistant organic residues (Boyle 1990) would have provided less energy for heterotrophs in the mine residues compared with the fresh GM. Thus, for the sludge treatments of the mine-soils, nutrient supplies were not as abundant as for the green manure-only treatments. Consequently, a lower proportion of microbial gums and hence a smaller aggregation effect on the soils would have been expected eventually from the sludge when applied without GM. Therefore, it is likely that, compared with the GM-only, the total 'decomposability' of the mine-soils treated with green manured sludge as a unit would have been decreased by its more resistant sludge fraction because the ratio of non-resistant to resistant manure would have then been substantially increased by its sewage sludge fraction. Therefore, it is not surprising that treatment effects on aggregate stability and HC occurred in the order of GM > SSGM > SS (Tables 5 and 6).

As mentioned earlier, when GM was added as a treatment to the sludge (i.e. as if the sludge were a soil to be treated, in the absence of mine tailings or leachate residue), the sludge proportion of stable macro-aggregates dramatically improved by >300 % (for >2 mm) compared with 0 % improvement for the sludge

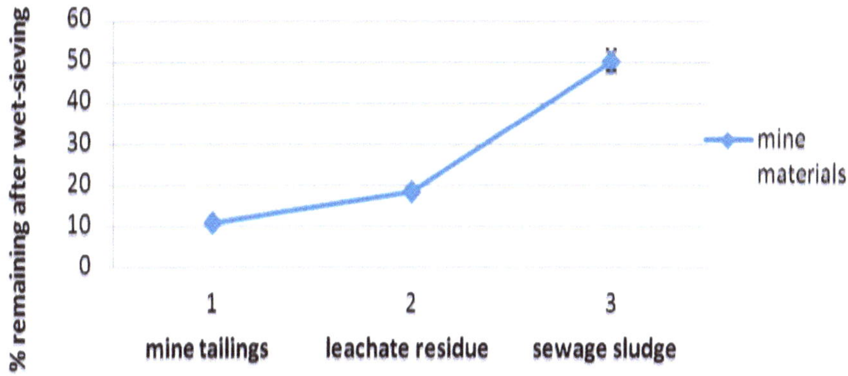

Fig. 11 Changes to aggregation of a sewage sludge and mine wastes after addition of green manure

controls (i.e. with no GM). There was at least three times the proportion of >2 mm aggregates of mine tailings or leachate residue stabilized by the GM treatment (i.e. 25 % increases; Table 5). This increase in physical properties of GM-treated sludge seems surprising because microbially toxic levels of the metals Cu, Cd, Cr, Pb and Zn are usually contained in sewage sludges, and such toxicity slows and stops microbial actions in soils (Fig. 11) (Kabata-Pendias and Pendias 1984). However, Seaker and Sopper (1988) point out that (among other criteria) effects of metals in sludge depend largely on sludge quality and the complexity of the sludge. Although the sludge of the present study was relatively low in organic matter, it was a complex sludge from a residential area, and contained a high colloid and lime fraction (Table 2). Hausenbuiller (1978) cites the metal-binding effects of crystalline lattices of clay particles (also of humic acids and proteins), and during its aerobic treatment prior to its use in the present study, 40 % clay (w/w) was added to the sewage sludge. Both of these materials would, therefore, have acted as sinks for the toxic metals inherent in the sludge, during its stabilization prior to use in this study. In addition, because a subsequent 0.1 M HCl extraction showed relatively low amounts of toxic heavy metals in the sludge (compared with that of other sludges used as amendments in other studies), it is likely that microbial activities would not have been greatly hindered in the sludge during incubation. Indeed, after 7 hours in water, the aggregates comprised of sludge and green manure oulasted all other combinations by at least 30 % (Table 8).

The 2.5 % soluble lime added to the sludge during its prior aerobic treatment would have provided extra bridging poly-cations for increased linkages between inorganic and potential organic colloids. Edwards and Bremner (1967) found that in soils containing soluble $CaSO_4$, $2H_2O$, or $CaCO_3$, and clays, opportunities exist for clay + polyvalent, metal + organic matter complexes, which may be represented as (C + P + OM)xy, where C = clay mineral particle, P = polyvalent metal, OM = organo-metallic complex: C + P + OM)x and C + P+ OM represent compound particles of clay size, and x and y are finite whole numbers with limits dictated by the size of the primary clay particles.

Table 8 Mean values for the index of the modified Emerson water dispersion test [MEWDT]. 0 No dispersion; 1 dispersion of some particles; 2 aggregate partly dispersed; 3 considerable dispersion; 4 total dispersion

Time					
Soil & treatment	0 min	2 min	15 min	360 min	720 min
Reference soil	0	0	0	0	0
LR	1	3	3	4	5
LRGM	1	3	3	4	5
LRGMS	0	1	2	2	3
LRSS	4	4	4	4	4
MT	4	4	4	4	5
MTGM	1	2	2	2	3
MTGMS	0	4	1	2	3
MTSS	4	4	4	4	5
SS	4	4	4	4	5
SSGM	0	0	1	2	2

LSD within treatments not significant at the $P = 0.05$ level

The mechanics of the above mentioned process as considered by Hayes and Mingelgrin (1991) show that when a single cation neutralizes two charges on adjacent molecular strands, the strands are pulled together, water is excluded, and a water insoluble gel or sol is formed. Large macromolecular flocculating structures are thus formed (Fig. 4). This possibly suggests that the high level of clay (40 % clay added), soluble lime (2.5 %) and the C from added readily decomposable decomposing plant material (green manure) in the sludge were major factors for aggregation, even though organic C measured before the start of the experiment was much lower than expected for a sludge (1.5 %).

4.5 Scanning Electron Microscopy of Water-Stable Aggregates

Scanning electron micrographs (SEM) of the sludge are presented in Fig. 7, where the increase in organically derived binding agents in the sludge is evident. These showed that green manure acted as a binding or stabilizing agent. It can be seen by eye estimation that individual soil particles are substantially larger after green manure treatment than before (Fig. 7). Therefore, green manure caused greater coalescence of soil particles. Individual strands of organic binding materials are seen under SEM as shown by its 'sticky string bag' effect on soil particles in Fig. 7b. In the context of water flow through soil, these agents would have stabilized the mesopore walls in the untreated sludge, thereby inhibiting the dislocation of clay particles, and thus increasing the hydraulic conductivity of the treated sludge.

4.6 Sludge Components

However, it should be emphasized that, despite the properties of the sludge that make it very responsive to GM treatment (thereby leading to its own aggregation), an ability of this sludge of itself to significantly increase mine tailings or leachate residue aggregation during its incubation with these was not evident. This was most likely because of its low C content caused by its long sub-aerial exposure leading to volatilization of its readily decomposable fraction prior to incubation.

As the added sludge brought some C to the mine materials during incubation, a change in their physical properties, however small or non-significant, was expected. This did not occur. Susanto (1992) found that mobile organic colloidal materials, ranging in size from a few hundred molecules to a diameter in the order of microns, move around in the soil solution. Goldberg et al. (1990) found that the positive edge charges of clay particles are offset by their specific adsorption of organic anions when those anions are not longer than the clay edges. The sludge used in this experiment had been sub-aerially stored for >2 years and, therefore, could have been classified as an aerobically aged sludge. After such a lengthy period of exposure to the weather, the anions would by then have reached a state of advanced de-polymerization. The fact that, in this study, a much lower microbial count was found in the sludge sub-aerially exposed for >2 years compared with those stabilized for shorter periods (experiment 1), suggests that there were minimal amounts of large decomposable organic units in the 2 year-old aerobically stabilized sludge of this study (experiment 2). Placed against clay particles, such small anions could have acted not only as a dis-aggregator for the sludge itself, but also as a deflocculant for the mine-spoils treated with the sludge. This point seems more likely because the values for spontaneous clay dispersion within the untreated sludge were just as high as those of the untreated mine tailings (Table 7). On the other hand, the dramatic increase in water stability and hydraulic conductivity of the sludge by GM (Tables 5 and 6) was caused by GM-derived, recently created organic acids and, hence, larger organic molecules than were contained in the sludge.

Because the sludge contained at least five times more native C than either the LS or mine tailings, priming of the old C in the sludge by the GM probably occurred, leading to a flush of microbial activity. As a result of there then being newly formed, larger organic anions than clay particles, flocculation of the clay particles in the sludge occurred, thereby reducing its dispersiveness and increasing its hydraulic conductivity constant (K_{sat}). This view seemed to have been confirmed by particle size analysis. Therefore, it seems that the main role of sludge was that of a supplier of degrading microorganisms, which accelerated decomposition of the green manure in the soil and the physicochemical surfaces conducive to aggregation.

5 Conclusion

The hypotheses were as follows: (1) green manure will improve the granulation of a mine tailings, (2) green manure will more effectively granulate mine tailings than aged sludge, and (3) green manure in combination with aged sludge will be at least as effective as the same quantity of green manure in the granulation of acid-sulphate mine tailings. In the foregoing short-term study, it was seen that parts (1) and (2) of the hypothesis were supported by the results. However, this was not the case for part (3) because the effect of sewage sludge applied with green manure on water stability of tailings aggregates was substantially smaller than that of green manure alone.

The implications of these results for some current conventions for sludge application to mine-spoils seem important because its incorporation may decrease rather than increase water infiltration and movement within the soil. Therefore, it is concluded that at least for such short-term studies, low C aerobically-aged sewage sludges added with green manures weaken the inherent ability of the green manures to improve the physical condition of mine-spoils.

Because of downward movements in soil materials of small colloids in solution, the application of such aged sludges as top-dressings may eventually cause similar problems. As volatilization and very long ageing periods lead to increased diminution of organic particles in sludge, which increases dispersion, the sooner the application of sewage sludge after completion of its stabilization the more effective a soil physical ameliorant it may be. Thus, it should be spread as quickly as possible after its stabilization has been completed.

6 Geobiotechnological Applications (After: Queensland Govt. 2015)

6.1 As shown above, sewage sludge inherently is weakly structured, as is the AMD leachate residue (Fig. 12), even when both are combined (Fig. 13). Therefore sewage sludge only marginally increased the physical properties of the equally weakly structured acid sulphate leachate residues (Fig. 12). But as green manure greatly improved the physical structure of sewage sludge, such improved sludges may increase the structure of leachate residues. More research is required to determine if such effects of green-manure-treated sewage sludge may also improve the aggregation of ASS soils. Activities that have the potential to disturb acid sulphate soils (ASS), either directly, or by affecting the elevation of the water table, need to be managed appropriately to avoid environmental harm (Queensland Govt. 2015). An acid sulphate soil management plan (ASSMP) should be prepared and implemented, following advice presented in this document, to effectively manage potential impacts of such activities. (Queensland Govt.

Fig. 12 Leachate residue ($CASO_4$ + contaminants) at Brukunga Mines, South Australia. Note structure-less consistency

Fig. 13 Leachate residue ($CASO_4$ + contaminants) intermixed with sewage sludge incubated with phyto-organics. Note improved aggregation

2015).

6.2 Under anaerobic conditions maintained by permanent groundwater, iron sulfides are stable and the pH is often weakly acid to alkaline. ASS only becomes a problem when they are disturbed and exposed to air. Typically, excavating or otherwise removing soils or sediments, extracting groundwater or filling land, causes disturbance of ASS. These activities can occur during construction of, for example, canal estates, golf courses, swimming pools, agricultural drainage systems, roads and other infrastructure. The extraction of sand and gravel can also result in the disturbance of ASS.

6.3 According to (Queensland Govt. 2015), in areas where *Special Acid Sulphate Soil Maps* are not available, the identification of areas with a high probability of containing ASS can be determined using the following geomorphic and site description criteria:

- land with elevation less than 2 m AHD (Australian height datum);
- soil and sediment of recent geological age (Holocene);
- marine or estuarine sediments and tidal lakes;
- low-lying coastal wetlands or back swamp areas, waterlogged or scalded areas, stranded beach ridges and adjacent swales, inter-dune swales or coastal sand dunes;
- coastal alluvial valleys5;
- areas where the dominant vegetation6 is tolerant of salt, acid and/or waterlogging conditions e.g. mangroves, salt couch, swamp-tolerant reeds, rushes, grasses (e.g. *Phragmites australis*), paperbarks (*Melaleuca* spp.) and swamp oak (*Casuarina* spp.); and
- areas identified in geological descriptions or in maps as:
- bearing sulfide minerals;
- coal deposits or marine shales/sediments (geological maps and accompanying descriptions may need to be checked); and
- deep older estuarine sediments below ground surface of either Holocene or pre-Holocene age.

6.4 Most ASS sediments are associated with coastal landforms that have surface elevations below 5 m AHD. Older estuarine sediments (Holocene or pre-Holocene age) have been found deep below ground surface with surface elevations up to 20 m AHD. These occurrences, such as those found in upper alluvial valleys, are mainly an issue if deep excavation or drainage is required. Areas with large tidal variations or tropical areas subject to cyclonic storm surges may also have ASS occurring above 5 m AHD.

6.5 *Soil sampling.* After assessing the available broad scale information and calculating the dimensions of the proposed disturbance, a site visit with soil sampling is required. The desktop assessment and site investigation for geomorphic and other indicators can be used to help define the location of boreholes for soil sampling. As many of the indicators for actual and potential ASS are quite different, the site inspection should investigate for the presence of both types of soil. Also note that it is common to have an actual ASS that also contains some un-oxidized iron sulfides or potential acidity.

6.6 Care should be taken especially on sites with more than one type of geomorphological unit, or clearly different land surface elevations, so that sampling is representative of the area.

6.7 The number of boreholes required is dependant on the volume of ASS disturbance or for disturbances greater than 1000 m^3, the area (m^2 or ha) to be disturbed. Table 2 summarizes the minimum number of boreholes to be drilled, described, field tested and sampled for non-linear disturbances.

6.8 Soil sampling involves going to the site, drilling or augering investigative boreholes to at least 2 m depth, or at least 1 m below the maximum depth of disturbance (whichever is the greater 14), describing and undertaking field soil tests 15 on the soil profiles retrieved and finally collecting and storing samples for laboratory analysis. The information gathered from this step will be required to assist in selecting appropriate samples for laboratory analysis and enable both the applicant and the reviewer of the application to interpret the results.

6.9 The information that will need to be collected and documented as part of the soil sampling procedure (DEC 2011) includes:

- the full grid reference of each borehole;
- the exact location of each borehole shown on an appropriately scaled map;
- an exact description of the vertical dimensions of the borehole relative to surface AHD;
- a brief description of the equipment and/or methods used to retrieve the samples;
- a field description for each soil profile including soil texture16, colour, mottling and other diagnostic features (e.g. jarosite, shell);
- results from field soil tests (field pH (pHF), pH after oxidation with hydrogen peroxide (pHFOX) and reaction with peroxide) at 0.25 m vertical intervals to the base of the soil profile; and
- collect samples at a maximum of 0.5 m intervals down the soil profile e.g. 0–0.5, 0.51–1.0, 1.01–1.5, 1.51–2.0 m etc., ensuring each horizon is sampled.

6.10 Once boreholes have been dug, the profile described and soil field tests conducted, soil samples must be collected from each profile at a vertical maximum of 0.5 m intervals. In deciding the appropriate sampling intervals, the operator should refer to the field description notes and identify any significant changes with depth down the profile in field description properties (such as a change in pH, colour, texture etc.). Samples with clearly different physical, visual or chemical properties should not be 'bulked' together, as this will reduce the precision of future laboratory results.

6.11 When collecting samples in the field it is important to prevent oxidation of the soil as much as possible. This can be achieved by immediately placing the sample in plastic bags or other suitable containers, excluding air, then placing in a field freezer or with ice in a portable ice-box. The samples should be carefully marked (using a waterproof pen) with bore-hole number and depth for easy identification, and be frozen or specially dried within 24 h of collection.

6.12 It is important to note that there are Occupational Health and Safety issues related to soil collection and field-testing, particularly in regard to handling hydrogen peroxide, digging soil inspection pits and dangers associated with hydrogen sulfide gas poisoning.

Acknowledgements The authors are grateful to Associate Professor Victor Gostin (Department of Geology and Geophysics), Professor Martin Williams (Foundation Professor in the Department of Geographical and Environmental Studies), Dr P. Rengasamy (Department of Soil Science), of the University of Adelaide, and to Dr N. Sethunathan, Visiting Scientist at the Commonwealth Scientific and Industrial Research Organisation CSIRO) for their constructive comments. The technical support of Adrian Beach, analytical chemist CSIRO) is appreciated. Many thanks to Dr Jenny Bourne of the Department of Geology and Geophysics for kind assistance with the references.

References

Abat T (2014) Mineralogy of Doña Amanda Prospect, Bayaguana, Dom. Rep.: a transitional epithermal-porphyry deposit in Los Ranchos Formation? macla nº 19. julio '14 revista de la sociedad española de mineralogía

Alexander M (1976) Symbiotic nitrogen fixation. In: Introduction to soil microbiology, Wiley, New York, pp 305–330

Alloway BJ, Morgan H (1986) The behavior and availability of Cd, Ni and Pb in polluted soils. In: Assink JW, vanden Brink WJ (eds) Contaminated soil, Martinus Nijhoff, Dordrecht, pp 101–113

Alperovitch N, Shainberg I, Keren R (1985) Effect of clay mineralogy and aluminium and iron oxides on the hydraulic conductivity of clay-sand mixtures. Clays Clay Min 33:443–450

Anderson WH, Cobb JC (2008) Foundation Problems and Pyrite Oxidation in the Chattanooga Shale, Estill County, Kentucky. Report of Investigations 18, Series XII, 2008. www.uky.edu/kgs

Aoyama M, Angers DA, N'Dayegamiye A, Bissonnette N (2000) Metabolism of 13 C-labeled glucose in aggregates from soils with manure application. Soil Biol Biochem 32:295–300

Baldock JA, Aoyama M, Oades JM, Grant CD (1994) Structural amelioration of a South Australian red-brown earth, using calcium and organic amendments. Aust J Soil Res 32:571–594

Bell LC, Jones CJ (1987) Rehabilitation of base metal tailings-Australian case studies. In: Farrell TP (ed) Mining rehabilitation'87, Australian Mining Industry Council, Canberra, pp 88–102

Bell LC, Ward SC, Kabay ED, Jones CJ (1989) Mine tailings reclamation in Australia—an overview. In: Walker DG, Power CB, Pole MW (eds) Proceedings of the conference: Reclamation, a global perspective, Calgary, Canada, 27–31 August 1989, Report no RRTAC 89-2, Alberta Land Conservation and Reclamation Council, Edmonton 2, pp 769–778

Bitton G, Boylan R (1985) Effect of acid precipitation on soil microbial activity. J Environ Qual 14(1):66–68

Boyle M (1990) Biodegradation of land applied sludge. J Environ Qual 19:640

Cerda A (1998) Soil aggregate stability under different Mediterranean vegetation types. Catena 32(2):73–86

Chaney K, Swift IRS (1984) The influence of organic matter on aggregate stability in some British soils. J Soil Sci 35:223–230

Chivenge P, Vanlauwe B, Gentile R, Six J (2011) Comparison of organic versus mineral resource effects on short-term aggregate carbon and nitrogen dynamics in a sandy soil versus a fine textured soil. Agric Ecosyst Environ 140:361–371

Daniels WL, Haering KC (1994) Use of sewage sludge for land reclamation in the central Appalachians. In: Clapp CE, Larson WE, Dowdy RH (eds) Sewage sludge: land utilization and the environment. Miscellaneous Publications, ASA, CSA, SSSA, Madison, WI, pp 105–121

Daniels RB, Buol SW, Kleiss HJ, Ditzler CA (1999) Soil systems in North Carolina. North Carolina State University Soil Science Department, NC

DEC (2011) Acid sulfate soils guideline seriestreatment and management of soils and water in acid sulfate soil landscapes. Department of Environment and Conservation. http://www.der.wa.gov.au/images/documents/your-environment/acid-sulfate-soils/guidelines/ass-management-guideline.pdf

Edwards AP, Bremner JM (1967) Micro-aggregates in soils. J Soil Sci 18:64–73

Emerson WW (1983) Interparticle bonding. In: Soils: an Australian viewpoint, CSIRO, Melbourne, Australia/Academic Press, London, pp 477–497

Epstein E (1975) Effects of sewage sludge on some soil physical properties. J Environ Qual 5:139–142

Epstein E, Grant WJ (1973) Soil crust formation as affected by raindrop impact. In: Ecological studies IV. Physical aspects of soil water and salts in ecosystems, Springer, Berlin Heidelberg, pp 195–201

Epstein E, Taylor JM, Chaney RL (1976) Effects of sewage sludge and sludge compost applied to soil on some soil physical and chemical properties. J Environ Qual 5:422–426

Fletcher GK (1978) Mineral sands—mining effects and rehabilitation. In: Flummery RA, Howes KMW (eds) Management of lands affected by mining, Division of Land Resources Management, Commonwealth Scientific and Industrial Research Organization, Australia, pp 129–139

Forster SM (1979) Microbial aggregation of sand in an embryo dune system. Soil Biol Biochem 11:537–543

Frenkel J, Goertzen JO, Rhoades JD (1978) Effects on clay type and content exchangeable sodium percentage and electrolyte concentration on clay dispersion and soil hydraulic conductivity. Soil Sci Soc Am J 42:32–39

Fresquez PR, Aldon EF (1986) Microbial re-establishment and the diversity of fungal genera in reclaimed coal mine spoils and soils. Reclam Reveget Res 4:245–258

Golchin A, Oades JM, Skjemstad JO, Clarke P (1994) Soil structure and carbon cycling. Aust J Soil Res 32:1043–1068

Goldberg S, Kapoor BS, Rhoades JD (1990) Effect of aluminium and iron oxides and organic matter on flocculation and dispersion of and zone soils. Soil Sci 150(3):588–593

Gulde S, Chung H, Amelung W, Chang C, Six J (2008) Soil carbon saturation controls labile and stable carbon pool dynamics. Soil Sci Soc Am J 72:605–612

Harris MA, Ragusa S (2000) Bacterial mitigation of pollutants in acid drainage using decomposable plant material and sludge. Environ Geol 40 (1–2):195

Hausenbuiller RL (1978) Soil science: principles and practices, 2nd edn. William C, Brown, Dubuque, Iowa

Hayes MHB (1980) The role of natural and synthetic polymers in stabilizing soil aggregates. In: Berkeley RCW (ed) Lynch

Hayes MHB, Mingelgrin U (1991) Interactions between small organic and soil colloidal constituents. In: Bolt H, DeBoot MF, Hayes MHB, McBride MB (eds) Interactions at the soil colloid—soil solution interface, Kluwer, Dordrecht, pp 323–407

Hayes MHB, Swift RS (1978) Chemistry of soil organic colloids. In: Greenland DJ, Hayes MHB (eds) The chemistry of soil constituents, Wiley, Chichester, pp 179–320

Hill RD, Montague PEA (1977) The potential for using sewage sludge and compost in mine reclamation. In: 3rd national conference on sludge management disposal and utilization, US Energy Research and Development Administration, Miami, Washington, DC, pp 14–16

Inbar Y, Chen Y, Hadar Y (1989) Solid-state carbon-13 nuclear magnetic resonance and infra-red spectroscopy of composted organic matter. Soil Sci Soc Am J 53:1695–1701
Janzen HH (2006) The soil carbon dilemma: shall we hoard it or use it? Soil Biol Biochem 38:419–424
Kabata-Pendias A, Pendias H (1984) Trace elements in soil and plants. CRC Press, Boca Raton
Kemper WD, Roseneau RC (1986) Aggregate stability and size distribution. In: Klute A (ed) Methods of soil analysis, 2nd edn. ASA, SSA, Madison, WI, pp 425–442
Khan SV (1970) Enzymatic activity in grey wooded soils as influenced by cropping systems and fertilizers. Soil Biol Biochem 2:137–139
Kladivko EJ, Nelson DW (1979) Changes in soil properties from application of anaerobic sludge. J Water Pollut Control Fed 51:325–332
Loch RJ, Cleary JL, Thomas EC, Glanville SF (1978) An evaluation of the use of size distributions of sediment in runoff as measure of aggregate breakdown in the surface of a cracking clay soil under rain. Earth Surf Process Land 13:37–44
Lunt HA (1959) Digested sewage sludge for soil improvement. Connecticut Agric Exp Stn Bull 622:30–31
Maartens DA, Frankenberger WT (1992) Decomposition of bacterial polymers in soil and their influence on soil structure. Biol Fertil Soils 13:65–73
MacRae RJ, Mehuys GR (1985) The effect of green manuring on the physical properties of temperate-area soils. Adv Soil Sci 3:1–94
Martin JP, Martin WP, Page JB, Raney WA, DeMent JD (1955) Soil aggregation. Adv Agron 7:1–37
McCalla TM (1944) Water-drop method for determining stability of soil structure. Soil Sci 58:117–121
McNeal BL, Layfield DA, Norvell WA, Rhoades JD (1968) Factors influencing hydraulic conductivity of soils in the presence of mixed-salt solutions. Soil Sci Soc Am Proc 32:187–190
Megharaj M, Singleton I, McClure NC (1998) Effect of pentachlorophenol pollution towards microalgae and microbial activities in soil from a timber processing facility. Bull Environ Contam Toxicol 61:108–115
Mehta NC, Strell H, Muller M, Deul H (1960) Role of polysaccharides in soil aggregation. J Sci Food Agric 11:40–47
Miller RM, Cameron RE (1978) Microbial ecology studies at two coal mine refuse sites in Illinois. Argonne National Laboratory Report ANL/LRP-3, National Technical Information Service US Department of Commerce, Springfield, VA
Min DH, Islam KR, Vough LR, Weil RR (2003) Dairy manure effects on soil quality properties and carbon sequestration in alfalfa-orchardgrass systems. Commun Soil Sci Plant Anal 34:781–799
Moe PG (1967) Nitrogen losses from urea as affected by altering soil urease activity. Soil Sci Soc Am Proc 31:380–382
NRCS (National Water Quality Handbook). National Resource Consevation (2003) Accessed September 4, 2015. www.nrcs.usda.gov/Internet/FSE_DOCUMENTS/stelprdb1044775.pdf
Oades JIM (1993) The role of biology in the formation, stabilization and degradation of soil structure. Geoderma 56:377–400
Oster JD, Shainberg I, Wood JID (1980) Flocculation value and gel structure of sodium/calcium montmorillonite and illite suspensions. Soil Sci Soc Am J 44:955–959
Pagliai M, Guidi G, La Marca M, Giachetti M, Lucamente G (1981) Effects of sewage sludges and composts on soil porosity and aggregation. J Environ Qual 10(4):556–561
Paulson KN, Kurtz LT (1969) Locus of urease activity in soils. Soil Sci Soc Am Proc 33:897–901
Pinck LA, Allison FE (1961) Adsorption and release of urease by and from clay minerals. Soil Sci 91:183–188
Ponnamperuma FM (1972) The chemistry of submerged soils. Adv Agron 24:29–96
Powell JL, Barnhisel RI, Craig FA, Armstrong JR, Ellis ML, Thom WO (1986) The use of organic amendments in the restoration of prime farmland. In: Harper J, Plass WT, (eds), New Horizons in Mined Land Reclamation, Proceedings of National Meeting of American Society of Surface Mining and Reclamation, pp 55–60

QGRM (2002a) Queensland Government Department of Natural Resources and Mines State Planning Policy 2/02 Guideline Acid Sulfate Soils Version 2. http://www.dip.qld.gov.au/resources/policy/spp-guidelines-oct-02-v2.pdf

QGRM (2002b) Treatment and management of soils and water in acid sulfate soil landscapes. Contaminated Sites Branch Environmental Regulation Division, Department of Environment and Conservation Acid sulfate Soils Guideline Series, Acid Sulfate Soil Planning Scheme Policy. http://www.brisbane.qld.gov.au/sites/default/files/Appendix2_AcidSulfate_PSP.pdf

Queensland Government (2015) http://www.qld.gov.au/environment/land/soil/soil-testing/soil-terms/

Rengasamy P (1983) Clay dispersion in relation to changes in the electrolyte composition of dialysed red-brown earths. J Soil Sci 34:723–732

Roberts JA, Daniels WL, Bell JC, Burger JA (1988) Early stages of mine soil genesis in a Southwest Virginia spoil lithosequence. Soil Sci Soc Am J 52:716–723

Robertson WK, Lutrick MC, Yuan TL (1982) Heavy applications of liquid-digested sludge on three ultisols: effects on soil chemistry. J Environ Qual 11(2):278

Seaker EM, Sopper WE (1984) Reclamation of bituminous strip mine spoil banks with municipal sewage sludge. Reclam Reveget Res 3:87–100

Seaker EM, Sopper WE (1988) Municipal sludge for minespoil rehabilitation: effects on microbial populations and activity. J Environ Qual 17(4):591–597

Segal W, Mancielli RL (1987) Extent of regeneration of the microbial community in reclaimed spent oil-shale land. J Environ Qual 16:44–48

Sekhon BS, Bajwa BS (1993) Effect of organic matter and gypsum in controlling soil sodicity in rice and wheat-maize system irrigated with sodic waters. Agric Water Manage 24:15–25

Sheoran V, Sheoran AS, Poonia P (2010) Soil Reclamation of Abandoned Mine Land by Revegetation. Int J Soi Sed Water 3(2) ISSN: 1940-3259

Smith HA, PR Stephenson (2011) Pueblo Viejo gold project Dominican Republic Technical report for Pueblo viejo dominicana corporation, Barrick gold corporation, Goldcorp inc. AMC Mining Consultants (Canada) Ltd

Strayer RF, Lin GJ, Alexander M (1981) Effect of simulated acid rain on nitrification and nitrogen mineralization in forest soils. J Environ Qual 10:547–551

Stroo HF, Jencks EM (1985) Enzyme activity and respiration in mine-spoils. Soil Sci Soc Am J 46:548–553

Susanto M (1992) Effect of gypsum, agricultural lime, and organic matter on the infiltration characteristics of a redbrown earth. Master of Agriculture Thesis, University of Adelaide, Australia

Tisdall JM, Oades JM (1982) Organic matter and water-stable aggregates in soils. J Soil Sci 33:141–163

Ussiri DAN, Lal R (2005) Carbon sequestration in reclaimed mine-soils. ISSN: 0735–2689 print/1549-7836, online DOI: 10.1080/07352680591002147. Crit Rev Plant Sci 24:151–165

Van Breemen N, Mulder J, Driscoll CT (1983) Acidification and alkalization of soils. Plant & Soil 75(3):283–308

Visser S (1985) Management of microbial processes in surface mined land reclamation in Western Canada. In: Tate RL, Klein DA (eds) Soil reclamation processes, Marcel Dekker, New York, pp 203–341

Walker JM (1994) Production use and creative design of sewage sludgebio-solids. In: Clapp CE, Larson WE, Dowdy RH (eds) Sewage sludge: land utilization and the environment, SSSA, Madison, WI, pp 67–74

Warkentin BP (1982) Clay soil structure related to soil management. Trop Agric Trinidad 59:167–177

Whalen JK, Hu Q, Lui A (2003) Compost applications increase water-stable aggregates in conventional and no-tillage systems. Soil Sci Soc Am J 67:1842–1847

Whyte RO, Sisam JWB (1949) The establishment of vegetation on industrial wasteland. Commonwealth Forestry Bureau, Joint Publication 14, Oxford, pp 1–2

Wilson HA (1965) The microbiology of strip mine spoil. West Virginia Univ Agric Exp Station Bull 506

Yu TR (1985) Physical chemistry of paddy soils. Springer, Berlin Heidelberg

Zantua MT, Bremner JM (1976) Production and persistence of urease activity in soils. Soil Biochem 8:369–374

Zhai Q, Barnhisel RI (1996) A reclamation approach for mined prime farmland by adding organic wastes and lime to the subsoil. In: Daniels WL (ed) Successes and Failures: Applying Research Results to Insure Reclamation Success. Proceedings of American Society of Surface Mining and Reclamation, pp 740–748, Princeton

Quenching of Phosphorus-Fixation in a Disturbed Caribbean Bauxite Mine Overburden Using Root Exudates: Implications for Acidic Tropical Soils

Mark Anglin Harris

Abstract Extreme phosphorus (P)-deficiency due to strong P sorption caused mainly by high levels of aluminium (Al) iron (Fe) oxides prevail in terra rossa soils of the humid tropics. Producing fixed P in the soil up to the fixing threshold (or saturation) is referred to as "quenching." The objective of this experiment was to determine the excess of phosphorus required to produce labile, plant-available P, and the conditions in stored bauxite overburdens which determine those thresholds. The capability of *Cajanus cajan* (pigeon pea) organic root exudates to quench phosphorus (P) fixation in a stored high phosphorus-fixing tropical bauxite overburden was examined in a field experiment. Growth and decomposition of roots in situ facilitated increased solubility and hence phosphorus fixation quenching in the soil. Normal plant-available P concentrations were exhibited by the overburden in the treated beds. Twenty-year-old, stored overburden treated with live and decaying pigeon pea roots exhibited P levels of 1.0 and 1.1 %, respectively. This represents a small excess over normal concentrations of free P in the overburdens. By releasing hitherto fixed P in the overburden, the capacity of the exudates to quench phosphorus was thus rapidly accomplished with an excess of P released in the overburden. Therefore, maintaining *C. cajanus* in rotations promise a long-term corrective for P-fixing in stored bauxite overburdens.

Keywords Acid soils • Ferralitic soils • Piscidic acid • Phosphate activation • Phosphorus deficiency

Highlights

- Root exudates quench phosphorus fixation in overburden.
- Maintaining *C. cajanus* in rotations promise a long-term corrective for P-fixing in stored bauxite overburdens.

M.A. Harris, *Geobiotechnological Solutions to Anthropogenic Disturbances*, Environmental Earth Sciences, DOI 10.1007/978-3-319-30465-6_14

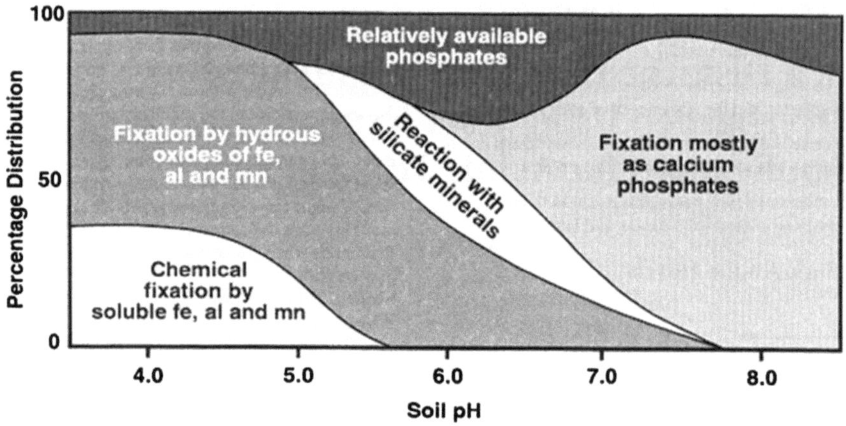

Fig. 1 General relationship between soil pH and phosphorus availability, and, by extension, P-diffusivity in soils. This is based on the kinds of P-compounds associated with pH. From http://passel.unl.edu. Adapted from the *"Soils Home Study Course," published in 1999 by the University of Nebraska Cooperative Extension*

1 Introduction

1.1 P-Deficiencies of Tropical Soils

As an essential element for plant and animal nutrition (FAO 2004), Phosphorus (P) is the main limiting macro-nutrient for crops and forage plants in acid tropical soils (Lal 1990; Balemi and Negisho 2012), and is the second-most limiting nutrient after nitrogen (N) for plant growth and crop production in many tropical locations (Kamprath 1984; Baligar and Fageria 1997; Bekunda et al. 1997). Hoagland and Aron (1941) showed that stunted plant growth was caused mainly by insufficient P, and plant growth is greatly restricted in P deficient soils. Decrease of P with distance along a transect is due to the weak diffusivity (Turner and Gilliam 1976; Singh et al. 2003) of soluble phosphorus (Fig. 1), and P has become the main factor limiting the rise of crop yield in these regions (Yuan et al. 2013).

P-fixation includes both P-sorption and P precipitation, both of which reduce phosphorus availability. Phosphorus deficiency is exacerbated in hot, humid environments (Ghose and Kumar 1997), especially in the highly weathered and leached ferralitic soils of the humid tropics, where primary minerals are generally absent except for quartz (Bekunda et al. 1997). The dominant secondary minerals here are a combination of kaolinite, gibbsite, goethite and hematite (Chesworth 2008), and the soils are enriched in iron and aluminum relative to silica and all major cations. Highly weathered and inherently infertile soils comprise extensive tracts of land in the tropical and subtropical regions of Africa, Asia and Latin America. Acid soils representing about 40 % of the total arable land area in the world, dominate the tropical and subtropical regions (Haug 1984). With 68 % of the soils having extreme P-deficiency, Tropical America's P-deficiency is far more serious than those of Tropical Asia (38 %) and Tropical Africa (27 %)

(Pandey and Gardener 1992). Their extreme P-deficiency is due largely to strong P sorption caused by high levels of aluminium (Al) and iron (Fe) oxides (Pandey and Gardener 1992). P-sorption occurs when the orthophosphates, H_2PO^{4-} and $HPO_4{}^2$, bind tightly to soil particles.

As soils weather, the Ca content tends to decrease and the proportion of the more reactive Al and Fe increase. These form complex phosphate compounds in the soil solution, which precipitate or are adsorbed in clay lattices or on amorphous oxides. Fox and Kamprath (1970) determined that amorphous oxides are the most efficient abiotic P-fixers in the soil. Under intense tropical weathering, but especially in highly weathered soils of warm humid regions, mineralogical structures consist of mainly either Fe or Al atoms coordinated with O atoms-the O atoms often associated with H ions to make hydroxyl groups. Some, such as gibbsite (Al-oxide) and goethite (Fe-oxide), form crystalline sheets while others form amorphous coatings on soil particles. Fox and Kamprath (1970) found that the intensity of P fixation by different minerals in soils is as follows: Amorphous oxides > crystalline oxides > 1:1 clays > 2:1 clays. As plants must absorb P in solution, these actions decrease plant-available P.

As its reactivity with air and other oxygen–containing substances is high, phosphorus is not found free in nature but is widely distributed in many different minerals in the soil. Since phosphate is an anion, particles that generate an anion exchange capacity will form strong bonds with phosphate. As stated above, these include aluminium and iron oxides, highly weathered kaolin clays (under acidic conditions), and amorphous materials. Thus phosphorus immobilization can be enhanced by the addition of amendments containing soluble forms of iron, aluminum or calcium such as aluminum sulphate (alum), iron sulphate, and the oxide and hydroxide of calcium (Sharpley 2000). Phosphate precipitation is a process in which phosphorus reacts with another substance to form a solid mineral. Hence, the physical positioning (and the oxidation potential and water status) of soil horizons can affect phosphorus fixation.

Phosphorus retention capacity in red ferralitic soils is therefore high, leading to high levels of plant-unavailable P (Pérez et al. 2014). Further, total P is not a true index of the soil P status because not all P in the soil becomes available to plants over the same time scale (Richter et al. 2006). Therefore, land degradation practices which increase the proportion of sesquioxides in the soil including deforestation, overgrazing, inappropriate farming practices and open-cast mining (O'Callaghan et al. 1998; Harris and Omoregie 2008), exacerbate P-fixation. Moreover, as its reactivity with air and other oxygen–containing substances is high, phosphorus is not found free in nature but is widely distributed in many different minerals in the soil (Robertson 2011). Phosphorus retention capacity in red ferralitic soils is therefore high. Thus Pérez et al. (2014) reported a high level of retention of (plant-unavailable) phosphorus in a ferralitic red soil (Pérez et al. 2014). Therefore, in addition to the problem of inherently low N levels in acid-tropical soils (Formoso and Cerri 1999), high inputs of P are required (Sanchez et al. 2003; Date et al. 1995) for optimum growth and adequate food and fibre production. However, while N inputs can be supplemented from processes such as biological nitrogen fixation, there is no natural biological replenishment of P. This means P-inputs need to be applied in order to improve the soil P status and ensure normal plant growth and adequate yields.

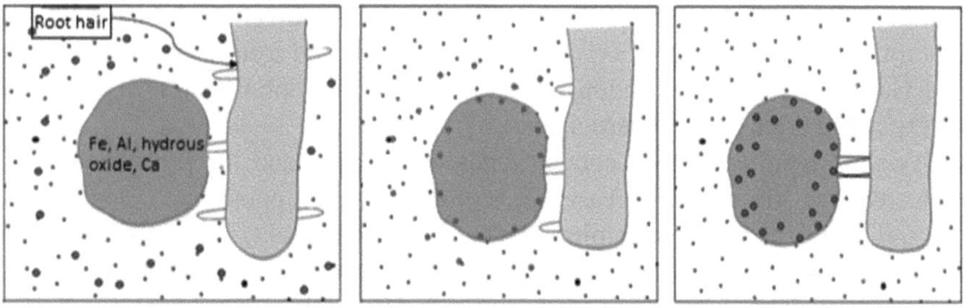

Fig. 2 Plant availability of P with time after application: within days of application (frame #1 at *left*), the oxide particle has fixed the P anions (frame #3 at *right*), thereby depleting labile P in the soil (After USDA 2012)

The factors affecting phosphorus availability thus include: proportions of Fe and Al oxides, clay types and amounts; and pH (Fig. 2). However, most nutrients in oxisol ecosystems are contained in the standing vegetation and decomposing plant material. Thus following deforestation, overgrazing, inappropriate farming practices, open-cast mining (Harris and Omoregie 2008; O'Callaghan et al. 1998), land degradation which can lead to low crop yields consequently follows. Therefore, for optimum growth and adequate food and fibre production, high inputs of P are required (Sanchez et al. 2003; Date et al. 1995). In the Caribbean, the locations of P deficiency are determined by soil texture and origin, topography, and moisture regime. Hence the problem of phosphorus fixation affects diverse locations in the Caribbean region including as follows.

1.2 Caribbean Red Soils

Typifying such a deficiency is the tropical humid karst, extensively found in Yucatan, Guatemala, Belize, Barbados, Antigua, Gaudalupe and Jamaica. These soils coexist spatially by zones in karst plains and highland karst environments (Ford and Williams 2007), such that ninety per cent of the Caribbean's 130,000 km^2 of karst is in the Greater Antilles, and the Cuban archipelago (Fig. 3) in particular hosts some of the most typical karst features in the Caribbean (Febles-Gonzalez et al. 2011).

The Jamaica bauxite soils are an example of an Oxisol which is developed from pure white limestone (Simpson 2013). But this claim is disputed. Others view this bauxite as having been developed from the weathering of volcanic ash deposited on the limestone during the Miocene (Comer 1975; Ahmad et al. 1966). The bauxite overlies limestone of middle Eocene to lower Miocene age and accumulates in karst depressions and fault-controlled valleys. Significant thicknesses of bentonitic limestone and bentonitic clay have been found in the middle and upper Miocene carbonate section along the north coast of Jamaica (Comer 1975). It is believed that volcanic ash alters to bauxite by desilication of glass, plagioclase, biotite, and ferromagnesians through either phyllosilicate intermediate phases or gel and

Fig. 3 Cuba: karst upland limestone adjacent to red P-deficient soils in valleys. *Source* www.fao.org

imogolite (Comer 1975). According to Comer (1975), vertical drainage of vadose water through faulted host limestone karst is extremely efficient in leaching silicon from overlying soils, thereby concentrating aluminium in the form of bauxite.

Non-bauxitic red soils of the Greater Antilles are located also in Jamaica, Cuba (Fig. 4) and the Dominican Republic, with the interior of Guyana in the southern Caribbean having the most extensive tracts (Simpson 2013). In Jamaica, the main production of bauxite comes from terra rossa karst sinkholes (Lyew-Ayee 1984). Most of these soils are characterized by extremely low native fertility, resulting from very low nutrient reserves, and low cation exchange capacity (CEC). The three major macronutrients, namely nitrogen, phosphorus and potassium are generally found to be deficient in overburden dumps (Sheoran et al. 2008).

1.3 Topography Effects: Puerto Rico

Total P is not a true index of the soil P status because not all P in the soil becomes available to plants over the same time scale (Richter et al. 2006). Mineral nutrients in soils are therefore not always available to plants. In addition to soil age and climate (as shown above), Mage and Porder (2012) found that other factors such as topography can cause phosphorus fixation in soils. Thus, elevation differences correlate with variations in soil redox potential and P availability (Silver et al. 1999), and extreme variations in topography typify Caribbean landscapes. As soil texture also affects P fixation (Raich et al. 1996), the extreme topographies producing widely differing soil textural regimes such as alluvial deposits of the Black River (Jamaica), Essequibo and Courantyne (Guyana); Yaque del Norte and Yuna

Fig. 4 Karst red soils within limestone region, Jamaica. These are residual soils remaining after desilication of Miocene volcanic ash. The desilication was facilitated by repeated drainage through the fissured limestone. *Source* Jamaica Bauxite Institute

(Dominican Republic); and clay-dominated intermontane depressions including the Cockpit Country of Jamaica, and the black clay soils of the lower reaches of the coral rocks (Harrison 2012) of Barbados indirectly influences P fixation.

The Luquillo Mountains, for example, located in Puerto Rico (Fig. 5), rise to elevations of 1075 m (3527 ft.) with precipitation inputs including rainfall and cloud drip of 5000 mm p.a. (Larsen and Concepcion 1998).

The Luquillo Mountains contain abundant silicon, iron, and aluminium. According to Mage and Porder (2012), intense weathering of minerals in the rocks formed the clays and reddish iron minerals that typify regional soils and stream sediments, yet, pockets of unfixed P exist there. In a matrix of sites in the Luquillo Mountains of Puerto Rico, Mage and Porder took readings from three topographic positions (ridge slope and valley) with an elevation gradient range of 350–800 m. They reported that though parent material more than any other factor determined the fraction of recalcitrant P, topographic position explained 10–15 % of the variance in those metrics. Soil-released P increased in valley floors compared to ridges and peaks. This was attributed to water logged valley floors releasing reduced forms of iron and aluminium which do not combine readily with P,

Fig. 5 The Luquillo Mountains, Puerto Rico. Note the juxtapositioning of rugged terrain, and several intermontane swales with opportunities for anoxic reducing soil conditions which alleviate P-fixation. *Source* http://eoimages.gsfc.nasa.gov/images

thereby increasing plant available P in such conditions. Indeed, Mage and Porter (2012) reported that total organic P is often correlated with total P in soils. Given that tropical environments often exhibit a variety of such variables, prediction of phosphorus status of tropical soils has remained elusive (Mage and Porder 2012).

1.4 Volcanic Materials of the Eastern Caribbean

Of all environments, volcanic soils tend to have greatest P-sorption because they develop from volcanic materials having a glassy non-crystalline structure which contain large amounts of amorphous material. The Lesser Antilles Volcanic Arc islands of the eastern Caribbean from the Virgin Islands in the north to the islands off the coast of Venezuela in the south lie on a subducting plate boundary zone (Fig. 6) containing seventeen active volcanoes including Soufrière Hills on Monsterrat, Mount Pelée on Martinique, La Grande Soufrière on Gaudeloupe and Soufrière Saint Vincent on St. Vincent (Feuillet et al. 2011).

Such materials, as mentioned above, generally have a great surface area. Three types of such amorphous minerals in volcanic soils include allophone, imogolite, and ferrihydride (Clermont-Dauphin et al. 2006). These are rich in Al and Si in the early stages.

On the other hand, due to their high anion exchange capacity (AEC), volcanic soils can efficiently adsorb organic colloids (Clermont-Dauphin et al. 2006)

Fig. 6 Volcanic mountains on the islands of the eastern Caribbean. Adapted from: earthquake.usgs.gov/

thereby alleviating P-fixation. When the same amount of fertilizer is applied to a volcanic soil and a moderately weathered grassland soil, the volcanic soil has less P available due to its greater P-fixation capacity. Hence on well-drained volcanic slopes of these islands (Figs. 7 and 8), maximum levels of P-fixation should be expected, whilst in the flatter localities with higher levels of organic colloids, more plant-available P should be available.

Again, on the volcanic, P-adsorbing soils of the island uplands of the Lesser Anitlles such as the Windward Islands and the French West Indies where banana mono-cropping has been practised since about 1980, large inputs of chemicals and replanting every three to five years after ploughing have become necessary (Clermont-Dauphin et al. 2006).

Fig. 7 Mount Pelee: part of the Lesser Antilles active volcanic system. P-sorption is high in amorphous volcanic materials. *Source* Smithsonian Institution Natural Museum of Natural History: Global Volcanism Program

1.5 Solubility of Phosphate Minerals

1.5.1 Reaction of Phosphorus

Compared with the other major nutrients, phosphorus is by far the least mobile and available to plants inmost soil conditions (Hinsinger 2001). The poor mobility of soil inorganic phosphorus is due to the large reactivity of phosphate (P) ions relative to numerous soil constituents and to the consequent strong retention of most of soil phosphorus onto those (Hinsinger 2001). Hinsinger (2001) observed that phosphate ions readily precipitate with metal cations, forming a range of P minerals. The type of mineral formed will depend on the soil pH in the first place as it governs the occurrence and abundance of those metal cations that are prone to precipitate with P ions in the soil solution, namely Ca, Fe and Al. Hence, in neutral to alkaline soils, P ions will rather precipitate as Ca phosphates (Hinsinger 2001).

In contrast to nitrate which remains highly soluble in the soil and moves with the water, phosphate leaches from soils with difficulty, and over a long period of time. The solubility of phosphate minerals is very dependent upon soil pH. The soil pH for optimum phosphorus availability is 6.5 (Fig. 1). At high or neutral pH, phosphate reacts with calcium to form minerals, such as apatite. Under acidic conditions, phosphorus may react with aluminum and iron to form minerals, such as strengite and varescite (Hinsinger 2001).

Fig. 8 Martinique, Lesser Antilles. Note the excellent radial drainage around two volcanic domes. Such efficient slope drainage combined with volcanic ash induces heavy P-fixation in the humid tropics. Adapted from: www.lib.utexas.edu/maps/world

1.5.2 Biological Role of Phosphorus

Large early growth rates are vital for high survival levels in mine-site re-vegetation programs. Phosphorus effects rapid plant growth, being responsible for energy storage in the form of ADP and ATP, thereby facilitating the translocation of nutrients across the cell wall. Though plant roots absorb phosphorus from the soil solution, in comparison to other macronutrients, the phosphorus concentration in the soil solution is much lower and ranges from 0.001 mg L^{-1} to 1 mg L^{-1} (Brady and Weil 2002). Highly weathered soils are P-depleted because all non-occluded P has been released, and the remaining mineral phosphorus is occluded. In these soils, organic phosphorus is the main source of plant nutrition (Brady and Weil 2002). Soil organic matter (SOM) is an important source of P for plant uptake because P comprises 1–3 % of SOM (Price 2006) and 50 % of total P in the soil (Price 2006). Increasing organic matter slows P-fixation in clay soils (Orcutt and Nilsen 2000) by decreasing the rate of mineralization, and by physically lowering the proportion of the inorganic fixing species in the soil.

Hence, soils that receive large applications of animal wastes and sewage sludge are vulnerable to releasing environmentally significant concentrations of dissolved P available to subsurface flow owing to the gradual saturation of the soil's P sorption capacity. (Siddique and Robinson 2003). Further, mining methods (including those for bauxite) which shift and misplace soil horizons containing organic matter will be inimical to the availability of plant-available soil phosphorus (Harris and Omoregie 2008).

Phosphorus also seems to improve the metabolism of other nutrients. For example, phosphorus is low in most Brazilian soils and when it has been applied it has been responsible for large root-yield increases. But research carried out by Gomes and Howeler (1980) in Brazil has shown that although nitrogen uptake by cassava is high, it does not always result in yield increases. In some cases productivity has been reduced. However, when applied as organic matter (which is normally high in P) very good responses have been obtained. This suggests that yield increases due to potassium and nitrogen occur more frequently in the presence of phosphorus.

1.6 Advantages and Drawbacks of Animal Waste

For quenching to be effective, the rate of P release from a soil additive must be higher than the soil P-fixation rate, and the more organic P in the soil, the higher the rate of mineralized P reaching the soil (Price 2006). Thus, heavy applications of organic manures arenormally required for soils which rapidly fix phosphorus. But this method has its drawbacks. Shrimp waste, for example, is considered a slow-release organic fertilizer derived from ground-up shrimp shells high in phosphorus and calcium with 5 % N, 8 % P, 15 % Ca, 18 % chitin and trace minerals (Ngoan et al. 2000). Land spreading of such waste has been discouraged due to its uniquely obnoxious odors (Hayes et al. 1994). Moreover, as the shrimp waste is recalcitrant and resists decomposition, grinding and ensiling is necessary to release macronutrients efficiently (Rao and Stevens 2006). This increases costs, especially as fossil fuels are often the energy source for the preparation phase of such additives.

1.7 Rationale for C. Cajan Organics

Low bioavailability of soil phosphorus (P) often limits N_2 (nitrogen) fixation and crop production in large parts of the tropics. Vesterager et al. (2008) studied the efficiency of P acquisition and P use by *C. cajan* in a pot experiment using two cowpea (*Vigna unguiculata*) genotypes as controls. They found that root surface (area) did not determine P acquisition of pigeon pea, yet absorbed 1.6 mg P cm^{-3} from 33 cm^3 soil compared with 0.17 mg P cm^{-3} from 387 cm^3 soil for cowpea at 66 d. This shows a > 10-fold level of P-absorption efficiency of *C. cajan* compared with cowpea, another legume.

Hinsinger (2001) found that low-molecular organic acids secretion induced by P deficiency including oxalic acid, malic acid and citric acid, etc., can activate and promote plants' utility of sparingly soluble phosphate in the soil by chelating metal ions associated with P and reducing pH of the rhizosphere. Research in India with rotation systems showed that pigeon peas not only increased the N status of the soils but that it also increased the amount of Phosphorus available for the follow-up crops in the rotation (Ae et al. 1990). (Raghothama and Karthikeyan 2005). Further, mycorrhizal interactions help to increase P availability around the roots of pigeon peas (Shibata and Yano 2003).

1.8 Aim

The objective of this study was to determine the effectiveness of phosphorus quenching by root exudates from *C. cajan* during growth in an aged (10-year-old) bauxite overburden. The aim was to not only exceed the P-sorption capacity of stored bauxite-mine overburden with *C. cajan*, but to determine how soil P-availability varies with the distance from the root of *C. cajan*.

1.9 Hypothesis

Increases of plant available (labile) P occur in soil after *C. cajan* has grown in it.

2 Method

Land preparation, seeding, and cultivation was done as follows: *C. cajan* was selected as a test plant based on its relatively high growth rate and its high requirement for soil P (Shibata and Yano 2003) and other nutrients (FAO 2012), and thus chosen to test for P-quenching of soil by root exudates. Seeds of *C. cajan* were sown at intervals of four metres apart in a soil bed in a terra rossa bauxite overburden. Sloping ground was chosen because *C. cajan* does not withstand waterlogged conditions (Özgül et al. 2007). Due to the great length of *C. cajan* roots (Fig. 9) the distance of 4 m between plants was chosen to avoid root-to-root contact. Plants were weeded in the early stages of growth. After two years, roots were dug up at

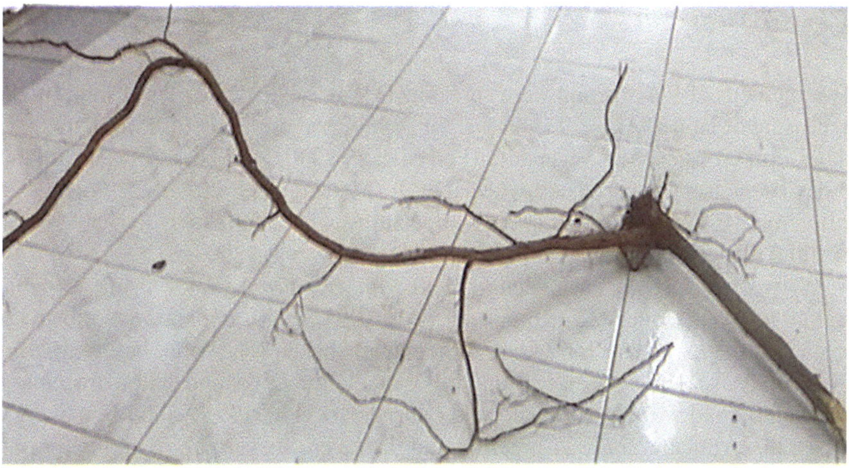

Fig. 9 Root of *C. cajan*, showing the great length, compared to shoot section (partly obscured) at right. Acidic root exudates which solubilize fixed P in soils directly vary with root length. Each floor tile is 250 mm in width (*Photo* Gad Onywere)

Table 1 Soil sample analysis of bauxite overburden before added fertilizers

Depth (cm)	pH	SOM (%)	Total N (%)	P	K	Zn (ppm)	Cu	Mn	Fe
0–20	6.98	2.7	0.15	38	29	5	4	19	66
15–30	7.92	3.0	0.17	39	39	8	5	4	10
30–45	7.77	4.5	2.1	182	185	9	4	30	58

a depth of at 5–15-cm depth. The 0–5-cm depth was excluded to avoid the P-rich organic matter in the A horizon. The 5–15 cm depth was adopted based on the growth habit of the *C. cajan* in the bauxite overburden. Thus the roots, though long, did not penetrate vertically but horizontally in the soil and at no greater depth than 15 cm. Topsoil contained greater amounts than deeper layers for all P fractions except residual-P. Loose soil from the roots was removed. At horizontal transect distances of 0-, 5-, 10-, and 50-cm from the roots, soil samples were taken. Further sampling was done below the root levels. Samples were collected in triplicate for contents such as organic matter (OM), nitrogen (N), potassium (K), phosphorus (P), zinc (Zn), manganese (Mn), Iron (Fe), and pH level (Table 1).

The amounts of K, Zn, Cu, Mn, and Fe present in the soil were determined by flame atomic absorption spectroscopy (FAAS).

Prior to crushing, all the obvious organic matter such as roots was removed. Air-dry samples from four locations were selected for analysis. Soil was hand-crushed and passed through a 2-mm-diameter sieve.

2.1 P-Extraction

Two methods of P-extraction were applied: (1) Water-extractable P is designed to be a measure of labile P (Castillo and Wright 2008), used primarily for P-availability of short-term crops such as vegetables (Forsee et al. 1940), and because water-soluble P has been strongly correlated with labile P (McCray et al. 2012). (2) For longer-term crops, 0.5 M acetic acid as a P extractant was used for slowly available forms of soil P related to crop availability as stated by Glaz et al. (2000).

Water-extractable P was determined with deionized water using a 4 cm^3 soil/50 mL extractant ratio (Korndorfer et al. 1995). Soil samples were allowed to remain in the extractant for 24 h and then were shaken for 50 min before filtering for P analysis. Using a 4 cm^3 soil/50 mL extractant ratio, acetic acid-extractable P was determined with 0.5 M acetic acid (Korndorfer et al. 1995). Soil samples were left in the extractant for a 24-h period and then were shaken for 50 min before filtering for P analysis. Settling, followed by gentle stirring provided opportunities for flocs to form.

The above specific acid extraction was applied because a determination of the *total* amount of P in the soil is not meaningful to a growing plant, a correlation between the amount of P extracted by a chemical removing total P and the amount of P taken up by the plant in question is necessary. This quantity of P is known as the *plant-available* P, and the better the correlation, the better the test. This choice was further justified because McCray et al. (2012) found that acetic acid-extractable P was correlated ($P < 0.05$) with residual P ($r = 0.47$) and negatively correlated with Fe–Al-bound P($r = -0.53$), and the objective of this study was not to determine P associated with hydroxides of Fe and Al, but plant-available P. Further, Yuan et al. (2013) showed that the greatest extraction capability of P from red soil was obtained by citric acid, followed by oxalic acid and malic acid and that of formic acid was the weakest. However, P extraction capabilities of acetic acid and succinic acid whose amounts were the most in root exudates were just 70 and 40 % of that of citric acid respectively (Yuan et al. 2013).

2.2 Study Area: A Former Mine-Site

The study site is located in southwestern Jamaica, (long. 35″1SE; lat. 33″Ol′N; alt. 500 m). The mean temperature of the Mandeville intermontane zone is 23 °C with a seasonal range of 5 °C (Morissey 1990), though the winter diurnal range is of the order of 5–15 °C. Annual rainfall totals average 2000-mm (Greenberg and Wilding 2007), occurring mainly as afternoon convectional thundershowers with two peaks between May and November. The dry season occurs in winter, where potential evapotranspiration exceeds precipitation (Greenberg and Wilding 2007). Here, mined-out bauxite lands totaling 10,000 ha are currently under mining leases (Jamaica Bauxite Institute 2007). The research plot was located in a former bauxite mine within the Northern Caribbean University property, Mandeville, Jamaica, mined and re-filled with its own overburden 10 years prior to this study. The soil is composed of a clayey oxisol replaced on a karst limestone basement. Using hexametaphosphate and ultrasonics to free clay bound to Fe oxides, Ahmad et al. (1966) found clay contents exceeding 60 % in all local bauxite soils, and up to 90 % in some replaced overburdens. The post-mined soil of this study, as is the case with other local post-mined soils (Greenberg and Wilding 2007), is higher in limestone fragments than the surrounding pre-mined soils, due, most likely, to mixing with the C-horizon on a limestone bedrock during the overburden removal process (Harris and Omoregie 2008).

2.3 Statistics

Differences of plant available P in relation to distance from root of C. cajan were examined using the Paired t-test.

3 Results

Results as indicated by the Paired t-test showed that in a terra rossa bauxite overburden soil of high phosphorus fixation, *C. cajan* produced an increase in the proportion of plant available phosphorus. The nutrient content of the bauxite overburden before and after treatment, at the 5–15-cm depths are depicted in Table 2. P-levels varied directly with distance from the roots of *C. cajan* (Table 2; Fig. 10). Twenty four months after treatment, plant available soil P close to the root of *C. cajan* was measured at 35 ppm, becoming exponentially less with distance from the root (Table 2; Fig. 10). This corroborates the observation of Yadvinder-Singh (2009) cited above, stressing the low diffusivity of P.

Greatest phosphorus (P) levels were found at the sampling locations 0–5-cm from the roots, i.e., in soil which adhered to the roots, while lowest P-levels levels were located at 50 cm from the roots (Table 2; Fig. 10).

Table 2 Concentration of labile (plant available) phosphorus compared with distance from root of *C cajan*

Horizontal distance (cm)	PO_4 (mg/L)	SOM (%)
0–20	35^a	0.15
100	31^a	0.16
200	15^b	0.15

For t-test: numbers followed by different letters are significantly different ($P < 0.05$)

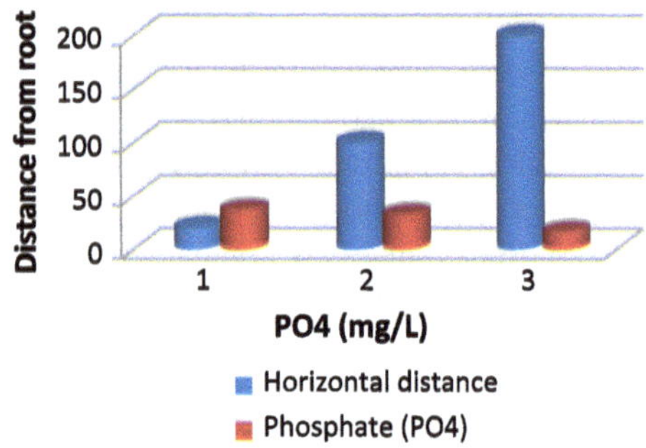

Fig. 10 Concentration of labile phosphorus in soil compared with distance from root of *C. cajan*

4 Discussion

In this study, *C. cajan* significantly increased plant-available phosphorus in the root zone in the karst limestone red soil. Sinclair (2004) similarly found that in a soil with a high phosphorus fixation rate, pigeon peas were better able to uptake P and to maintain adequate growth while other crops such as corn and soybeans were not able to survive under such low P conditions. These findings are supported by Barber and Navarro (1994) who found that the amount of nutrients released from root decomposition amounted to over 40 kg ha^{-1} (40 lb Ac^{-1}) of nitrogen and over 80 kg ha^{-1} of phosphorus, representing a potential valuable pool of nutrients for the following crops in the rotation. According to Shibata and Yano (2003), the roots of pigeon pea excrete organic acids such as citric, piscidic, and tartaric acid, which help to mobilize phosphorus in the soil. In this study, de-ionized water extracted plant-available P from this soil in contrast to 0.5 M acetic acidic which failed to extract P. This was against expectations because the prototype anaerobic equation for biologic phosphorus removal (BPR) includes acetate (Randall and Chapin 1997) as follows:

$$\text{Acetate} + \text{ATP} + \text{NAD} \rightarrow \text{HB} + \text{ADP} + \text{Pi} + \text{NADH} \quad (1)$$

However, Randall and Chapin (1997) found that industrial waste water was considerably more inhibitory to biological phosphorus removal (BPR) than equivalent amounts of sodium acetate. They ascribed this to the probable presence of other organic chemicals in the industrial waste water, such as isopropyl alcohol and acetone, although high concentrations of inorganic chemicals such as magnesium and sulphate were also present. In the current experiment, due to the small aperture of the filtering medium, the extraction process for each sample entailed several hours of prolonged filtering of each soil sample. The samples waiting to be filtered were thus stored at room temperature for periods exceeding 24 h. This permitted unintended prolonged anaerobic conditions within the soils. The 0.5 M acetic acid could therefore have reacted with organically formed alcohols. On the other hand, the relatively weak extract—deionized water—removed plant-available P in comparatively large quantities. This suggests that active plant available P present in the high P-fixing soil was either released by the roots, and quenched the fixation process after the optimum fixation capability of the soil was exceeded in the soil particles adjacent to the roots.

In an earlier study (Brown 2007), sixteen weeks after treatment and harvest, soil phosphorus from chicken manure-only and shrimp silage-only treatments (500-g/hole each) was measured at 1000 and 1100 ppm, respectively. However, in that study, after 32 weeks, terra rossa overburden fertilized with shrimp silage, and chicken manure, exhibited P levels of only 55 and 65 ppm, respectively. Those are a more than an 18-fold decrease from 1000 ppm. Nevertheless, they were approximate increases of 50 and 30 % higher than the already depleted control (unfertilized soil at 38 ppm. Price (2006) found that plant-available phosphorus levels in most soils are in the order of 0.1 % (1000 ppm). Thus at 8 months after application, both chicken manure-only and shrimp waste-only beds retained a small but ineffective increasing trend of soil P over the controls. Therefore, such residual soil P levels in the terra rossa soil were still inadequate for high-yielding crop growth. Further, in the study of Brown (2007), there was no inorganic fertilizer added with the organic treatment, therefore there was no initial luxury absorption of inorganic P by micro-organisms as occurs on such occasions (Hue 1992). Soil microorganisms therefore would have immediately attacked the well-watered and aerated organic manures. But, as the P-fixing species in that study had already been substantially quenched at 16 weeks, excess organic P from the post-harvest stubble left in the ground was expected to release P by slow microbial degradation during the ensuing 16 weeks, and hence become slowly fixed. However, this did not happen, possibly as a result of the slow degradation rate of the plant stubble, compared with that of the manures. At some stage, the P mineralization rate became lower than the P immobilization rate. Had P release from the stubble been higher than that of P immobilization rates in the soil (i.e., had the stubble degradation rate been sufficiently high), the level of soil P would have been higher after week 16. But without human intervention, this is impossible. Hence the advantage of exudates over added organic matter can be seen: dissolution of fixed P continues as long as there are living roots in the soil. On the other hand, quenching of P fixation using organic additives require perpetual addition of such materials to the soil. This is unfeasible where organic matter in large quantities are unavailable.

Had the treatment of Brown (2007) consisted of inorganic rather than organic additives, the effect could not have lasted long enough to produce a continuous growth in week 8 through to week 16. Conversely, the manure-treated overburden still contained adequate P when growing plants were harvested from the soil. It is the balance between the organic and inorganic that determines the availability of P in the soil. The gradual incremental mineralization of organic P in the manures by microbial activity caused the prolonged availability of P in the soil. Such prolonged P availability would not have resulted from the application of inorganic fertilizers. Similarly, without the acidic exudates from the root of *C. cajan* to continually produce labile P from fixed P in the soil, an increase in labile P would not have occurred. Each study therefore showed an increase in soil P. However, Brown (2007) reported seemingly high P in that study exceeding 1000 ppm compared to that of the present study of less than 100 ppm. However, whereas Brown (2007) extracted total P, that of the present study was of plant-available P as it was done with de-ionized water. Further, there are three pools of P in soil: (1) solution P (2) active P, and (3) fixed P. As the ratio of fixed to unfixed P is normally in the order of 1: 500 (worse in tropical ferralitic soils), it is reasonable to conclude that the treatment of Brown (2007) increased soil P from unavailable levels to plant available P in the karst terra rossa soil.

Yadvinder-Singh (2009) found that at low and medium P levels, soil solution concentration is more important in determining P uptake than the ability of the roots to absorb P. Vice versa, at high P level, the ability of the roots to absorb P is more important than P supplying characteristics of the soil. At all P levels the influence of root growth rate (k) for P influx was found to be very scarce (Yadvinder-Singh 2009). Nevertheless, the large percentage of root mass observed for *C. cajan* (Fig. 8) in this study, in addition to the effects of the exudates, indicate a strong plant available P effect of *C. cajan* in acidic tropical soils. Moreover, root exudates enhanced the mobilization of red soil plant-unavailable phosphorus by 100–211 % (Yuan et al. 2013). Thus root exudates collected by solution method were employed to extract P in soil and sparingly soluble phosphate. Soil P extraction test revealed that the dissolving of soil P by exudates in Changsha, Changde and Chenzhou was remarkably higher than that by water ($P < 0.05$) and P concentration extracted in these three samples were 100,157 and 211 % higher than that by water respectively (Yuan et al. 2013).

Under the condition of low P, kinds and amounts of organic acids secreted by plants changed, such that secretion and amounts of organic acids were noticeably higher in P-free and low P cultures than those in high P culture (Yuan et al. 2013). Also, they found that 4 acids (oxalic acid, fromic acid, malic acid and citric acid) can be influenced significantly by low P induction, and that this coincides with findings in other plants.

Hoffland et al. (2006a) found that Rape (*Brassica napus* L.) plants, for example, are efficient users of rock phosphates. This they ascribed to the excretion of malic and citric acids from the roots during phosphorus deficiency, resulting in a decrease of the pH in the rhizosphere and a consequent solubilization of the rock phosphate. Malic acid which is prevailing in the exudate is probably newly–synthesized within the excreting root segment. When no phosphate was supplied, this excretion zone was located 1–2 cm behind the root tips. However, when rock phosphate was applied locally this zone shifted along the root to that part which was in direct contact with the rock phosphate particles. As stated above, this facility of some plants to solubilize phosphate is species-selective. Thus Hoffland et al. (2006b) observed no accumulation and excretion of organic acids with the roots of P–deficient *Sisymbrium.* On the other hand, pigeon peas exude piscidic acid which enhances the availability of phosphate from iron-phosphate and rock-phosphate (Ae et al. 1990). Thus, unlike other legumes, pigeon pea is one of the few crop species that can utilize iron bound P efficiently (Subbarao et al. 1997) making it capable of producing appreciable yields even under P limiting conditions.

On the assumption that *C. cajan* increased P in the soil of this study mainly with its root exudates, uncertainties still remained as to the source of the increased soil P. This is because *C. cajan*, due to its extensive tap-root (Fig. 1) and dense canopy, adds substantial amounts of organic matter to the soil in the form of roots and leaves, which in turn improves the physical condition of the soil (Yadvinder-Singh 2009). It is therefore possible that organic parts, and not exudates, caused increased P in the 5–15-cm soil layer contiguous to roots (Table 2). In Bolivia, for example, the root biomass of pigeon peas at a 15 cm (6 in.) depth was determined to be over 5 t ha^{-1}. The amount of nutrients released from root decomposition amounted to over 40 kg ha^{-1} (i.e. 40 lb Ac^{-1}) of nitrogen and over 80 kg ha^{-1} of phosphorus, representing a potential valuable pool of nutrients for the following crops in the rotation (Barber and Navarro 1994).

Moreover, pigeon peas are considered to have greater N fixation rates, compared to other legume species (Chikowo et al. 2004), and develop effective mycorrhizal associations, improving nutrient uptake efficiency (Chikowo et al. 2004). Mycorrhizal associations enhanced the ability of pigeon peas to uptake phosphorus by a rate of 10× compared to the 1× uptake without root inoculation (Shibata and Yano 2003). To exclude the effects of organic matter in the present study such as rotten shed leaves, no soil samples were taken from the A-horizon.

Thus Fohjse et al. (1988), who observed P efficiency as being related to uptake efficiency of the plant (which in turn is determined by both root-shoot ratio and absorption rate per unit of root), is applicable to this study. They found that species of low efficiency such as onion, tomato and bean, had low influx rates and low root-shoot ratios, whereas species of high efficiency had either high influx rates (rape and spinach) or high root-shoot ratios (ryegrass and wheat). The combination of high root-shoot rate and high influx of P was not found in any of the species studied (Fohjse et al. 1988). On the assumption that this "one or the other" principle also applies to many other crops, the high root/shoot ratio of *C. cajan*, and the presence of piscidic acid in *C. cajan* exudates, suggest that root exudates, micorrhyza and high root/shoot ration combination causes efficient quenching of phosphorus fixation in the root zone of terra rossa mine overburden.

5 Conclusion

The main P-fixing problem-sites in the Caribbean are exposed ferralitic and volcanic slopes, displaced mine overburdens, and karst plains. Permanent cropping of such locations in low-income areas is very difficult due to low cation exchange capacities and high phosphorus fixation on iron and aluminium oxides. Maintenance of soil P at an adequate target value in such locations can be accomplished by growth of *C.cajan* in the soils, rather than costly chemical P fertilizers or through periodic incorporation of crop residue or of organic manures. This process operates on a continuous basis thereby causing more plant-exchangeable P in the soil for a longer period of time.

But despite the apparent effectiveness of *C cajan* to release P in the above soils, the effect was detected in a narrow zone around the root. This may vary with soil texture because the P-fixing sites are located mainly in sesquioxides, which in turn largely comprise the clay particles in such soils. Wehr et al. (2006) found that Jamaica bauxite waste contains, on average, 75 % clay, at least twice as much clay as any of the other bauxites wastes. Therefore, such a high concentration of P-fixing entities may help to explain the short distances of availability from the root of *C. cajanus* in such soils.

Based on the short distances of P availability from the root zone of *C. cajanus*, intensive growth of *C. cajanus* in a crop rotation may be an effective corrective for P-fixation in terra rossa soils.

6 Geobiotechnological Applications

- In addition to the role of root exudates in the above study, a better understanding of the influence of anthropogenic interventions on such ecosystems is necessary to guide the management to enhance phosphorus availability for age production, conservation and recreation. As stated above, anaerobic conditions enhance the presence of plant-available phosphorus in soils.
- Before P becomes fixed in soil, certain anaerobic organisms can increase phosphorus availability (Oehmen et al. 2005) through storage in their biological tissues. Hence the question of soil water status in this context applies not just to terra rossa topography but to that of other vulnerable soil types in the Caribbean.
- Waste water can, for example, accumulate in depressions. According to Oehmen et al. (2005), enhanced biological phosphorus removal (EBPR) is one of the most commonly used and environmentally sound processes for phosphorus (P) removal from wastewater. Thus EBPR can be used to increase P availability, where wastewater accumulates.
- The special conditions favouring EBPR are the operation of an activated sludge system with a regime of anaerobic and aerobic conditions in a sequence which promotes the growth of polyphosphate-accumulating organisms (PAOs) (Oehmen et al. 2005). Under anaerobic conditions PAOs take up volatile fatty acids (VFAs), thereafter converting them to intracellular poly-b-hydroxyalkanoates (PHAs) (Oehmen et al. 2005). Mino et al. (1998) attributes the energy for these anaerobic transformations as being generated through hydrolysis of their internally stored polyphosphate (poly-P) and glycogen. Glycogen glycolysis also produces the reducing power necessary to maintain the cell redox balance (Mino et al. 1998). When aerobic conditions prevail, PAOs oxidize PHA to gain energy for growth, for glycogen replenishment, and phosphorus uptake. Sludge with a high poly-P content supplies the phosphorus so removed, a group of bacteria known as Candidatus Accumulibacter phosphatis being the only identified PAO (Oehmen et al. 2005).
- In the above-mentioned acid soils, phosphate is fixed by iron and aluminum. Hence phosphate fixation can be minimized by adjusting soil pH to an optimum

level for phosphate availability. Phosphate fixation is minimized relative to soil pH when the pH is just slightly acidic, between 6.0 and 7.0. This can be maintained by liming acid soils.

- For ferralitic soils in the tropics, 2–3 year rotations of *C. cajan* can release fixed phosphorus. However close planting to the extent of achieving closed canopies could substantially increase labile P levels for subsequent crops. Further research is required to quantify rates of release by *C. cajan* in such soil conditions.

References

Ae N, Arihara J, Okada K, Yoshihara T, Johansen C (1990) Phosphorus uptake by pigeonpea and its role in cropping systems of the Indian subcontinent. Science 248:477–480

Ahmad N, Jones RL, Beavers AH (1966) Genesis, mineralogy and related properties of West Indian soils. 1: Bauxite soils of Jamaica. Soil Sci Soc Am Proc 305:719–722

Balemi T, Negisho K (2012) Management of *soil* phosphorus and *plant* adaptation mechanisms *to* phosphorus stress for sustainable crop production: a review. J. Soil Sci. Plant Nutr 12(3) Temuco set. 2012 Epub 31-Ago-2012http://dx.doi.org/10.4067/S0718-95162012005000015

Baligar VC, Fageria NK (1997) Nutrient use efficiency in acid soils: nutrient management and plant use effi ciency pp. 75–93. In: Monitz AC, Furlani AMC, Fageria NK, Rosolem CA, Cantarells H (eds) Plant-soil interactions at low ph: sustainable agriculture and forestry production. Brazilian Soil Science Society Compinas, Brazil

Barber RG, Navarro F (1994) Evaluation of the characteristics of 14 cover crops used in a soil rehabilitation trial. Land Degrad Rehabil 5:201–214

Bekunda MA, Bationo A, Ssali H (1997) Soil fertility management in Africa: a review of selected research trials. SSSA Special Publication no 51, pp 63–79. American Society of Agronomy and Soil Science, Madison, USA

Brady NC, Weil RR (2002) The nature and properties of soils, thirteenth edition. Upper Saddle River, NJ. Prentice Hall

Brown B (2007) The utilization of shrimp waste to form ration for poultry and fish and the formulation of organic fertilizer. MSc. Thesis Northern Caribbean University.

Castillo MS, AL Wright (2008) Soil phosphorus pools for Histosols under sugarcane and pasture in the Everglades, USA. Geoderma. 145:130Y135

Chesworth W (2008) Encyclopaedia of soils, 2008, pp 237 Encyclopedia *of* soil science. In: Carlota GP, Teresa TR, Behan-Pelletier VM (eds). ISBN:978-1-4020-3994-2 (Print) 978-1-4020-3995-9 (Online). http://www.ctahr.hawaii.edu/oc/freepubs/pdf/pnm9.pdf

Chikowo R, Mapfumo P, Nyamugafata P, Giller KE (2004) Woody legume fallow productivity, biological N2-fixation and residual benefits to two successive maize crops in Zimbabwe. Plant Soil 262:303–315

Clermont-Dauphin C, Cabidocheand YM, Meynard JM (2006) Effects of intensive monocropping of bananas on properties of volcanic soils in the uplands of the French West Indies. Soil Use Manage 20(2):105–113. doi:10.1111/j.1475-2743.2004.tb00345.x

Comer JB (1975) Genesis of jamaican bauxite. Econ Geol 70(8):1486–1487. doi:10.2113/gsecongeo.70.8.1486

Date RA, Grundon NJ, Rayment GE and Probert ME (1995) Plant-soil interactions at low pH: principles and management. Developments. Plant and Soil Sciences, 64. Kluwer Academic Publishers, Dordrecht, The Netherlands, vol 64, 822 pp

FAO (2004) Use of phosphate rocks for sustainable agriculture. Food and Agriculture Organization, Rome, Italy. http://www.fao.org/docrep/007/y5053e/y5053e00.htm

FAO (2012) Grassland species index. Cajanus cajan. http://www.fao.org/ag/AGP/AGPC/doc/Gbase/DATA/PF000150.HTM. Accessed 6 June 2012.

Febles-Gonzalez JM, Vega-Carreno MB, Tolo N-Becerra A and Lastra-Bravo X (2011) Assessment of soil erosion in karst regions of Havana, Cuba. Land Degrad Develop Land Degrad Develop, Published online in Wiley Online Library (wileyonlinelibrary.com). doi:10.1002/ldr.1089

Feuillet N, Beauducel F, Tapponnier P (2011) Tectonic context of moderate to large historical earthquakes in the Lesser Antilles and mechanical coupling with volcanoes. J Geophys Res 116: B10308. Bibcode:2011JGRB.11610308F. doi:10.1029/2011JB008443

Fohjse D, Claassen N, Jungk A (1988) Phosphorus efficiency of plants. 1. External and internal P requirement and P uptake efficiency of different plant species. Plant Soil 110:101–109

Ford D, Williams PW (2007) Karst hydrogeology and geomorphology. Wiley, 562 pp

Formoso MLL, Cerri CC (1999) Workshop on Tropical Soils. Rio de Janeiro, Brazil, Brazilian Academy of Sciences, 192 pp

Forsee WT (1940) Recent plant responses to some micronutrient deficiencies on Everglades peat. Soil Crop Sci Soc Fla Proc 2:53–58

Fox RL, Kamprath EJ (1970) Phosphate sorption isotherms for evaluating the phosphorous requirements of soils. Soil Sci Soc Am Proc 34:902–907

Ghose MK, Kumar A (1997) Management of topsoil for geoenvironmental reclamation of coal. Environ Geol 40, 11/12, 1405–1410

Glaz B, Powell G, Perdomo R, Ulloa MF (2000) Sugarcane response to phosphorus fertilizer in relation to soil test recommendations on Everglades Histosols. Agron J 92:375Y380

Gomes J de C, Howeler RH, Weber EJ, Toro MJC, Graham M (eds) Journal conference title cassava cultural practices. 1980 pp. 93–102 Record number 19831976360

Greenberg WA, Wilding LP (2007) Pre and post-mined bauxite soils of Jamaica. Physical and chemical properties. Soil Sci Soc Am J 71:181–189

Harris MA, Omoregie SN (2008) Post-mining deterioration of bauxite overburdens in Jamaica: storage methods or subsoil dilution? Environ Geol 54(1):111–115

Harrison JB (2012) Geology of Barbados. Outlook Verlag 2012

Haug A (1984) Molecular aspects of aluminium toxicity. CRC Critical Reviews in Plant Sciences 1:345–373

Hayes LA, Richards R, Mathur SP (1994) Eccentric viability of commercial complexity of fisheries waste by passive aeration. In: The compost council of Canada symposium. Shigawake Organics Ltd, Canada

Hinsinger P (2001) Bioavailability of soil inorganic P in the rhizosphere as affected by root-induced chemical changes: a review. Plant and Soil 237:173–195

Hoagland DR, Aron DI (1941) Physiological aspects of availability of nutrients for plant growth. Soil Sci 51:431

Hoffland E, van den Boogaard R, Nelemans J, Findenegg G (2006a) Biosynthesis and root exudation of citric and malic acids in phosphate-starved rape plants. New phytolo 122(4). doi:10.1111/j.1469-8137.1992.tb00096.x

Hoffland E, Wei C, Wissuwa M (2006b) Organic anion exudation by lowland rice (Oryza sativa L.) at zinc and phosphorus deficiency. Plant Soil 283:155–162. doi:10.1007/s11104-005-3937-1

Hue NV (1992) Correcting soil acidity of a highly weathered ultisol with chicken manure and sewage sludge. Commun Soil Sci Plant Ann 23:241–264

Jamaica Bauxite Institute (2007) Sale/lease and use of pre-mined and mined out lands, vol 6, sect 3

Kamprath EJ (1984) Crop response to acid soils in the tropics. In: Adam F (ed) Soil acidity and liming, 2nd edn. Agronomy Monograph 9. American Society of Agronomy and Soil Science Society of America, Madison, Wisconsin, USA, pp 643–698

Korndorfer GH, DL Anderson, Portier KM, Hanlon EA (1995) Phosphorus soil test correlation to sugarcane grown on Histosols in the Everglades. Soil Sci Soc Am J 59:1655Y1661

Lal R (1990) Soil erosion in the tropics: principles and management. McGraw- Hill Inc, New York, USA

Larsen, MC, Concepción IM (1998) Water budgets of forested and agriculturally developed water-sheds in Puerto Rico. In: Segarra García RI (ed) Proceedings, tropical hydrology and caribbean water resources. American Water Resources Association, San Juan, Puerto Rico, 12–16 July 1998, pp 199–204

Lyew-Ayee P (1984) Bauxites in the Caribbean. In: Jacob L. Jr. (ed) Bauxite symposium. American Institute of mining, Metallurgical and Petroleum Engineers. Proceedings, pp 262–296

Mage SM, Porder S (2012) Parent material and topography determine soil phosphorus status in the Luquillo Mountains of Puerto Rico. Ecosystems. 2012 Springer Science & Business Media New York. doi:10.1007/s10021.012.9612.5

McCray JM, Wright AL, Luo Y, Ji S (2012) Soil phosphorus forms related to extractable phosphorus in the Everglades Agricultural Area. Soil Sci 177:31Y38. doi:10.1097/SS.0b013e31823782da. ISSN: 0038-075X

Mino T, Van Loosdrecht MCM, Heijnen JJ (1998) Microbiology and biochemistry of the enhanced biological phosphate removal process. Water Res 32(11):3193–3207

Morissey M (1990) Our land, Jamaica. Longmans Caribbean Ltd, San Juan, Trinidad

Ngoan LD, An LV, Ogle B, Lindberg JE (2000) Ensiling techniques for shrimp by-products and their nutritive value. Asian Aust J Animal Sci 14:216–223

O'Callaghan WB, McDonald SC, Richards DM, Reid RE (1998) topsoil-free vegetative cover on a former red mud disposal site. Alcan Jamaica Rehabilitation Project Paper

Oehmen A, Zeng RJ, Yuan Z, Keller J (2005) Anaerobic metabolism of propionate by polyphosphate-accumulating organisms in enhanced biological phosphorus removal systems. Advanced Wastewater Management Centre (AWMC), The University of Queensland. Wiley InterScience (www.interscience.wiley.com). doi:10.1002/bit.20480

Orcutt DM, Nilsen ET (2000) The physiology of plants under stress: soil and biotic factors. Wiley, New York, p 30

Özgül M, Turan M, Ketterings QM (2007) Short- and long-term phosphorus availability in four soil orders under indigenous vegetation in Turkey Acta Agriculturae Scandinavica, Section B. Soil Plant Sci 57(4), 357–364. doi:10.1080/09064710601029695

Pandey S, Gardener CO (1992) Recurrent selection for population, variety and hybrid improvement in tropical maize. Adv Agron 48:81–87

Pérez MM, Bossens J, Rosa E, Tack FM (2014) Water Sci Technol 70(9):1561–1568. doi:10.2166/wst.2014.410

Price G (2006) Australian soil fertility manual; fertilizer industry federation (Australia) Inc, and CSIRO, 3rd edn. Blackwell-Synergy, Oxford, pp 39–45

Raghothama KG, Karthikeyan AS (2005) Phosphate acquisition. Plant Soil 274:37–49

Raich JW, Russell AE, Crews TE, Farrington H, Vitousek PM (1996) Both nitrogen and phosphorus limit plant production on young Hawaiian lava flows. Biogeochemistry 32:1–14

Randall CW, Chapin RW (1997) Acetic acid inhibition of biological phosphorus removal. Water Environ Res 69(5):955–60

Rao MS, Stevens WF (2006) Fermentation of shrimp bio-waste under different salt concentrations with amylolytic and non-amylolytic lactobacillus strains for chitin production. Food Technol Biotechnol 44:83–87

Richter DD, Allen HL, Li J, Markewitz D, Markewitz D, Raikes J (2006) Bioavailability of slowly cycling soil phosphorus: major restructuring of soil phosphorus fractions over four decades in an aggrading forest. Oecologia 12/2006 150(2):259–271. doi:10.1007/s00442-006-0510-4

Robertson Murray 2011 www.rsc.org/periodic-table/element/15/phosphorusWater Sci Technol. 2014;70(9):1561–8. doi:10.2166/wst.2014.410

Sanchez P, Palm C, Buol S (2003) Fertility capability soil classification: a tool to help assess soil quality in the tropics. Geoderma 114(3–4):157–185. doi:10.1016/S0016-7061(03)00040-5

Sharpley Andrew N (2000) Agriculture & phosphorus management. Lewis Publishers/CRC Press, Boca Raton, Florida

Sheoran AS, Sheoran V, Poonia P (2008) Rehabilitation of mine degraded land by metallophytes. Min Eng J 10(3):11–16

Shibata R, Yano K (2003) Phosphorus acquisition from non-labile sources in peanut and pigeon pea with mycorrhizal interaction. Applied Soil Ecology 24:133–141

Siddique MT, Robinson JS (2003) Phosphorus sorption and availability in soils amended with animal manures and sewage sludge. J Environ Qual 32(4):1114–1121. doi:10.2134/jeq2003.1114)

Silver WL, Lugo AE, Keller M (1999) Soil oxygen availability and biogeochemistry along rainfall and topographic gradients in upland wet tropical forest soils. Biogeochemistry 44:301–328

Simpson LA (2013) Status of soil resources in the Caribbean and experiences of its sustainable management. Workshop to launch the Global Soil Partnership: towards prevention and restoration of degraded soils in Central America and The Caribbean. FAO 2013: Global Soil Partnerships

Sinclair TR (2004) Increasing yield potential of legume crops—similarities and contrasts with cereals. New directions for a diverse planet. In: Proceedings of the 4th international crop science congress, 26 Sep–1 Oct 2004, Brisbane, Australia

Singh B, Odeh IOA, McBratney AB (2003) Acid buffering capacity and potential acidification of cotton soils in northern New South Wales. Australian Journal of Soil Research, 41, 875–888

Subbarao GV, Ae N, Otani T (1997) Genetic variation in acquisition, and utilization of phosphorus from iron-bound phosphorus in pigeonpea. Soil Sci Plant Nutr 43:511–519

Turner FT, Gilliam JW (1976) Increased P diffusion as an explanation of increased P availability in flooded rice soils. Plant and Soil 45(2):365–377. doi:10.1007/BF00011700

Vesterager Jens M, Neilsen Neils E, Henning Hogh-Jensen (2008) Effects of cropping history and phosphorus source on yield and nitrogen fixation in sole and intercropped cowpea–maize systems. Nutr Cycling Agroecosyst 80(1):61–73. doi:10.1007/s10705-007-9121-7

USDA (2012) Inherent factors affecting soil phosphorus. www.nrcs.usda.gov/Internet/FSE_DOCUMENTS/nrcs142p2_053254.pdf

Wehr JB, Fulton I, Menzies NW (2006) Revegetation strategies for bauxite refinery residue: a case study of Alcan Gove in Northern Territory, Australia. Environ Manag 37:297–306

Yadvinder-Singh (2009) Poultry litter as a nitrogen and phosphorous source for the rice–wheat cropping system. Biology & Fertility of Soils 45(7): 701–710

Yuan J, Tan X, Ye S, Zhou N, Shi B (2013) The organic acids in root exudates of oiltea and its role in mobilization of sparingly soluble phosphate in red soil. J Chem Pharm Res 5(11):572–577. ISSN: 0975-7384

Cold-Water Starch Dissolution Bioreactors: Role in Treating Heap-Leached Wastewater from Precious Metal and REE Ores

Mark Anglin Harris

Abstract Sodium cyanide, like other soluble cyanide salts, is among the most rapidly acting of all known poisons. Acting on mitochondrial cytochrome oxidase and hence blocking electron transport, NaCN is a potent inhibitor of respiration. This causes decreased oxidative metabolism and oxygen utilization followed by Lactic acidosis resulting from anaerobic metabolism. An oral dosage as small as 200–300 mg can be fatal. At abandoned mines, it is postulated that the alkaline cyanide wastewater from heap leaching of precious metals can be microbially degraded if alkaliphilic bacteria and a suitable microbial substrate can provided. The facultative alkaliphile *pseudoalcaligenes* CECT5344 was identified as being capable of degrading cyanide and metal-complexed cyanides in wastewater with a pH of up to 11. The aim was to provide a self-generating continuous source of a suitable, low cost microbial substrate in anaerobic conditions. "Bitter" *Bambusa vulgaris* (bamboo) chips from freshly harvested 1 and 2-year-old plants were submerged in neutral (pH 7) cold water for 15 days. This process dissolved starch which was then leached from the bamboo. The system acted as a cold water anaerobic bioreactor because (1) anaerobes produced alcohol using the leached bamboo sugars, (2) starch dissolution and release was perpetuated. The starch dissolution was attributed to reactions between alkalis from the intrinsic alakloids in the cells of "bitter" bamboo, and the alcohols produced by yeast fermentation. This slow-release, continuous source of starch is a potential substrate for degradative anaerobes exposed to the cyanide in wastewater from heap leaching. Moreover, this nutrient source can be applied to heavy metal detoxification of acidic wastewater such as those from REE heap-leaching, after neutralization by SRB's.

1 Introduction

The most commonly used process for gold extraction is cyanidation by heap leaching, which extracts gold from low-grade ore by converting the gold to a water soluble coordination complex (Rubo et al. 2006), and from higher grade ore by agitation leaching (Zhang et al. 2016). At low gold concentrations and in the

M.A. Harris, *Geobiotechnological Solutions to Anthropogenic Disturbances*, Environmental Earth Sciences, DOI 10.1007/978-3-319-30465-6_15

absence of sufficient difference in density, physical extraction processes alone are neither economical nor quantitative (Australia Government 2008). When physical separation is not achievable, gold is usually separated from the other constituents of the ore by chemical dissolution in cyanide (Minerals Council of Australia 2005). This process is generically referred to as leaching, and as cyanidation with specific reference to cyanide. The process occurs in conjunction with physical processing (crushing, milling, gravity separation and flocculation) (Mineral Councils of Australia 2005).

Cyanide is a naturally occurring chemical that is found in low concentrations throughout nature including in fruits, nuts, plants, and insects. It has been used by the mining industry to separate gold and silver particles from ore for over 120 years (Lottermoser 2003).

1.1 Occupational Dangers of Cyanide

Cyanide extraction of gold through milling of high-grade ores and heap leaching of extremely large heaps of ores containing low concentrations of gold, requires cycling of millions of liters of alkaline water containing high concentrations of potentially toxic sodium cyanide (NaCN), free cyanide, and metal-cyanide complexes, with some milling operations resulting in tailings ponds of 150 ha and larger (Eisler and Wiemeyer 2004). Surface discharge of excess mine dewatering water and other waters to main waterways may contain excess quantities of arsenic, total dissolved solids, boron, copper, fluoride, and zinc (Eisler and Wiemeyer 2004). These activities have created a number of serious environmental problems affecting wildlife and water management (Eisler and Wiemeyer 2004; Zhang et al. 2016).

Cyanides do not cause cancer, and do not build up or "biomagnify" in the food chain (Laberge Environmental Services 2001). Cyanide does not bioconcentrate as it undergoes rapid metabolism in exposed animals (Donato and Smith 2007; Mudder 1997). They do not persist in the environment, and are quickly broken down into less toxic chemicals by sunlight and air. But, in comparison to acid mine drainage, cyanide has a higher toxicity, but is shorter lived. Acid mine drainage can last for thousands of years, whereas cyanide breaks down within, at most, a few years (Canada Health Canada 2008). Nevertheless, its high toxicity opens the possibility of even a relatively small spill having major consequences (Canada Health Canada 2008). Further, when cyanide salts are used to extract gold, mercury cyanide complexes that form enhance mercury mobility and increase mercury concentrations in groundwater (Coles and Kimberly 2006). In one case this caused mercury contamination of drinking water supplies where removal to an acceptable level could only be achieved at an exorbitant cost (Coles and Kimberly 2006).

Additionally, cyanide is toxic to aquatic life, especially fish which are one thousand times more sensitive to cyanide than humans. Aquatic organisms show a range of sensitivities to cyanide, but fish are generally the most sensitive aquatic organisms, with 24-hour LC50 concentrations (concentrations at which kill 50 %

of the individuals) as low as 40 μg/L free cyanide for some species. For humans, a maximum concentration of <0.01 ppm is allowed for potable water, which requires effective chemical or biological treatment of cyanide-containing water or waste water (Conn 1969). Exposure to cyanide in solution through consumption of surface water is the main exposure route for most animals affected by cyanide poisoning (Lottermoser 2003). In August 1995, Guyana, more than 3.2 billion litres of cyanide-laden tailings were released into Essequibo river in Guyana when a dam collapsed at the Omai gold mine (Jodah 1995; Rainforest Information Centre 2006). Studies by the Pan American Health Organization have shown that all aquatic life in the 4 km long creek that runs from the mine to the Essequibo was killed (Ishmael 1995). Communities downstream from the Omai gold mine have reported a "reddish, silty" discharge in the Essequibo making the river, which is the main or only source of water for many villages, unfit for use (www.rainforestinfo.org.au/gold/spills.htm).

In Jamaica, Serju (2011) reports sodium cyanide concentration in the water tanks at a local abandoned gold mine reportedly containing 4500 mg per litre before being brought down to 38 mg per litre. But an attempt to drill through the walls in an effort to pour out the liquid failed when the specialised equipment broke. An examination of the tank revealed that some of its liquid content had adhered to the walls, becoming 'concretised'. Neither the solidified material nor an appropriate treatment strategy had been identified after the liquid had been removed. Though gold mining operations in Jamaica ceased in 2004, a cow reportedly died in 2011 after the animal drank from a spring near the property (Cunningham 2014). Cyanide had reportedly leaked from a rotting container (containing sodium cyanide), which found its way into the water supply. It can therefore be seen that water monitoring and management at such used (and disused) sites are very important.

1.2 Chemistry of Heap Leaching

Sodium cyanide (NaCN) is a highly soluble, white deliquescent crystalline powder forming a liquid solution when exposed to air. In the form of a very dilute sodium cyanide solution, it dissolves and separates gold from ore without dissolving many other ore components (copper, zinc, mercury and iron are the most common soluble impurities) (Kappes 1998). As the chemistry of leaching gold and silver from their ores is essentially the same for both metals, silver is also removed (Kappes 1998). Cyanide forms complexes with gold, mercury, cobalt and iron that are very stable even under mildly acidic conditions (ICMI 2012). Due to their high toxicity, cyanide compounds must be destroyed or removed from wastewaters prior to discharge (USP Technologies 2016).

The cyanide-rich solution is maintained at an alkaline pH of 9.5–11(Eisler and Wiemeyer 2004). Below a pH of 9.5, cyanide consumption is high. Above a pH of 11, metal recovery decreases (Kappes 1998). The pH of the resulting slurry is raised by adding lime or another alkali to ensure that cyanide ions do not change into toxic cyanide gas (HCN) and become lost from the system (Adams 2001). The free cyanide ions combine selectively with gold (Australian Government 2008) and oxygen is added to complete the reaction (Adams 2001).

The end result equation is:

$$4Au + 8NaCN + O_2 + 2H_2O \rightarrow 4Na\left[Au(CN)_2\right] + 4NaOH \quad (1)$$

In this redox process (called cyanidation), oxygen removes, via a two step reaction, one electron from each gold atom to form the complex Au $(CN)^{-2}$ ion.

In aqueous, alkaline cyanide solution gold is oxidized and dissolves to form Au(I) cyanide. Moreover, cyanidation is also adversely affected by the presence of free sulfur or sulfide minerals in the ore (ICMI 2012). Cyanide will preferentially leach sulfide minerals and will react with sulfur to produce thiocyanate. These reactions will also enhance the oxidation of reduced sulfur species, increasing the requirement for lime addition to control the pH at a sufficient level to avoid the volatilization of hydrogen cyanide (HCN). Further, several operations use cement (instead of lime) for agglomeration (Kappes 1998).

The above processes in tandem can push the pH of the leaching liquid up to a value of 10.

At higher pH levels, disinfectant by-products in the water increasing the formation of trihalomethanes, which are carcinogenic (Nieminski et al. 1993; Diehl et al. 2000; Virtual Amrita Laboratories 2012) are an added danger. There are thus two environmental concerns: (1) cyanide toxicity and (2) soil and water alkalinity, particularly for contiguous streams and ground water reservoirs.

1.3 Remediation

Chemical and physical methods of cyanide breakdown have been used. Applying acidification, Riveros et al. (1996) treated a high-cyanide waste solution for cyanide and metal recovery. Gonen et al. (2004) achieved acidification with H_2SO_4, followed by volatilization of hydrogen cyanide (HCN) with air stripping. Absorption of the HCN occurred when basic solution stages were applied, and under optimum conditions, 100 % of free cyanide and 48 % of complex cyanide and consequently 70 % of the total cyanide in the liquid phase of gold leach effluent were recovered. But, each of these methods have serious drawbacks (Raybuck 1992), being often complex and costly (Akcil 2003).

Akcil (2003), and Siller and Winter (1998) note that the most accepted and most feasible form of biological treatment of cyanide in modern metal purification systems is through the use of bacteria. Hundreds of plant and microbial species (bacteria, fungi and algae) can detoxify cyanide quickly to environmentally acceptable levels and into less harmful by-products (Akcil and Mudder 2003). Full-scale bacterial processes have been used effectively for many years in commercial applications in North America. Several species of bacteria can convert cyanide (HCN) under both aerobic and anaerobic conditions using it as a primary source of nitrogen and carbon (Akcil and Mudder 2003). Some of the more notable bacteria utilized are *Pseudomonas, Achromobacter, Flavobacterium, Nocardia, Bdellovibrio, Mycobacterium*, and the tried and true nitrifiers, *Nitrosomonas* and *Nitrobacter*. Biological methods of waste treatment are feasible because a number of fungi and other microorganisms have methods of producing cyanide. Several workers have reported on instances of cyanide biodegradation (Barclay et al. 1998a, b; Dhillion and Shivaraman 1999). Bacteria are used to naturally biodegrade both free and metal-complexed cyanides to bicarbonate and ammonia (Akcil 2003). Padmaja (1995) found that fermentation and ensiling are efficient techniques for removing cyanide from cassava peels.

Siller and Winter (1998) found that the optimal pH for cyanide anaerobic degradation was 6–7.5, (and the optimal temperature 25–37 °C). They removed up to 4 g potassium cyanide L^{-1} of wastewater at a hydraulic retention time (tHR) of 4 days, equivalent to a maximal cyanide space loading of 400 mg CN-L($^{-1}$) day^{-1}. The original concentration went from 200 mg L^{-1} to the residual cyanide concentration of 0.2–0.5 mg L^{-1}. But though biological degradation of cyanide is probably more sustainable than chemical methods (Siller and Winter 1998), microbial activity requires organic substrates for the success of bacterial technologies. Thus, the need for an efficient carbon source (e.g. starch) is necessary to keep the population active enough to cope with any possible inhibiting effect of cyanide (Siller and Winter 1998). Using a cold-water anaerobic methanogenic bioreactor which produced organic acids, Harris and Koomson (2015) dissolved and extracted starch from soaked *Bambusa vulgaris* (bamboo) logs. *Bambusa vulgaris* is a rhizomatous plant. It is postulated that such starch could provide the carbon source required for ongoing cyanide degradation by facultative and obligate anaerobes.

However, in high pH aqueous liquids (such as sodium cyanide wastewater environments) decomposition rates are generally low. But Luque-Almagro et al. (2005a, b) conducted a study identifying *P. pseudoalcaligenes* CECT5344 as being capable of degrading cyanide and metal-complexed cyanides. Moreover, this microorganism is able to degrade cyanide in wastewaters with as much as 30 mM free cyanide at pH 11.5. They further observed *P. pseudoalcaligenes* CECT5344 growing in jewelry wastewater containing heavy metals (Baxter and Cummings 2006).

1.4 Location of Potential Study

Gold mining activities in Jamaica are located at the Main Ridge Gold Mine, in the central section of the island, in the Parish of Clarendon (Fig. 1), approximately 5 km west of Pennants on Highway B3.

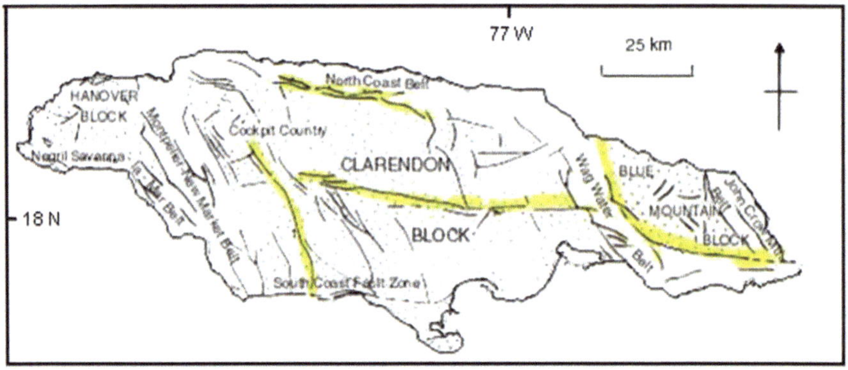

Fig. 1 Location of Main Ridge abandoned gold mine is in the Clarendon Block. *Source* Exploration Management Services

1.5 Geological Setting

The mine is located within the Central Inlier (Figs. 1 and 2) consisting of an Upper Cretaceous volcano-sedimentary sequence surrounded by Tertiary limestone (Exploration Management Services 2016). Gold is found in low-sulphidation epithermal quartz-carbonate Au-Ag setting (Exploration Management Services 2016)

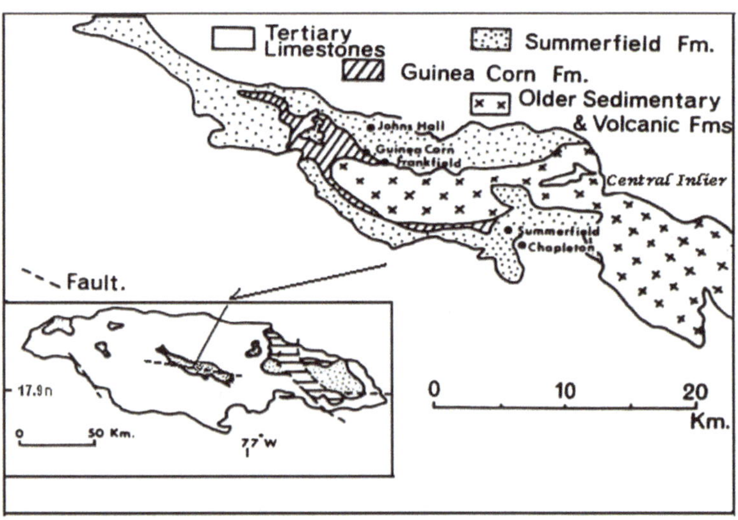

Fig. 2 Geological setting of the abandoned gold mine. An east-west trending fault runs through the central inlier. *Source* Exploration Management Services

1.6 Mineralization Setting

The mine sequence is within the Bullhead Formation (calc-alkaline andesite ± dacite) to basalt composition with minor flows and fine sediments (Exploration Management Services 2016). Mineralization is controlled by the east-west, steeply south dipping Bennett Fault, and gold is associated with stockwork quartz supporting volcaniclastic breccias (Exploration Management Services 2016). Stockwork is defined as metalliferous deposits laying not in strata or veins, but in solid masses, so as to be worked in chambers or stories or characterized by the impregnation of the mass of rock with many small veins or nests irregularly grouped.

1.7 Deposit Grade and Size

According to Exploration Management Services (2016), typical grades are from 5 to 8 g t^{-1} gold, with average mine head grade of 7 g t^{-1}. The west zone has more erratic grades (to 120 g t^{-1} gold) and head grade varying from 5 to 15 g t^{-1}.

1.8 Exploration History

Prior to 1997, the following mining operations ensued: (1) BHP Minerals International Exploration, Inc (BHP) 1989–1993 (2) Orvana from 1993 to 1996: 95,725 tonnes at 9.01 g t^{-1} containing 27,739 oz gold to a pit floor at 310 m RL (reduced level).

1.9 Aim and Hypothesis

Reporting on 28 operations, Kappes (1998) notes that the level of cyanide in heap onflow solutions ranged from 100 to 600 ppm NaCN, and averaged 240 ppm for 28 reporting operations. Heap discharge solution (pregnant solution) averaged 110 ppm. As noted above, the onflow in the Jamaica mines was recorded initially at 4500 ppm. For microbial decomposition of cyanide in waste water, the need is of a continuous-supply adequate, low-cost carbon source for anaerobic microorganisms. The goal of this investigation is to produce a qualitatively slow-release cold-water starch producing mechanism as a carbon source for cyanide- and other degrading microorganisms in mine wastewater.

2 Methodology

Freshly harvested bamboo logs were chopped with a band-saw into <2 cm diameter particles and placed into a polyacrylate horizontal cylindrical bioreactor. To extricate soluble carbohydrates from bamboo for microbial nutrients during their exposure to cyanide, bamboos were first submerged in water (Fig. 3) for thirty days (after Harris and Koomson 2015). Samples of the leachate were refrigerated and evaporated until only granular deposits remained.

3 Results and Discussion

Based on the strong odor of decaying organic matter detected after just 4 days of submergence, carbohydrates in solution were anaerobically converted to alcohols and organic acids.

Not only had sugars been dissolved from the bamboo, but starch had been solubilized during soaking. Thus, evaporation of some of the leachate left a substantial mass (2 % w/w) of white, granular crystalline residue identified as starch by a blue-black reaction (Fig. 3) with Iodine Biostain C4194 g-stain #2 (Post-Apple Scientific Inc.TM).

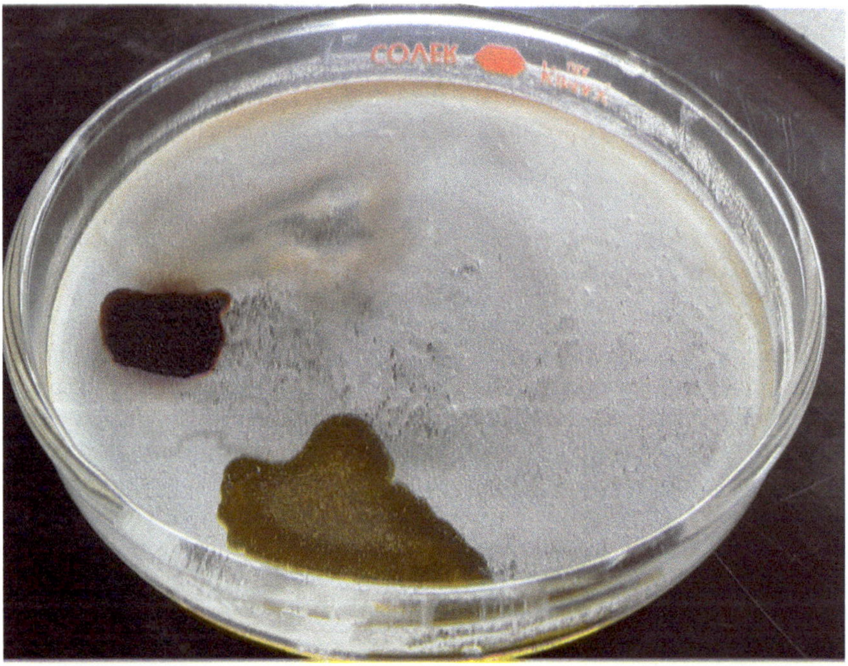

Fig. 3 Granular starch observed after evaporation of water in an anaerobic bioreactor containing bamboo. Iodine biostain caused colour change at *left side above*

How did such starch dissolution occur in cold water? As stated above, sugar alcohols such as inositol, sorbitol, and mannitol, and mono- and disaccharides comprise (largely) the carbohydrates in bamboo. In such a nutrient-rich, submerged, placid environment, fermentation over several weeks occurred. The overall process of fermentation is to convert glucose sugar ($C_6H_{12}O_6$) to alcohol (CH_3CH_2OH) and carbon dioxide gas (CO_2). Fermentation is the process of culturing yeast under favorable thermal conditions to produce alcohol.

This process is normally carried out at around 35–40 °C to produce ethanol from starchy materials after conversion into sugars. But, in the cold water, with sugar but no starch available, sugar was converted to alcohol, and starch may well have been conserved under such conditions as the monohydric alcohols abundant in bamboo such as inositol, sorbitol, and mannitol would have been metabolized first.

Using alcoholic alkali treatments, Jane et al. (1986) reported the dissolution of starch in cold water. The absence of gelatinization in that, and in this, study, proved that starch had been removed in cold water. Alkaloids are basic and occur in a limited number of plants, one of which is bamboo (Yakuba and Bukoye 2009), which contains alkaloids, tannins, phenolics, glycosides, saponins, flavonoids and anthraquinones (Yakuba and Bukoye 2009).

As alkaloid salts are soluble in water and dilute alcohols, dissolving the alkaloids may have released free OH-ions. Anghore et al. (2014) also showed that ethyl acetate removed alkaloids from the leaves of *B. vulgaris*, but in their study, water did not remove alkaloids. However, on the contrary, despite using ethyl alcohol as an extractant, Coffie (Pers.Com.) detected no alkaloids in the leaves of *B. vulgaris*. This apparent contradiction may be due to the age of the leaves used. Whereas Anghore et al. (2014) gave no indication of the condition of the leaves used, Coffie (Pers.Com.) state that leaves were harvested "before the end of April at a time non-structural carbohydrate content (TNC) levels were expected to be at a critical low and when the plants were most likely to be hostile to herbivory." Therefore the high cellulose content of leaves used by Coffie (Pers.Com.) indicates in their original samples: (1) lowered proportion of constituents such as alkaloids, and (2) increased sample age. Nitrogen is a major constituent in alkaloids, and nitrogen becomes depleted with increasing leaf age. Thus the precise selection of leaves with very low levels of non-structural carbohydrates indicates an intrinsic deficiency of alkaloids in that study.

While Harris and Koomson (2010) used heat to gelatinize starch in cassava at 50 °C, Jane et al. (1986) observed their cold-water (at <30 °C) solubilized starch as a granular structure. The starch in the present study also had not gelatinized because the granular crystals could be seen after samples of the fermenting liquor evaporated. Therefore, compared to the high cellulose leaves of Coffie (Pers.Com.) reactions of plant material containing higher levels of alkaloids in a fermenting liquor with in situ parenchymal starch over several weeks could have changed the starch status from insoluble to soluble.

The blue-black coloration in this investigation shows that the starch had not cleaved, thereby indicating that enzymatic breakdown by amylase did not occur even during fermentation. The effect of fermentation on alkaloids was demonstrated by Sarangthem and Singh (2013), who decreased alkaloid content in fresh bamboo shoots (0.98 mg 100 g^{-1} and 0.87 mg 100 g^{-1} dry wt.) by fermenting both the samples (0.63 mg 100 g^{-1} and 0.34 mg 100 g^{-1} dry wt.). Such a "decrease" indicates that the substance was used up in reactions. This is how, in this study, cold water (originally non-alkaline) treatment could have dissolved the starch in a fermenting environment.

4 Conclusions

Anaerobic degradation of toxins in wastewater is a slow process. Consequently, this study aimed to produce a qualitatively slow-release cold-water starch producing mechanism as a carbon source for cyanide- and other degrading microorganisms in mine wastewater. In this study, anaerobes produced alcohol using the leached bamboo sugars, and starch was also released and dissolved. The starch dissolution was attributed to reactions between alkalis from the intrinsic alkaloids in the cells of bamboo, and the alcohols produced by yeast fermentation. This slow-release, continuous source of starch is a potential substrate for degradative anaerobes in wastewater from heap leaching.

5 Applications

5.1 Field Applications

Bacterial mitigation of cyanide-rich wastewater from heap-leaching of precious metals can be applied in gold mine tailings in the Guianas. Biological degradation of cyanide in which CN^- is converted to CO_2, NH_3, and OH^- by bacteria, when appropriate, is considered the most cost-effective method in cyanide detoxification and has been used in cyanide detoxification of heap leaches containing more than 1.2 million t (Mosher and Figueroa 1996).

In biotechnologically treating cyanide wastewater extreme acidity is not an insurmountable obstacle. Applying a particular combination of sewage sludge and ryegrass, Harris and Ragusa (2000) observed pH increases of >4 units in an AMD of pH 2.8 after 100 days. Therefore, as many species of bacteria are pH extremophiles, it is hypothesized, that, likewise, bacterial can acclimatizate to, and similarly reduce high pH values in an alkaline waste water. It is therefore hypothesized that, given sufficient acclimatization time in cyanide waste water, alkaline pH of heap leached wastewater will be reduced and cyanide will be biodegraded.

The following sequence of activities are entailed in heap leaching (after Lottermoser 2003):

1. Mine the ore
2. Crush the ore (if necessary)

3. Agglomerate the ore (if necessary)
4. Place the ore on a lined pad
5. Irrigate the ore with the appropriate lixiviant to dissolve the metals (leachate)
6. Collect the leachate in a pond or tank (pregnant or value bearing solution)
7. Process the pregnant solution to recover the metals
8. Recycle the barren solution (with additional lixiviant) back to the heap.

If the barren solution is not to be recycled, as for example, where a mine has been abandoned, treatment of the wastewater may be necessary, as depicted in Figs. 4, 5 and 6.

Regarding the above arrangement Kappes (1998) notes the following:

> At the simplest level, the ore is hard but contains a large percentage of fines. Agglomeration means simply wetting the ore with water so the fines stick to the coarse particles, and do not segregate as the heap is built. A drainage base of crushed rock and embedded perforated pipes is installed above the plastic leach pad and below the ore heap. The importance of this drainage base cannot be overemphasized. Solution should percolate vertically downward through the entire ore column, and then enter a solution removal system with zero hydraulic head. If the drainage base cannot take the entire flow then solution builds up in a stagnant zone within the heap, and leaching within this stagnant zone can be very slow… Flow is carried in pipes within the base.

From bamboo 7:

Juice and spent process water are valuable fertilizers disposed of by land spreading (International Starch Institute)

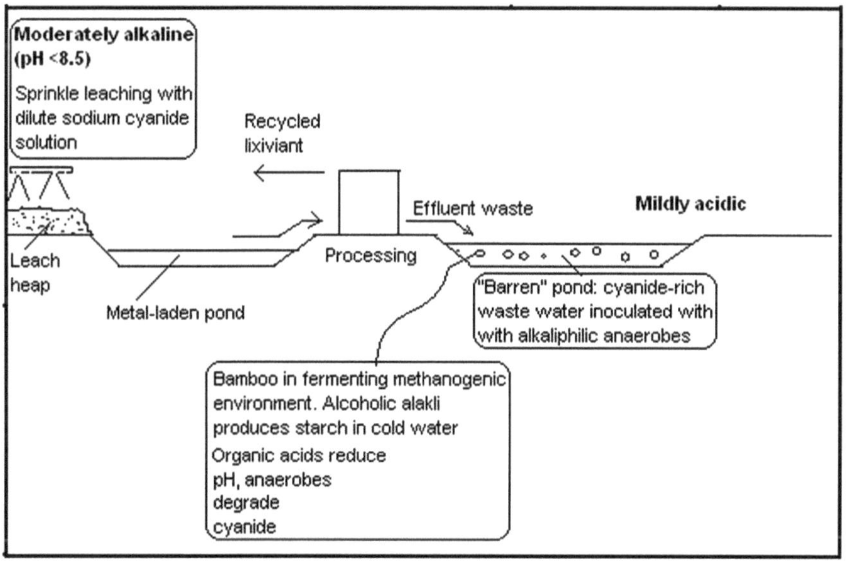

Fig. 4 Biodegradation of mildly alkaline heap-leached wastewater in starch-fueled bioreactor. *Drawn by MA Harris*

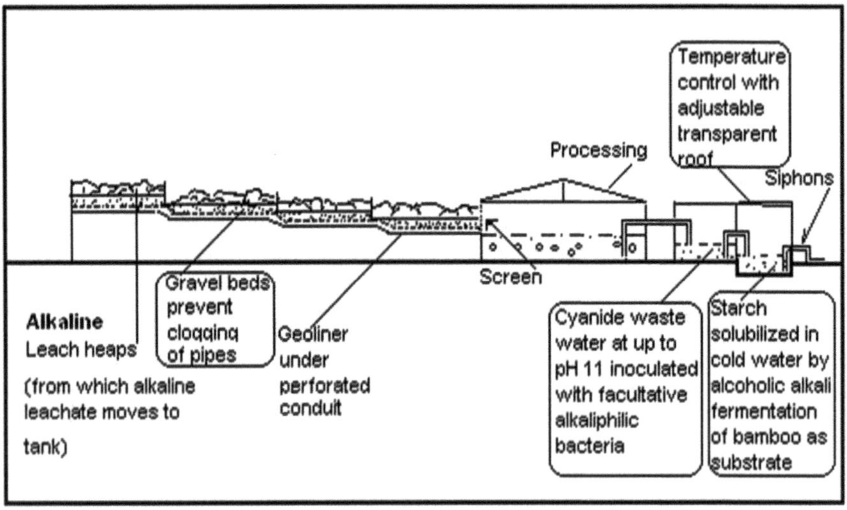

Fig. 5 Treatment of alkaline heap-leach wastewater in starch-fueled bioreactor. Covering the ponds prevent evaporation, and the resulting recirculating solution in the fermenting pond can be attain temperatures above 30 °C. *Drawn by MA Harris*

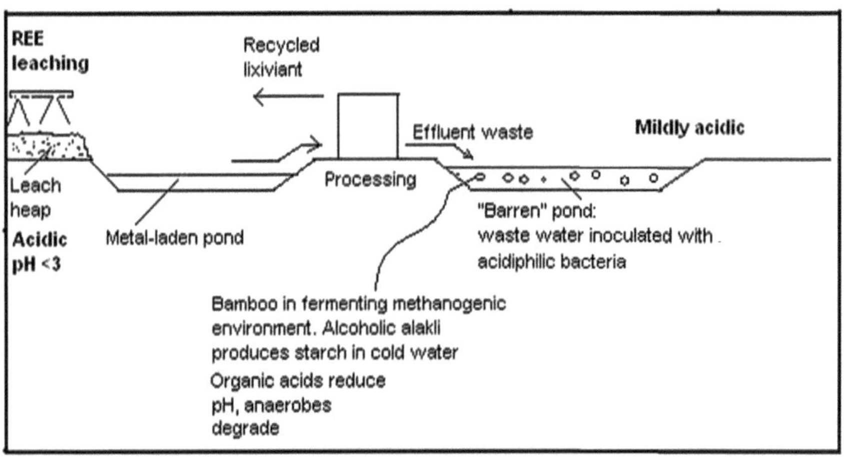

Fig. 6 Treatment of REE heap-leach wastewater in starch-fueled bioreactor. The REE ore can be of variable pH, including highly alkaline bauxite red mud waste treated with acidic solvents. *Drawn by MA Harris*

References

Adams MD (2001) A methodology for determining the deportment of cyanide losses in gold plants. Miner Eng 14(4):383–390

Akcil A (2003) Destruction of cyanide in gold mill effluents: biological versus chemical treatments. Biotechnol Adv 21:501–511

Akcil A, Mudder T (2003) Microbial destruction of cyanide wastes in gold mining: process review. Biotechnol Lett 25(6):445–450, doi:10.1023/A:1022608213814

Anghore D, Kulkarni GT, Subramanian AR (2014) Preliminary phytochemical investigation of Bambusa vulgaris var. striata Holtum. J Chronotherapy Drug Deliv 5(1):27–31 (2014)

Australia Government, Department of Resources, Energy & Tourism. Cyanide Management, 2008. Commonwealth of Australia

Barclay M, Hart A, Knowles CJ, Meeussen JCL, Tett VA (1998a) Biodegradation of metal cyanide by mixed and pure cultures of fungi. Enzyme Microb Technol 22:223–231

Barclay M, Tett VA, Knowles CJ (1998b) Metabolism and enzymology of cyanide/metallocyanide biodegradation by *Fusarium solani* under neutral and acidic conditions. Enzyme Microb Technol 23:321–330

Baxter J, Cummings SP (2006) The current and future applications of microorganism in the bioremediation of cyanide contamination. Antonie Van Leeuwenhoek 90:1–17

Canada Health, Canada (2008) Cyanide. Environmental and Workplace Health. http://hc-sc.gc.ca/ewh-sempt/pubs/water-eau/cyanide-cyanure/index-engine.php. Accessed 10 May 2016

Coles CA, Cochrane K (2006) Mercury cyanide contamination of groundwater from gold mining and prospects for removal. Sea Sky Geotechnique 1118–1122. www.engr.mun.ca/~ccoles/Publications/0227-231.pdf

Conn EE (1969) Cyanogenic glucosides. J Agric Food Chem 17:519–526

Cunningham Anastasia (2014) Mining water instead of gold. Gleaner 12 May 2014

Diehl AC, Speitel GE, Symons JM, Krasner SW, Hwang CJ, Barrett SE (2000) DBP formation during chloramination. J Am Water Works Assoc 92(4):76–90

Dhillion JK, Shivaraman N (1999) Biodegradation of cyanide compounds by a *Pseudomonas* species (S1). Can J Microbiol 45:201–208

Donato DB, Smith GB (2007) Summary of findings: ACMER Project 58, Sunrise Dam gold mine sponsor's report, AngloGold Ashanti Australia, Donato Environmental Services SA, p 27

Eisler R, Wiemeyer SN (2004) Cyanide hazards to plants and animals from gold mining and related water issues. Rev Environ Contam Toxicol 2004(183):21–54

Exploration management services (2016) Main Ridge Gold Mine Jamaica. ABN 66 092 158 686 http://www.explorems.com.au. Accessed 13 May 2016

Gonen N, Kabasakal OS, Ozdil G (2004) Recovery of cyanide in gold leach waste solution by volatilization and absorption. J Hazard Mater 113(1–3):231–236

Harris MA, Koomson CK (2010) Moisture-pressure combination treatments for cyanide reduction in grated cassava. J Food Sci. 76(1):T20–4. doi:10.1111/j.1750-3841.2010.01942.x. Epub 6 Jan 2011

Harris MA, Koomson CK (2015) Cold-water in situ dissolution of bamboo starch in a bioreactor: inducing forest recovery from an anthropogenic disturbance. Int J Environ Sci Technol 12:2503–2510. doi:10.1007/s13762-014-0646-4

Harris MA , Ragusa S (2000) Bacterial mitigation of pollutants in acid drainage using decomposable plant material and sludge. Bioremediation of acid mine drainage using decomposable organic matter and sludge. Environ Geol 40(1):195–215

ICMI (International Cyanide Management Institute) (2012) Cyanide Facts. http://cyanidecode.org/cyanidefacts.php

Ishmael O (1995) Statement by ambassador Odeen Ishmael, permanent representative of Guyana to the Organization of American States (OAS), to the Permanent Council of the OAS. Washington DC, 30 Aug 1995

Jane J, Craig SAS, Seib PA, Hoseney RC (1986) Characterization of granular cold water-soluble starch. Starch/Sta¨rke 38:258–263. doi:10.1002/star.19860380803
Jodah DK (1995) Courting disaster in Guyana. Multinational Monitor 16(11):9–13
Kappes D (1998). Heap leach or mill? Economic considerations in a period of stable gold prices. Randol Gold & Silver Forum '98. Golden, Colorado: Randol International)
LaBerge Environmental Services (2001) Cyanide. The Facts. Accessed 13 May 2016. Available from http://www.geology.gov.yk.ca/pdf/MPERG_2001_2.pdf
Lottermoser BG (2003) ISBN: 3-540-00-526-9. Springer. Berlin, pp 3–11
Luque-Almagro VM, Huertas MJ, Martínez-Luque M, Moreno-Vivian C, Roldán MD, García-Gil LJ, Castillo F, Blasco R (2005a) Bacterial degradation of cyanide and its metal complexes under alkaline conditions. Appl Environ Microbiol 71:940–947
Luque-Almagro VM, Blasco R, Huertas MJ, Martínez-Luque M, Moreno-Vivián C, Castillo F, Roldán MD (2005b) Alkaline cyanide biodegradation by Pseudomonas pseudoalcaligenes CECT5344. Biochem Soc Trans 33:168–169
Minerals Council of Australia (2005) Fact Sheet—cyanide and its use by the minerals industry
Mosher JB, Figueroa L (1996) Biological Oxidation of Cyanide: A Viable Treatment Option for the Minerals Processing Industry? Minerals Engineering 9(5):573–581
Mudder TI (1997) 'The Sources and Environmental Significance of Low Levels of Cyanide', in short course notes on Management of Cyanide in Mining, Perth, WA, April 1997, Australian Centre for Mining Environmental Research, Kenmore, Queensland
Nieminski EC, Chaudhuri S, Lamoreaux T (1993) The occurrence of DBPS in Utah drinking waters. J Am Water Works Assoc 85(9):98–105
Padmaja G (1995) Cyanide detoxification in cassava for food and feed uses. Crit Rev Food Sci Nutr 35(4):299–339
Rainforest information centre (2006) Cyanide spills and leaks. www.rainforestinfo.org.au/gold/spills.htm. Accessed 19 May 2016 www.rainforestinfo.org.au/gold/spills.htm
Raybuck Scott A (1992) Microbes and microbial enzymes for cyanide degradation. Biodegradation 3:3–18
Riveros PA, Molnar R, Basa F (1996) Treatment for a high-cyanide waste solution for cyanide and metal recovery. Canadian Institute of Mining, Metallurgy and Petroleum, Montréal, PQ, Canada (1974) (Revue), 89(998), 153–156. ISSN: 0317-0926
Rubo A, Kellens R, Reddy J, Steier N, Hasenpusch W (2006) Alkali metal cyanides, in Ullmann's encyclopedia of industrial chemistry, Wiley-VCH, Wein-heim, Germany [online]. http://onlinelibrary.wiley.com/, doi:10.1002/14356007.i01_i01/abstract
Sarangthem K, Singh T (2013) Fermentation decreases the antinutritional content in bamboo shoots. Int J Curr Microbiol Appl Sci 2(11):361–369. ISSN: 2319-7706
Serju C (2011) Cyanide contamination still a concern at Kraal Gold mine. The Gleaner, 22 Oct 2011
Siller H, Winter J (1998) Degradation of cyanide in agroindustrial or industrial wastewater in an acidification reactor or in a single-step methane reactor by bacteria enriched from soil and peels of cassava. Appl Microbiol Biotechnol. Sep;50(3):384–9
Virtual Amrita Laboratories (2012) Determination of pH of waste water sample. Retrieved 12 May 2016, http://vlab.amrita.edu/?sub=3&brch=272&sim=1414&cnt=1
USP Technologies (2016) Cyanide treatment with H_2O_2. Accessed 11 May 2016. www.h2o2.com/industrial/applications.aspx?pid=106&
Yakuba MT & Bukoye BB (2009) Abortifacient potentials of the aqueous extract of Bambusa vulgaris leaves in pregnant Dutch rabbits. J Adolesc. Health 80(3):231–322. doi: 10.1016/j.contraception.2009.03.003
Zhang J, Baodong Z, Bryan S (2016) Separation hydrometallurgy of rare earth elements. Springer, Berlin

Air, Land, and Water Contamination from Industrial Processing of Bitter Cassava: Implications for Pre-emptive Moisture-Pressure Combination Treatments

Mark Anglin Harris and Charles Kofi Koomson

Abstract Commercial processing of cassava produces vast quantities of cyanide-laced waste which can adversely infiltrate water supplies and air breathed by factory workers. This study aimed to determine (1) the comparative concentration of cyanogens in the cassava peel as opposed to that of the pith and (2) the effect of the moisture-pressure combination treatments on cyanide concentration. A semi-quantitative test using the picrate-spectrophotometer method was applied, where, at room temperature in a closed vial, reactions caused liberation of HCN which reacts with a picrate paper. The results showed an at least 25 % higher level of cyanogen concentration in casssava peels compared to that of blended peels and pith. To assess the effects of moisture-pressure combination applications, six treatments were applied: confining pressure for 12 h, wetting for 4 h at 25 °C, 2 h at 25 °C, 2 h at 40 °C, and 2 h at 50 °C, or each of the above followed by pressure for 12 h. Treatments released cyanide from samples in the order: 2-hr wet at 50 °C + pressing > 4-h wet at 25 °C + pressing = 2-h wet at 40 °C + pressing > 2-h wet at 25 °C + pressing = 4-h wet at 25 °C > 12-h pressing. Wetting for 2 h at 50 °C followed by pressure for 12 h reduced cyanide levels by at least 20 % more than that of any other treatment. The combination of moisture and pressure enhanced the contact time between linamarin and linamarase to increase the release of HCN. Physiological cyanide overload in organisms from cassava processing occurs in water, land, and air. Therefore the reduction in concentration observed in this study, if applied at an early stage of the cassava processing should reduce the rate of morbidity in environments at risk.

Keywords Cassava beer · Cassava cyanide · Cyanogen · Linamarin · Gari

Highlights

- Higher level of cyanogen concentration in casssava peels compared to that of blended peels and pith.
- Wetting for 2 h at 50 °C followed by pressure for 12 h reduced cyanide levels by at least 20 % more than that of any other treatment.

M.A. Harris, *Geobiotechnological Solutions to Anthropogenic Disturbances*, Environmental Earth Sciences, DOI 10.1007/978-3-319-30465-6_16

1 Introduction

Starch finds uses in fast food, sweets, sausages, tablets, paper, corrugated board etc. and plays a prominent part in everyday life (International Starch Institute 2014) including, in recent years, in parts of Africa, followed by Jamaica, the making of beer. However, according to International Starch Institute (2014), the many varieties of cassava fall into two main categories, namely bitter and sweet cassava (*Manihot palmata* and *Manihot aipi*), depending on their content of cyanohydrin (CAADP 2010). For industrial purposes bitter varieties are most often used because of their higher starch content. Sweet cassava is preferred for food due to its taste and dough forming facility (Huang 2001).

Cassava wastewater contains toxic materials that can endanger humans, as well as other living organisms if they are not properly treated before disposal (Okafor and Maduagwu 2000). Cyanogenic concentrations in cassava roots range from 10 to >500 mg hydrogen cyanide (HCN) equivalent/kg dry weight (Siritunga and Sayre 2004), fifty times greater than the recommended safe levels of 10 ppm for human consumable food products (FAO/WHO 1991) and levels exceeding 100 ppm are a health danger (Ernesto et al. 2002). During cassava starch production, large amounts of cyanoglycosides are often released and hydrolyzed by plant-borne enzymes, leading to cyanide concentrations in wastewater as high as 200 mg L^{-1} (Siller and Winter 1998). Akinrele (1986) reported that large scale cassava processing could be hazardous, not by consuming residual cyanide in food, but the discharge of hydrocyanic acid into the air. In this context, the hydrocyanic acid contamination of the atmospheric air (Okafor and Maduagwu 2000) and natural water sources (Okafor et al. 2001) in areas near large scale "gari" processing as well as possible occupational exposures of humans to cyanide poisoning during large scale cassava processing (Okafor et al. 2002) have been reported. At such locations, about 5–15 metric tons of cassava is processed per working day at each centres where they work. Blood and urine samples were collected between the hours of 5.00–7.00 p.m. after workers had been exposed to hydrocyanic acid discharged into the environment during processing.

Determining urine thiocyanate based on the quantitative oxidation of thiocyanate in acid permanganate at room temperature in a closed vial with liberation of HCN which reacts with a picrate paper, their study showed high exposures to hydrocyanic acid among cassava processing workers and cigarette smokers in Nigeria. From the criterion of urine cyanide overload results mainly from gari processing rather than ingestion of cassava foods. They found a statistically significant difference ($p < 0.05$) between the mean thiocyanate excretion of the processors and the consumers, such that gari processing is the highest source of cyanide exposure among Nigerian communities dependent on cassava as their major staple.

They attributed the high level of urine thiocyanate to inhalation and skin absorption of hydrocyanic acid during processing, in addition to ingestion of cassava foods. Absorbed from the skin or inhaled during roasting of gari, hydrogen cyanide readily passes through the lungs into the blood stream and is converted to SCN in the liver and kidneys (Okafor 2004).

Most of the cassava processing workers who are mainly women complained of dizziness, headache, pains and other symptoms, consistent with signs of sublethal acute cyanide intoxication. Many of them take off a lot of time due to sickness which could be attributed to inhaling hydrocyanic acid gas. Cynanide overload, in cassava consuming populations of Nigeria results mainly from cassava processing rather than ingestion of cassava foods.

1.1 Usefulness of Thiocyanate as a Marker

Cyanide ion is readily absorbed by the gastrointestinal tract and is rapidly converted into thiocyanate by the enzyme rhodanese (WHO 1996). Oral and subcutaneous doses of cyanide in rats are excreted as thiocyanate, primarily in the urine (Okoh and Pitt 1982; Okoh 1983).

Plasma proteins (especially albumin) are known to be involved in cyanide detoxification via its conversion to thiocyanate (Manahan 2009). Nevertheless, the amounts of sulphur needed to detoxify ingested cyanide of cassava is very small compared with the daily intake of sulphur containing amino acids and therefore cannot affect levels of protein energy malnutrition (Bradbury and Denton 2010). In other words, physiological detoxification of cyanide does not cause protein deficiency. The corollary is that even an adequate level of dietary protein cannot confer protection against the effects of exceeding maximum limits.

1.2 Commercial/Industrial Advantages of Cassava

In the tropics, large-scale cassava inputs for beer-making and soft drinks are a commercially less costly alternative to the importation of barley malt and corn syrup, based on the following reasons:

1. It is affordable not only as a nutrition source but as a commercial and industrial source of starch.
2. Cassava is one of the most drought-tolerant crops, growing successfully in marginal soils.
3. Yields are reasonable where many other crops do not grow well.
4. Cassava is well adapted within latitudes 30° north and south of the equator, at altitudes up to 2000 m (6600 ft) above sea level in widely ranging rainfall regimes, and to poor soils with a pH ranging from acidic to alkaline.

The USDA Foreign Agricultural Service (2014) reports the following:

1. China is poised for a 6 % increase in the manufacture of biofuels. The government is encouraging development of non-food grain feed stocks, such as cassava and sweet sorghum.
2. However, these crops still compete with food crops for land, and only one cassava and one sweet sorghum ethanol plant are approved for production by the government. Currently 8 percent of fuel ethanol is produced using cassava.
3. The 11th Five-Year Plan (2006–2010) set goals for expanding non-grain based ethanol production, targeting cassava and sweet sorghum. The world's first cassava ethanol plant was built in Guangxi in 2007 with an annual production capacity of 200,000 tons.

Dai et al. (2006) noted that cassava fuel ethanol is more energy efficient than gasoline, diesel fuel and corn fuel ethanol but less efficient than biodiesel (Dai et al. 2006). Through fuel ethanol production, one Joule of petroleum fuel, plus other forms of energy inputs such as coal, can produce 9.8 J of fuel ethanol (Dai et al. 2006). Biodiesel from cassava was expected to be to 200 thousand tonnes in 2010. This is equivalent to 10 million tonnes of petroleum. Cassava starch production, therefore is potentially a growing activity in the tropics.

1.3 Respiratory, Dermal, and Oral Exposure

Adding to the advantages of choosing bitter cassava cited above, the difficulty of avoiding its use is exacerbated because farmers often prefer the bitter (most dangerous) varieties because they deter pests, animals, and thieves (Otuu et al. 2013) because they deter pests, animals, and thieves. Urine SCN is a useful biomarker of exposure to cyanide from cassava foods (Oluwole and Oludiran 2013). There is strong ecological association of exposure to cyanide and endemicity of ataxic polyneuropathy.

Addition of potassium cyanide at a concentration of 200 mg/L to the drinking-water of a group of seven male Sprague-Dawley rats produced a slight elevation in liver weight but had no effect on body weight gain after 21 days. Pregnant golden hamsters were exposed to sodium cyanide (0.126–0.1295 mmol/kg per hour) on days 6–9 of gestation by infusion via subcutaneously implanted osmotic minipumps. High incidences of resorptions and malformations were seen in the offspring, the most common abnormalities observed being neural tube defects.

Three groups of fish (25 in each group) were treated with different concentrations of 0 (control), 0.1 and 0.2 mg/L KCN for 2 weeks Manjunatha et al. (2015). Blood samples were drawn after 2 weeks of exposure and serum biochemical analysis. KCN exposure at both concentrations caused significant ($P < 0.05$) elevation of AST and LDH activities and creatinine concentration as compared to the control. Indeed, ALT, ALP and glucose levels in the fish treated with 0.2 mg/L KCN were significantly higher than those levels from controls. On the other hand, the biochemical parameters including GGT, triglyceride, cholesterol, total protein, and albumin did not change significantly following cyanide exposure. Nevertheless, in Nile Tilapia (*Oreochromis niloticus*), a cichlid fish, tissue damages occurred, especially in liver and kidney, due to chronic cyanide intoxication (ref).

1.4 Cassava Wastewater and Cassava Peels

1.4.1 Study Area

St. Thomas-Kingston sub-region.

Cassava plantations in the southeast section of Jamaica for supplying a beer brewery in the same sub-region have been established. Being a rain shadow zone (Fig. 1), calm conditions often prevail. Diurnal temperatures range from 20 to 35 °C (April–September) and 17–35 °C (October–March) with average monthly temperatures varying by 2.5 °C (4.5 °F). The annual average temperature is 27.1 °C (80.8 °F). There is no cold season.

Kingston's climate can be classified as a tropical wet and dry/savanna climate (Köppen-Geiger classification: Aw) with a pronounced dry season in the months having a low-sun-altitude (December–April) months. The wet season occurs in months dominated by a high sun-angle (June–November) with total annual Precipitation averages 811 mm (31.9 in.). On the other hand, northeastern Jamaica receives more than three times as much rain mainly from the North-East Trade Winds (Fig. 1).

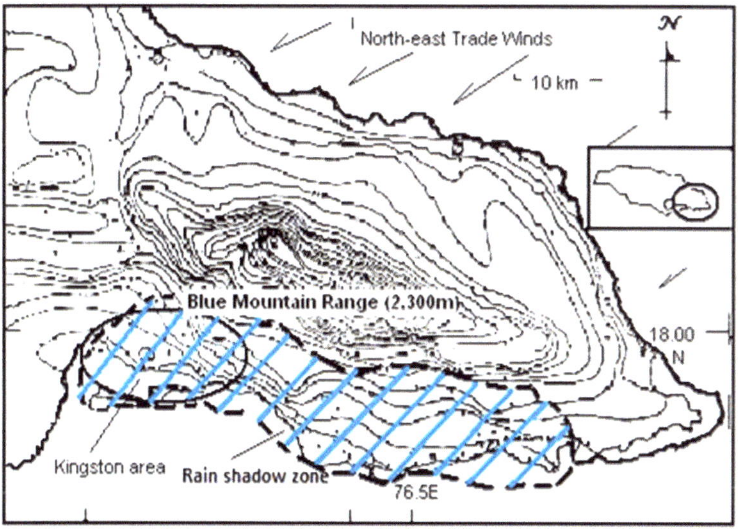

Fig. 1 The location of bitter cassava processing in southeastern Jamaica. The beer brewery is located here. Adapted after: National Land Agency of Jamaica

1.5 Dangers of Cassava Processing in Study Area

According to Okafor et al. (2001, 2002), occupational exposure remains a major problem associated with all phases of cassava processing. Otuu et al. (2013) state that: "In addition to manual peeling of cassava tuberous roots, several other processing operations involve semi-mechanized systems. The grater pulverizes the tuber into homogenous moist paste; the presser squeezes the milky cyanide—rich juice out; the sieve removes fibrous particles while the roaster gelatinizes the juice-free cassava paste into garri. In these stages of production, hydrocyanic acid is released, either as hydrogen cyanide gas during gelatinization, or as cyanogenic glycoside in the effluent (Juice extract) into the environment. While the gas joins atmospheric air, the cyanogenic glycosides make way to the soil water bodies, depending on the topography and geology of the soil environment."

The diurnal/nocturnal winds in the study area (Fig. 1) are weak, compared with those of the Trade Wind system of the northern section of the island. Such mainly calm conditions and insufficient air movements (Fig. 2) should fail to dilute and disperse toxic atmospheric gases. Under very similar climatic conditions (south-central Nigeria) to those of the study area in Jamaica, Otuu et al. (2013) found cassava peels as waste products discarded in like manner to the effluents which on decomposition release hydrogen cyanide

(HCN) into the environment. In Akwuke and Abakpa-Nike areas of Enugu metropolis (Nigeria), Otuu et al. (2013) refer to the clustering location of cassava processing plants without any designated site for waste disposal. To improve

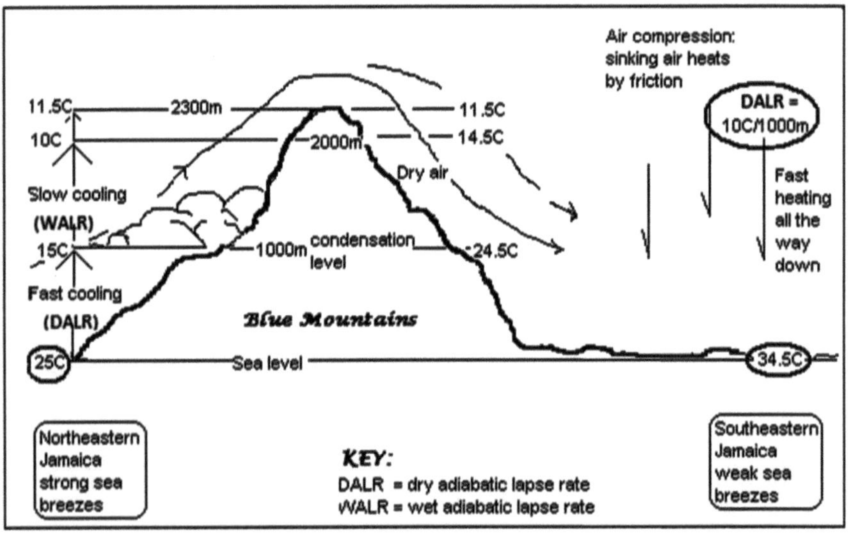

Fig. 2 Calm atmospheric conditions of southeastern Jamaica are inimical to the dilution and dispersion of environmentally toxic gases. *Drawn by MA Harris*

public health of the inhabitants, their study was expected to guide relevant government agencies on relocation of existing cassava plants to minimize effluent infiltration into wells to avoid goiterogenesis and related health hazards. In this context, buildings without adequate designated sites for disposal of cassava solid peels and liquid wastes remain a health concern.

Further corroboration of such effects are provided by the location of water wells tested for cyanide contamination. Thus of 11 water samples collected from hand-dug wells located around cassava processing plants, as well as effluent discharge from the plant, Otuu et al. (2013) found cyanide concentrations which indicated contamination of the well water and proceeding plant environs. They discovered a progressive decrease in cyanide content as the distance of the well increased from the plant site.

Otuu et al. (2013) note that the location with regards to slope influences cyanide content to a greater degree than the linear distance. Thus two sites at varying distances from the processing plant had the same cyanide content because the closer site was located on a slope, which enhanced the flow of the effluent into the well water. Well water bodies in their study areas were polluted with cyanide, and therefore unsafe for human consumption. The toxicity of cyanide is dose and time dependent (as is the case with most toxicants). Therefore, even relatively low doses, if continually absorbed, as for factory workers or residents close to waste heaps and/or drinking contaminated water, can be physiologically cumulative. This is made worse because cassava harvest can take place most of the year with less manufacturing and reduced storage costs (International Starch Institute 2014). These results of tapioca starch give a competitive advantage over the use of potato starch in many applications (International Starch Institute 2014).

1.6 Previous Processing Methods

Very little research except Bolhuis (1953) has been published on the effectiveness of the pressing method for releasing HCN from freshly grated cassava root. However, Cumbana et al. (2007) found that wetting fresh shredded cassava with added linamarase in water for 5 h decreased cyanide levels by up to 84 % in 5 g samples. Nevertheless, as many cassavas contain low levels of linamarase (Bokanga 1994), such favourable results were often obtained when the enzyme linamarase was deliberately added to the sample (Bradbury and Denton 2010), and the cost of such enzymes is often prohibitive. Further, the wetting treatment would be insufficient for cassavas containing >300 ppm of cyanide because >50 ppm cyanide (a safety limit) would remain in the product. Therefore Harris and Koomson (2011) studied the effects of moisture-pressure combination treatments on the removal of cyanide from bitter cassava. They found that pressure for a specified time, when preceded by moisture applications, decreased the concentration of cyanogens as compared with any wetting treatments. Harris and Koomson reported that traditional 12-h pressing of grated bland or bitter cassava roots containing sufficient reactants rarely reduced cyanide concentration to safe levels. They found that although long wetting was more efficient at releasing HCN than pressing, pressing following wetting was more efficient at releasing HCN than either wetting or pressing acting alone. Thus, of all their treatments, pressing for 12 h was easily the least effective, but the combination of moisture and pressure increased contact between linamarin and linamarase to increase the release of HCN.

1.7 Aim

The peels of cassava contain starch, and starch is the ingredient sought by many cassava processers. Therefore, commercial operations should save substantial costs by using unpeeled cassava tubers. However, only a few, if any, studies on relative toxicity of cassava peels have been observed in the literature. This project therefore aims to study (a) the differences in cyanogen concentration between peeled and unpeeled cassava roots and (b) the effects of moisture-pressure combination treatments on the whole tuber inclusive of pith and peel.

1.8 Hypothesis #1

Cyanogens in plants protect against destruction by predators seeking food (Minerals Council of Australia 2016). Therefore it is reasonable to hypothesize that highest levels of cyanogens are located in the outer section (peel, or, rind) of the cassava tuber.

1.9 Hypothesis #2

Moisture-pressure treatments are more efficient than moisture and/or heat in removing cyanogens from whole cassava tubers (pith + peel).

2 Method

Cassava tubers were thoroughly washed in de-ionized water, and dried. The gratings of (1) peeled, and (2) unpeeled (rind + pith) bitter cassava roots were soaked overnight after Harris and Koomson (2011) because wastewater from cassava processing contains, among insoluble constituents such as starch grains, cellulose fibres and peels, cyanide as a toxic substance (Siller and Winter 1998).

2.1 *Experiment #2: Moisture-Pressure Treatments of Grated Cassava*

Both peeled and unpeeled, grated cassava was subjected to soaking and pressure according to the method of Harris and Koomson (2011) as follows.

Some of the fresh, grated material was divided up and pressed for 12 h (P). For pressing, twenty grams of grated cassava were placed in an 8 × 12-cm cotton cloth bag and the bag placed on a concrete surface inclined at 30 degrees from the horizontal. A 30 × 15 × 10-cm concrete block weighing 10 kg was placed on the sample and left in that position for 12 h. The inclined surface was evaluated as a necessary step to facilitate the potential outward flow of liquid from the porous sample bag. All other samples were wet in the proportion 1:1.25 cassava: water (w/w) for 2 h at 50 °C. This is because Harris and Koomson (2011) found that wetting for 2 h at 50 °C removed more cyanogens than wetting at temperatures below 50 °C. This was followed by pressing (P) for 12 h. For total cyanide content analysis, linamarase/buffer papers were placed in plastic vials. Samples of 100 mg were added followed by picrate paper and a lid. The vials were kept at room temperature for 18 h, the picrate paper was removed from the plastic strip and the paper eluted for 30 min with 5 mL of distilled water. The absorbance (A) of the solution was measured using 10 mm cuvettes in a Genesys 20 spectrophotometer in the direct reading mode, against a blank solution prepared from a 4 cm^2 picrate paper not exposed to HCN and eluted with 5 mL distilled water. The total cyanogen content in mg HCN equivalents/kg sample (ppm) was calculated by the equation (Bradbury et al. 1999):

$$\mathrm{ppm} = 396 * \mathrm{A} * 100/\mathrm{z}\,(\mathrm{mg}) \tag{1}$$

where, (A) = absorbance, and (z) is the mass of the sample

2.2 *Statistical Method*

Three replicates were compiled for each treatment. Statistical analyses were conducted on treatment means using the t-test procedure (Instat Software® Version 3.0, GraphPad, San Diego, CA).

3 Results and Discussion

3.1 *Experiment #1: Cyanide in Peel Versus Pith*

Table 1 and Fig. 3 depicts the concentration of cyanide in freshly grated peeled cassava or unpeeled cassava. It can be seen that concentration in peel + pith

Table 1 Concentration of cyanide in peeled versus unpeeled + peeled cassava

Samples	Replicates	Mean	S.D.
Pith + peel	294, 266, 289	284	9.9
Pith	241, 250, 236	242	7.1

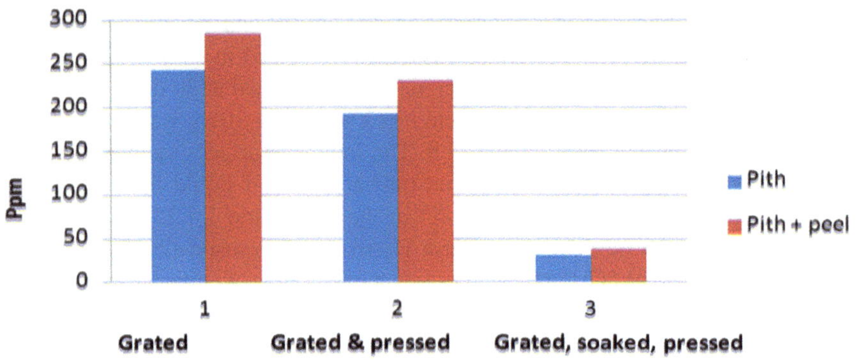

Fig. 3 Cyanogen remaining in cassava tuber samples after treatment

exceed that of the pith by >25 %. Hence, as the concentration in the peel had been diluted by that of pith, the difference is even greater than shown by the calculations. Nevertheless, as the samples in this case were within the same cassava variety, more studies may be required.

3.2 Moisture-Pressure Treatments

In this study, the most effective treatment is pressing for 12 h after wetting (without added linamarase) for 2 h at 50 °C, which reduced cyanide in a bitter cassava from 284 ppm cyanide to 38 ppm, an 87.2 % drop. The amount of cyanide remaining in samples after that treatment was therefore only 12.8 %. But, using the same cassava varieties and applying moisture-pressure combination treatments, Harris and Koomson (2011) reduced similarly high cyanide concentration in the ambient surroundings. They observed the greatest reduction of cyanide from the 2W50 treatment in the following order: 2WP50 > 4WP25 = 2WP40 > 2WP25 > 4W25 > 2W25 > 12P. These results show that even more bitter cassavas could be brought to much safer levels with this treatment. Therefore, even for cassava peels, the method of Harris and Koomson (2011) could reduce this high cyanide concentration in the ambient surroundings (Fig. 4).

Also, these results, like those of Harris and Koomson (2011) again exceeded those of Bradbury and Denton (2010) who achieved a maximum of 16 % of cyanide remaining in their samples when they applied wetting only for 2 h at 50 °C.

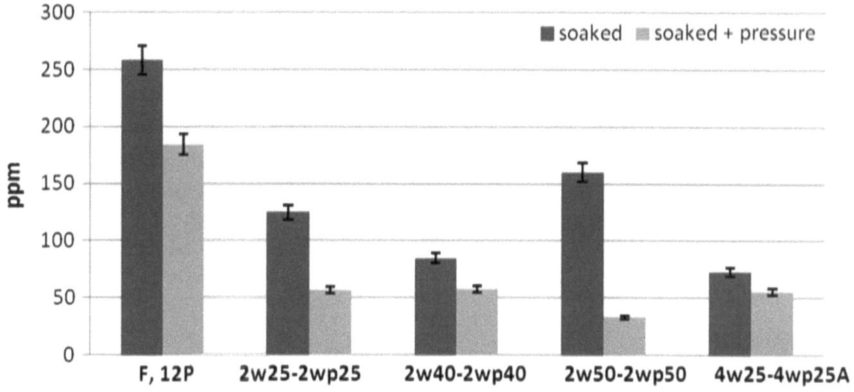

Fig. 4 Concentration of cyanide in cassava before and after treatments. Graph of data from Harris and Koomson (2011)

Linamarin breakdown varies directly with the availability of linamarase, and the amount of linamarase lost through its catalysis of linamarin breakdown varies directly with temperature and duration of wetting. In this study, linamarase acted efficiently due to increased rupturing of cells. This was also observed by Harris and Koomson (2011), where the high viscosity amylopectin with molecules (up to 80,000,000 in mass with many branch points) mainly comprising the cassava gel (WHO 1974) could have sealed off internal conduits, thereby trapping cyanide gas in internal pores of the gel mass of their treated samples. This is because gelatinized starch traps gases in its microstructure (Gan et al. 1990). Such a process could have restricted internal movement by 'locking in' gas molecules.

As release of reactants for the production of HCN occurs only after the rupture of cells, this implies the rupturing of cells during pressing. A longer time-period of contact between enzyme and substrate combined with a higher concentration of reactants contributed to the increased release of HCN. By restricting external escape of HCN gas (until subsequent pressure was applied), they acquired an anomalously high build-up of cyanide level for the 2W50 samples. However, though some rupturing must have occurred, there is no proof that rupturing is the only cause of the high level of cyanide release. Yet, on the assumption that pressing merely expelled pre-trapped HCN gas from gelatinized samples, the effectiveness of the treatment is even more convincing.

4 Conclusions

Higher cyanide levels exist in cassava peels compared with the pith. Removing the peels before processing bitter cassava can reduce the concentration of cyanide in the atmosphere of factories which process bitter cassava. However, environmentally safer disposal of unwanted bitter cassava peels requires further research.

4.1 Hypotheses

Hypothesis #1 postulated that highest levels of cyanogens are located in the outer section (peel, or, rind) of the cassava tuber. Hypothesis #2 above claims that the additive effect of pressure and moisture is more effective than combinations of wetting and heating in removing cyanide from freshly grated cassava whole roots (pith + peel). Both hypotheses have been supported by the results.

5 Applications

Determination of thiocyanate content of urine (Bradbury et al. 1999) can be used to check the cyanogens overload of populations associated with commercial quantities of cassava including factory workers, and individuals in close proximity to waste waters affected by cassava products (Okafor and Maduagwu 2000). A simple kit method for determination of thiocyanate in urine could be used to monitor the cyanide overload in cassava consuming populations was developed by Haque and Bradbury (1999). Just as this simple kit method was successfully used to assess the exposures of cassava processors (mainly gari processors, i.e., those involved in the roasting of grated cassava), and cigarette smokers in Nigeria to hydrocyanic acid, it can be used to evaluate the same exposures in other factories such as beer-making, etc.
Processing precautions may include:

(a) Deep peeling of cassava prior to further processing.
(b) Excluding the peels from further production.
(c) Applying an increase of linamarase during the pressing phase.

References

Akinrele IA (1986) Hydrocyanic acid hazard during large scale cassava processing. Trop. Sci. 26:59–65

Bokanga M (1994) Distribution of cyanogenic potential in cassava germplasm. ISHS Acta Horticulturae Int'l Workshop on Cassava Safety

Bolhuis GG (1953) A survey of some attempts to breed cassava varieties with a high content of protein in the roots. Euphytica 2:107–12

Bradbury JH, Denton IC (2010) Rapid wetting to reduce cyanogen content of cassava flour. Food Chem 121:591–594

Bradbury MG, Egan SV, Bradbury HJ (1999) Picrate paper kits for determination of total cyanogens in cassava roots and all forms of cyanogens in cassava products. J Sci Food Agric 79:593–601

Cumbana A, Mirioneb E, Cliff J, Bradbury JH (2007) Reduction of cyanide content of cassava flour in Mozambique by the wetting method. Food Chem 101(3):894–897

Dai D, Zhyuan H, Gengqiang P, He L, Chengtao W (2006) Energy efficiency and potentials of cassava fuel ethanol in Guangxi region of China. Energy Convers Manage 47(13–14):1686–1699. doi:10.1016/j.enconman.2005.10.019

Ernesto M, Cardoso PA, Domingos N, Mirione E, Massaza F, Cliff J, Haque MR, Bradbury JH (2002) Persistent konzo and cyanogen toxicity from cassava in northern Mozambique. Acta Tropica 82:357–362

FAO/WHO (1991) Joint FAO/WHO food standards programme, Codex Alimentarius Commission Xll. FAO, Rome, Italy (Supplement 4)

Gan Z, Angold RE, Williams MR, Ellis PR, Vaughan JG, Galliard T (1990) The microstructure and gas retention of bread dough. J Cereal Sci 12:15–24

Harris MA, Koomson CK (2011) Moisture-pressure combination treatments for cyanide reduction in grated cassava. J Food Sci 76(1), T20–T24. doi:10.1111/j.1750-3841.2010.01942.x

Haque MR, Bradbury JH (1999) Simple method for determination of thiocyanate in urine. Clin Chem 45(9):1459–1464

Huang J (2001) Analysis of post-harvest deterioration in tuberous roots of cassava. Published PHD Thesis, Wageningen University. ISBN 90-5808-548-1

International Starch Institute (2014) Technological memorandum on cassava starch. www.starch.dk—international@starch.dk. Accessed 19 May 2016

Manahan SE (2009) Environmental chemistry. CRC Press, USA

Manjunatha B, Tirado JO, Selvanayagam M (2015) Sub-lethal toxicity of potassium cyanide on Nile tilapia (orechromis niloticus): biochemical response. Int J Pharmacol Pharm Sci 3(7). ISSN:0975–1491

Minerals Council of Australia (2016) Resources Australia's gold industry. Cyanide Management. www.minerals.org.au/resources/gold/cyanide_management

Nhassico D, Muquingue H, Cliff J, Bradbury CA, Howard J (2008) Rising African cassava production, diseases due to high cyanide intake and control measures. J Sci Food Agric 88:20–43

Okoh PN (1983) Excretion of 14C-labelled cyanide in rats exposed to chronic intake of potassium cyanide. Toxicol Appl Pharmacol 1983(70):335–339

Okoh PN, Pitt GAJ (1982) The metabolism of cyanide and the gastrointestinal circulation of the resulting thiocyanate under conditions of chronic cyanide intake in the rat. Can J Physiol Pharmacol 60:381–386

Okafor PN, Maduagwu EN (2000) Cyanide contamination of the atmospheric air during large scale "gari" processing and the toxicity effects of such cyanide equivalent on rats. Afr J Biomed Res 3:19–23

Okafor PN, Okoronkwo CO, Alaneme FO, Maduagwu EN (2001) Cyanide contamination of natural water sources during large scale cassava processing. Afr J Biomed Res 4:25–27

Okafor PN, Okoronkwo CO, Maduagwu EN (2002) Occupational and dietary exposures of humans to cyanide poisoning from large scale cassava processing and ingestion of cassava food. Food Chem Toxicol 40:1001–1005

Oluwole OSA, Oludiran AO (2013) Normative concentrations of urine thiocyanate in cassava eating communities in Nigeria. Int J Food Sci Nutr 64(8):1036–1041. doi:10.3109/09637486.2013.825697

Otuu F, Inya-Agha S, Nnamani PO, Kenechukwu FC, Attama AA (2013) Cyanide content of well water round-about cassava processing plants in Enugu, south–eastern Nigeria. Int J Environ Biol 2014; 4(1):10–12. ISSN 2277–386X

Poulton C, Berhanu K, Chinsinga B, Cooksey B, Golooba-Mutebi F, Loada A (2010) The comprehensive Africa agriculture development programme (CAADP): political incentives, value added and ways forward. Working Paper 077. www.future-agricultures.org. February 2014

Siller H, Winter J (1998) Degradation of cyanide in agro-industrial or industrial wastewater in an acidification reactor or in a single-step methane reactor by bacteria enriched from soil and peels of cassava. Appl Microbiol Biotechnol 50(3):384–389

Siritunga Dimuth, Sayre Richard (2004) Engineering cyanogen synthesis and turnover in cassava (Manihot esculenta). Plant Mol Biol 56:661–669

USDA Foreign Agricultural Service (2014) China's 2014 fuel ethanol production is forecast to increase six percent. Biofuels Ann. GAIN report # CH14038

WHO (1974) Seventeenth report of the joint FAO/WHO 423 expert committee on food additives, World health organization technical report series no 539; FAO nutrition meetings report series

WHO (1996) Cyanide in drinking water guidelines for drinking-water quality, vol 2, 2nd edn. Health criteria and other supporting information. World Health Organization, Geneva

A Multiphase Outflow Purification System (MOPS) for Long-Term Sustainability of Acid Mine Drainage Detoxification and Metal Recovery

Mark Anglin Harris

Abstract Treatment of acid mine drainage (AMD) produces large quantities of solid wastes suffused with heavy metal contaminants. An AMD purification system comprising a pH-buffered anaerobic bioreactor was powered by sulphate-reducing bacteria. This was followed by removal of heavy metals from the sludge using organic acids before applications as a soil amendment. The most recalcitrant pollutant in SRB bioreactor systems is SO_4, which persisted in the AMD supernatant above acceptable levels. Excess SO_4 in the treated AMD was removed by lime neutralization, as: $Ca(OH)_2 + H_2SO_4 = CaSO_4 + 2H_2O$. The pure gypsum is environmentally safe and hence can be commercially utilized. This contrasts with the method of full-time lime-neutralization of AMD which produces polluted gypsum.

Keywords Hydroxide precipitation • Lime treatment • Organic acids • sulphate reducing bacteria

Highlights

- Gypsum AMD process: not as a waste but as a sustainable raw material.
- Sewage sludge cleans AMD and then is to decontaminate for land applications.

1 Introduction

The chemical weathering of an individual mineral within a poly-mineralogical rock can produce acidic (H^+ producing), acid buffering (H^+ consuming), or neutral (e.g. from quartz) conditions (Lottermoser 2010). To meet federal government- or sub-federal thresholds, acidic mine effluents are treated to neutralize acidity, remove suspended solids, and decrease base metals such as As, Cd, Cu, Fe, Ni, Pb, and Zn (Dinardo et al. 1991).

M.A. Harris, *Geobiotechnological Solutions to Anthropogenic Disturbances*, Environmental Earth Sciences, DOI 10.1007/978-3-319-30465-6_17

As metals cannot be destroyed, without recovery, they remain where they were left, most often embedded in lime sludge. Consequently, several treatment techniques have shifted to recovery of metal contaminants from AMD, the advantage being the potential economic gain of metals recovery as an incentive (Dinardo et al. 1991). Harris and Ragusa (2000) verified the potential of primary sewage sludge (PSS) plus finely comminuted ryegrass within a specific proportional range to provide the electron donor/carbon source for sulphate reduction in AMD wastewater. The contaminants removed from the AMD waste water therefore remain within the ameliorant, in this case the sewage sludge. To subsequently remove the metals from sewage sludge, acid dissolution is normally applied.

1.1 Limestone/Lime Treatment: Pros and Cons

The low cost of lime and the precipitation of many species are the main advantages of lime neutralization (Gionet et al. 1989). However, the high level of maintenance, the instability of the large volume of sludge waste produced, and the enactment of stricter pollution laws require alternative approaches to effluent treatment (Dinardo et al. 1991).

In reaction with AMD waters, limestone consumes H^+ ions to produce bicarbonate ions and sparingly soluble sulphates, carbonates and hydroxides as follows (Lottermoser 2010):

$$CaCO_3(s) + H^+(aq) + SO_4{}^{2-}(aq) + Pb^{2+}(aq) \rightarrow PbSO_4(s) + HCO_3{}^-(aq)$$

$$CaCO_3(s) + Pb^{2+}(aq) \rightarrow PbCO_3(s) + Ca^{2+}(aq)$$

$$CaCO_3(s) + Zn^{2+}(aq) + 2H_2O \rightarrow Zn(OH)_2(s) + Ca^+(aq) + H_2CO_3(aq)$$

Lime neutralization proceeds to completion if enough sulphate is in solution, as follows for hydrated lime (Lottermoser 2010):

$$Ca(OH)_2(s) + 2H + (aq) \rightarrow Ca^{2+}(aq) + 2H_2O(1)$$

$$Ca(OH)_2 + Metal^{2+} \text{ or } Metal^{3+} \rightarrow Metal(OH)_2 + Metal(OH)_3 + Ca^{2+}(aq)$$

$$Ca^{2+}(aq) + SO_4{}^{2-}(aq) + 2H_2O(1) \rightarrow CaSO_4 \cdot 2H_2O(s)$$

Though lime neutralization is effective for removing such metals as cadmium, copper, iron, lead and zinc from AMD, metals differ in the pH which removes them most efficiently, and the lowest amount of metal remaining in AMD is not achieved for all metals at the same pH (Matlock et al. 2002). For example, manganese does not precipitate below pH 9. Therefore, for CaO treatments, various additives are required to increase levels of alkalinity. Caustic soda (NaOH), for instance, is effective for treating AMD waters with a high manganese content, as it raises pH to >10 (Lottermoser 2010), but extreme pH values of waste water present risks in contiguous ecosystems. Moreover, NaOH is very costly, and dangerous to handle, producing highly alkaline sludges (Lottermoser 2010). Neutralization is therefore best applied only up to the levels necessary for precipitating and adsorbing metals.

Although hydrated lime $Ca(OH)_2$ is safe and easy to use, effective, and relatively inexpensive (Lottermoser 2010), its chief disadvantages are (1) It is bulky and hence costly to transport (2) very large quantities are required (3) It generates voluminous waste as metal-rich gypsum (Figs. 1 and 2) with a wide range of metals. In Canada, the average acidity of four mines with an approximate pH of 2 from widely disparate locations varied between 6 and 44 g L^{-1} of $CaCO_3$ (Dinardo et al. 1991). Thus for an effluent leaking 50 ML/pa, the 6 g/l requirement of $CaCO_3$ needed for neutralization would exceed 300,000 kg/year of $CaCO_3$. Taylor and Cox (2003) state similar quantities for the Brukunga pyrites mine of South Australia. Hence comparable rates of $CaCO_3$ are expected for the AMD seepages of Pueblo Viejo in the Dominican Republic and within the Camaguey region of Cuba. On the other hand, the neutralization efficiency of $CaCO_3$ compared to CaO or $Ca(OH)_2$ is 30 % (Skousen 1996). However lime production requires large inputs of energy for crushing and heating, and increases air pollution.

Fig. 1 A decommissioned pyrites mine. Note the contaminated gypsum in foreground after neutralization treatment of the AMD with hydrated lime

Fig. 2 Residue from hydrated lime treatment of AMD. These entities contain metal hydroxides including those of some heavy metals

2 Method

2.1 Treatment Selection

Lime CaO therefore can efficiently detoxify AMD, but as the toxic waste generated (metal-rich gypsum) is not easily disposed of, the process is not environmentally sustainable (Marchioretto et al. 2002). But sewage sludge used to detoxify AMD can itself be subsequently detoxified with acid treatments. The detoxified sewage sludge can then be safely applied as a soil fertilizer. A major advantage of this approach is that two contaminants, sewage sludge and AMD, are decontaminated rather than just one, when lime treatment is applied to AMD.

The procedure is postulated in seven phases (Fig. 3):

1. Extraction of metals from AMD and decanting of purified water (except >SO_4 concentration)
2. Acidification (cleansing) of sludge to solubilize metals
3. Cleaning of sludge effluent by precipitation of dissolved metals
4. Precipitation of SO_4 from sludge supernatant
5. Application of detoxified sewage sludge as bio-solids to improve physical, chemical and biological status of soil
6. Use of uncontaminated gypsum as industrial and agricultural raw materials.

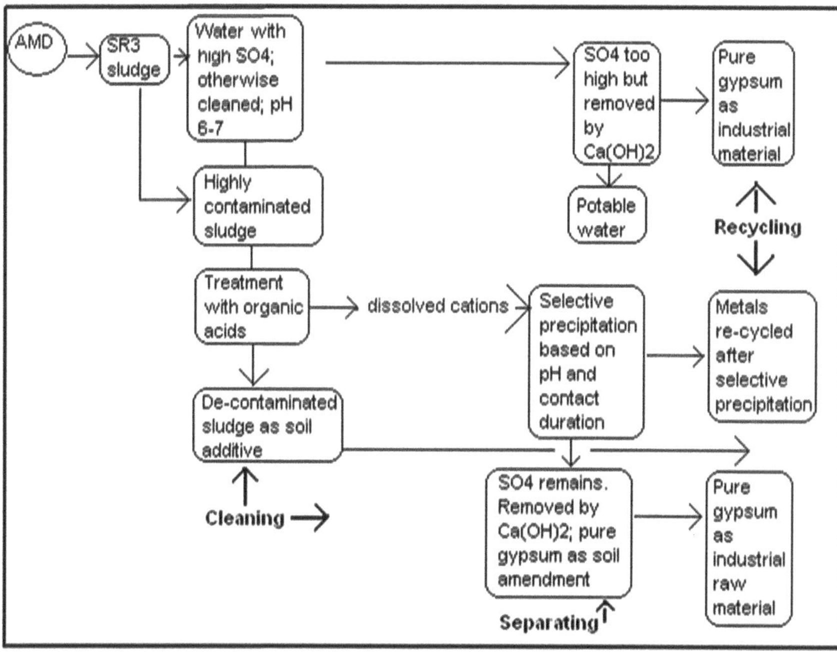

Fig. 3 Flow chart for multiphase outflow purification system (MOPS) of acidic mine drainage. SR3 treatment = lime-treated aerobically digested sewage sludge + 5 % ground ryegrass (<2 mm). *Drawn by MA Harris*

2.2 *Phase 1: AMD bioreactor (After Harris and Ragusa 2000)*

Though research has shown the effects of sewage sludge (primary and secondary) to provide the electron donor/carbon source for sulphate reduction (Harris and Ragusa 2000; Wong and Henry 1988), Harris and Ragusa (2001) found that sewage sludge, when pre-buffered and combined with phyto-organics, produced far greater detoxification in less than half the time (Harris and Ragusa 2000, 2001).

Harris and Ragusa (2000) found that the mass of the most successful treatment (S3/R) was 95 g L^{-1}, consisting of 90 g of sewage sludge and 5 g of chopped ryegrass (<2 mm diameter). As already stated, this treatment had the largest pH increases, such increases occurred in two phases: (1) days 0–50, (2) days 60–100, the latter increase being sudden, the most substantial so far, and coinciding with the appearance of SRB. Though this treatment removed the most sulfur, the remaining concentration in the AMD was far higher than the ANZECC threshold (Table 1). This is accomplished in Fig. 4.

Table 1 Decontamination levels in AMD after sludge-ryegrass treatment (Harris and Ragusa 2000)

Treatment	Al	Co	Cu	Fe	Mn	Ni	S	Zn
S3/R	0.1	0	<0.1	0.6	<0.1	0.1	649	<0.1
AD control	58	0.2	0.4	2160	127	0.4	3055	6.1

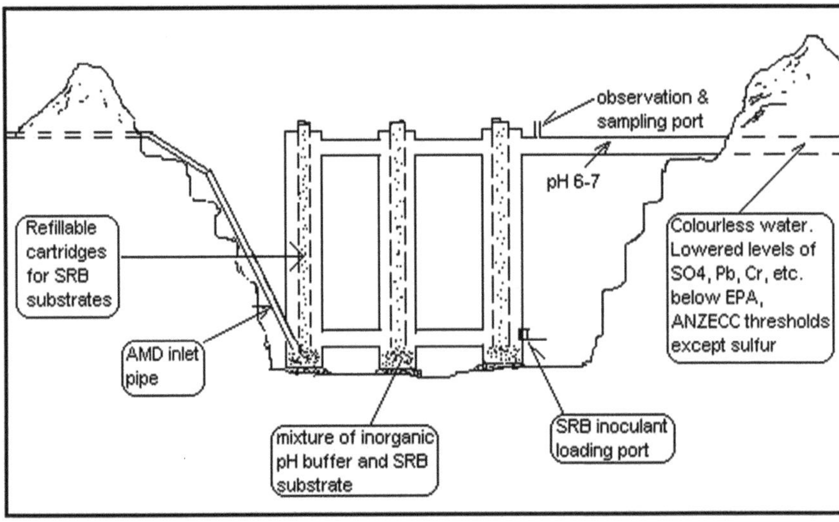

Fig. 4 Phase #1(Extraction from AMD): Multiphase Outflow Purification System (MOPS) for acid mine drainage (AMD). Progressively larger cartridge volumes towards effluent outlet enhance maximum cation removal. *Drawn by MA Harris*

2.3 Phase 2: Organic- or Mineral Acids?

Removing contaminants from an AMD increases the pollutant load of the ameliorant. The extraction of heavy metals from the sludge before land application is thus necessary, to achieve a greater sustainability (Veeken and Hamelers 1999). Mineral acids, especially HNO_3, are very efficient releasers of metals from sewage sludge (Deng et al. 2009) at low pH values. Studying sewage sludges, Gaber et al. (2011) found that solubilization of metals using inorganic acids (HNO_3, HCl) started at pH values around 2, and achieved its maximum extraction efficiency (Cr-88 %, Cu-82 %, Ni-71 %, Pb-94 %, Zn-89 %) at pH value around 1. However when they applied a complexing agent like citric acid or oxalic acid, metals dissolution began at a higher pH value (3–5) than when a strong acid such as HNO_3 and HCl is applied at the same pH. Organic acids are attractive extracting agents because extraction can be performed at mildly acidic conditions and they are biologically degradable (Gaber et al. 2011). Further, the removal efficiencies of heavy metals are drastically improved as the extraction time increased from 1 h to 10 days (Gaber et al. 2011).

For example, Gaber et al. (2011) report that Cr underwent maximum removal (90 %) achieved at pH 2 after 5 days of contact with citric acid while at pH around

3 maximum removal of Cr (66 %) attained at one day of contact. For Cu, they observed that a one day of extraction duration is the optimum condition that achieve higher removal efficiency at pH 2 (86 %) and at pH 3 (48 %). For Ni, at pH 2 an increase in Ni extraction efficiency starts after one day duration to achieve higher removal efficiency (96 %) after 5 days of extraction. For Pb, maximum removal efficiency (85 %) occurred after 10 days contact with citric acid at pH around 2, while at pH 3 maximum efficiency of lead removal (66 %) attained after 1 day of extraction. For Zn, optimum extraction time was one day duration, giving a maximum efficiency (88 %) at pH around 2 followed by pH 3 (68 %).

2.4 Sources of Organic Acids

Fungi are among the most efficient organic acid producers. *Aspergillus niger* in aerated stirred-tank-reactors convert glucose to citric acid with greater than 80 % efficiency and at final concentrations in hundreds of grams per liter (Magnuson and Lasure 2003), a phenomenal rate (Magnuson and Lasure 2003) of productivity. Fermentation using the substrates glucose or sucrose lead to the massive accumulation of organic acids by such filamentous fungi, particularly *Aspergillus niger* under industrially comparable conditions (Kubicek et al. 2010).

2.5 Phase 3: Selective Removal of Metals from Sludge

Dacera and Babel (2006) concluded that citric acid is a promising extractant for the removal of heavy metals from sewage sludge. The citric acid extraction efficiencies for Cr, Cu, Ni, Pb, Zn, are high enough to reduce the heavy metal content in sludge to levels below the legal standards. Temporal separation of metal groups occur at various time periods, and fewer metal species in a solution facilitate lower recovery costs. Gaber et al. (2011) report that pH 2.43 citric acid seemed to be highly effective in extracting Cu, Zn, mostly after 1 day of extraction time, Cr, Ni at 5 days leaching time, while Pb removal at the same pH was also high but at a longer leaching time of 10 days. At pH 3, citric acid seemed to be also highly effective in extracting Cr, Cu, Pb, Zn at 1 day leaching time. A relatively higher removal was also attained for Ni at only 4 h leaching time (Gaber et al. 2011).

Optimum pH values for precipitation are: Fe 3, Cr 5, Cu 6.5, Zn 6.75, Ni 8.2, Cd 8.2 (USEPA 1973). Therefore, effluents to be removed in pH phases for metal precipitation can therefore be grouped thus: Group 1 = Fe, Group 2 = Cr, Group 3 = Cu, Zn, Group 4 = Ni, Cd. Trial runs on small batches should characterize each particular AMD according to the following parameters:

1. No. of days, pH, height in column
2. Processing time: process until time of lowest amount left in AMD
3. Establish beforehand, levels of extraction required

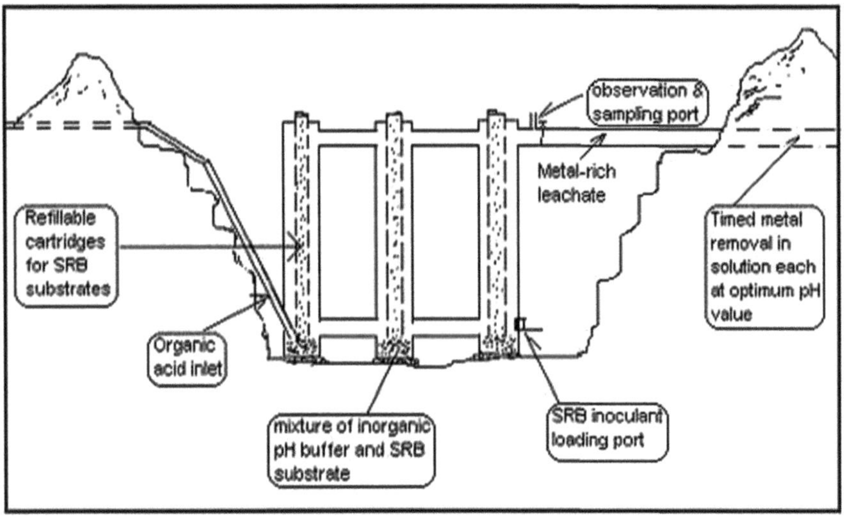

Fig. 5 Role of organic acids (e.g. citric acid) in solubilizing and releasing metals in sewage sludge

Phase 4

For sulphate precipitation, hydrated lime [$Ca(OH)_2$] is added to sulfur-rich AMD (Fig. 5).

2.6 *Phase 4*

2.6.1 Precipitation

The liquid containing the solubilized metals is separated from the suspended solids fraction by a physical separation step, e.g., centrifugation (Gaber et al. 2011). Each batch of supernatant would have contained a maximum quantity of soluble metallic ions solubilized and decanted at a particular pH. This is, ideally, converted to insoluble metallic forms by chemical precipitation. The precipitate formed is subsequently removed from the liquid by a physical separation process, e.g. sedimentation, flotation, or membrane filtration (Gaber et al. 2011).

2.7 *Phase 5*

Precipitation of SO_4 from sludge supernatant (Fig. 6).

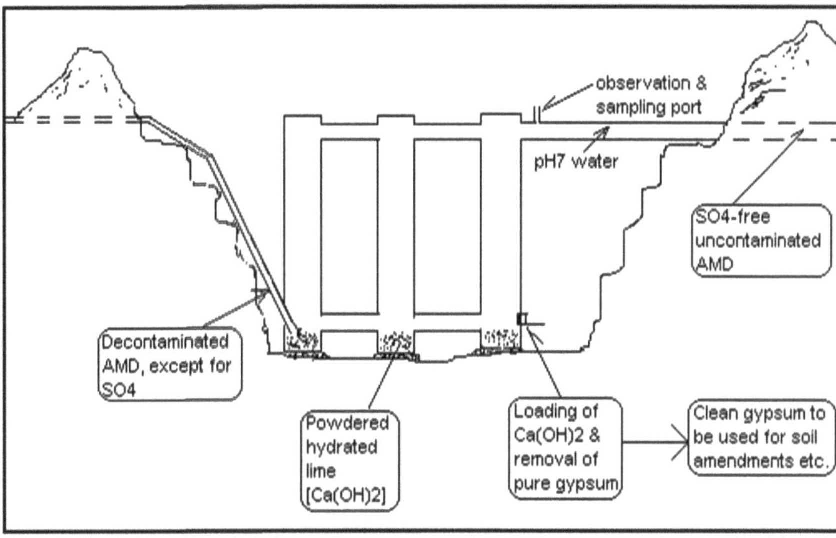

Fig. 6 The role of hydrated lime in SO_4 precipitation

2.8 Phase 6

Application of detoxified sewage sludge as bio-solids to improve physical, chemical and biological status of soil.

2.9 Phase 7

Use of uncontaminated gypsum as industrial and agricultural raw materials.

References

Dacera D, Babel S (2006) Use of citric acid for heavy metals extraction from contaminated sewage sludge for land application. Water Sci Technol 54(9):129–135

Deng J, Feng X, Qiu X (2009) Extraction of heavy metal from sewage sludge using ultrasound-assisted nitric acid. Chem Eng J 152(1):177–182. doi:10.1016/j.cej.2009.04.031

Dinardo O, Kondos PD, McKinnon DJ, McKready RGL, Riveros PA, Skaff M (1991) Study on metals recovery from acid mine drainage. MEND Treatment Committee, Mineral Sciences Laboratories, 555 Booth St., Ottawa, Ontario

Gaber Seleem E, Rizk Mahmou S, Yehia Moh M (2011) Extraction of certain heavy metals from sewage sludge using different types of acids. Biokemistri 23(1):41–48

Gionet M, Liebich Associates Ltd (1989) Minimization/utilization of mineral industry effluent treatment sludges. Report for CANMET, DSS contract no. 23440–7-9212/15-SQ, June 1989

Harris MA, Ragusa S (2000) Bioremediation of acid mine drainage using decomposable organic matter and sludge. Environ Geol 40(10):1192–1204

Harris MA, Ragusa S (2001) Bioremediation of acid mine drainage using decomposable organic matter in a constant-flow bioreactor. Environ Geol 40:1192–1240

Kubicek Christian P, Punt Peter, Visser Jaap (2010) Industrial applications. Production of organic acids by filamentous fungi. Volume 10 of the series The Mycota, pp 215–23. Online ISBN:978-3-642-11458-8. doi:10.1007/978-3-642-11458-8_10

Marchioretto M, Bruning H, Loan NTP, Rulkens H (2002) Heavy metals extraction from anaerobically digested sludge. Wat. Sci. Technol. 46(10):1–8

Lottermoser BG (2010) Mine wastes: characterization, treatment and environmental impacts. Springer, Berlin

Magnuson J, Lasure L (2003) Organic acid production by filamentous fungi.http://www.pnl.gov/biobased/docs/organic_acids_ch_12.pdf. Accessed 25 July 2016

Matlock MM, Brock SH, Atwood DA (2002) Chemical precipitation of heavy metals from acid mine drainage. Water Res 36(19):4757–4764. doi:10.1016/S0043-1354(02)00149-5

Shanmuganathan RT, Oades JM (1984) Influence of anions on dispersion and physical properties of the a horizon on a red brown earth. Geoderma 29:257–277

Skousen Jeffrey G (1996) Acid mine drainage: sources and treatment in the United States. West Virginia University, Morgantown, West Virginia. doi:10.1002/047147844X.gw3

Taylor G, Cox R (2003) The Brukunga pyrite mine—a field laboratory for acid rock drainage studies 6th ICARD Cairns, QLD, 12–18 July 2003

USEPA: Penn Environmental Consultants (1973) Processes, procedures and methods to control pollution from mining activities. U.S. Environmental Protection Agency. USEPA-430/9-73-011 Washington DC October 1973

Veeken M, Hamelers M (1999) Removal of heavy metals from sewage sludge by extraction with organic acids. Wat Sci Technol 40(1):129–136

Wong K, Henry G (1988) Bacterial leaching of heavy metals from anaerobically digested sludge, vol III. Wise, D.L. (ed.). CRC, Boca Raton, USA

MIX
Papier aus verantwortungsvollen Quellen
Paper from responsible sources
FSC® C105338

If you have any concerns about our products,
you can contact us on
ProductSafety@springernature.com

In case Publisher is established outside the EU,
the EU authorized representative is:
Springer Nature Customer Service Center GmbH
Europaplatz 3, 69115 Heidelberg, Germany

Printed by Libri Plureos GmbH
in Hamburg, Germany